Drought and Groundwater Development

Drought and Groundwater Development

Editors

Sang Yong Chung
Gyoo-Bum Kim
Venkatramanan Senapathi

MDPI • Basel • Beijing • Wuhan • Barcelona • Belgrade • Manchester • Tokyo • Cluj • Tianjin

Editors
Sang Yong Chung
Earth and Environmental Sciences
Pukyong National University
Busan
Korea, South

Gyoo-Bum Kim
Department of Construction Safety and Disaster Prevention Engineering
Daejeon University
Daejeon
Korea, South

Venkatramanan Senapathi
Disaster Management
Alagappa University
Karaikudi
India

Editorial Office
MDPI
St. Alban-Anlage 66
4052 Basel, Switzerland

This is a reprint of articles from the Special Issue published online in the open access journal *Water* (ISSN 2073-4441) (available at: www.mdpi.com/journal/water/special_issues/drought_groundwater).

For citation purposes, cite each article independently as indicated on the article page online and as indicated below:

LastName, A.A.; LastName, B.B.; LastName, C.C. Article Title. *Journal Name* **Year**, *Volume Number*, Page Range.

ISBN 978-3-0365-7859-0 (Hbk)
ISBN 978-3-0365-7858-3 (PDF)

Contents

About the Editors

Sang Yong Chung

Dr. Sang Yong Chung completed his Ph.D. program in the Department of Geological Sciences, University of Nevada, Reno, U.S.A., and his M.Sc. and B.Sc. degrees in the Department of Geological Sciences, Seoul National University, Republic of Korea. He is interested in hydrogeology, geostatistical and artificial intelligence applications to groundwater, and the assessment of groundwater contamination vulnerability. He worked in the Department of Earth and Environmental Sciences at Pukyong National University, Republic of Korea, for 35 years, and he is now an Emeritus Professor. He served as President of the Korean Society of Soil and Groundwater Environment (KOSSGE) 20 years ago and is currently an advisor to the KOSSGE. He has published over 110 articles, including SCI international and domestic publications, and has presented his research at more than 150 international and domestic academic conferences.

Gyoo-Bum Kim

Dr. Gyoo-Bum Kim completed his B.Sc., M.Sc., and Ph.D. in the Department of Earth and Environmental Sciences at Seoul National University in South Korea. From January 1991 to October 2015, he worked in various fields, including water resource policies, groundwater databases, groundwater investigations, and research at the Korea Water Resources Corporation. Since October 2015, he has been a professor of Disaster Safety Engineering at Daejeon University, where he is responsible for research and education in fields such as hydrogeology, geostatistics, artificial neural networks, artificial recharge, and environmental monitoring. He has served as the head of two large government-funded research projects, the "Riverside Groundwater Research Project" and the "Drought Response Groundwater Research Project (GW-SMART Project)", for 10 years and has published over 110 SCI international and domestic research papers to date.

Venkatramanan Senapathi

Dr. Venkatramanan Senapathi completed his Ph.D. in the field of Environmental Geochemistry at Annamalai University, India, in 2013. He was a postdoctoral fellow (Brain Korea, BK21) at the School of Earth Environmental Hazard System, Pukyong National University, Busan, Republic of Korea, from 2013 to 2017. He also worked as a visiting research faculty member in the Department for Management of Science and Technology Development, Faculty of Applied Sciences, Ton Duc Thang University, Ho Chi Minh City, Vietnam, from 2016 to 2020. He is currently working in the Department of Disaster Management, Alagappa University, Karaikudi, Tamil Nadu, India. He has published more than 200 papers, which are indexed in Thomson Reuters and Scopus (https://www.researchgate.net/profile/Venkatramanan-Senapathi). He was also an alternate faculty member in the Department of Earth and Environmental Sciences at Pukyong National University during the 2015–2016 academic year. He has more than 10 years of experience in research and teaching. Recently, he edited two books on "GIS and Geo-statistical Techniques for Groundwater Science" and "Groundwater Contamination in Coastal Aquifers", published by Elsevier in 2019 and 2022. He is also currently editing books on "Meso- and Microplastic Risk Assessment in Marine Environments: New Threats" with Elsevier. His research mainly focuses on the environmental geochemistry of water and sediments with respect to new contaminants and their sustainable management.

Preface to "Drought and Groundwater Development"

This reprint covers the important and current aspects of describing drought and groundwater development. The reprint also brings together various current techniques for assessing groundwater quality and its relationship to different fields such as science, engineering, planning, and management practices. Recent applications of groundwater assessment and management also address groundwater quality in different regions and management tactics. The authors have emphasized integrated approaches such as monitoring techniques, analysis, and remediation of groundwater management during drought. The authors have also addressed groundwater management issues considering climate change. The goal of this reprint is to bridge the gap between groundwater assessment and management. The articles presented here were written by researchers who have presented solutions to groundwater problems from a variety of perspectives. The articles are divided into modern assessment and management topics and include papers on drought and groundwater management. This reprint focuses on various technologies to solve water scarcity and pollution caused by abnormal droughts in Asia and Africa. It includes research on the use of groundwater level and precipitation data, drought prediction and diagnosis, and assessment technologies to safeguard water resources. The content of this reprint will be useful for young researchers as it provides basic information as well as details on advanced technologies in hydrology, hydrogeology, and hydrogeochemistry. The improved advanced methods are discussed, and the case studies, most of which are for illustrative purposes, present techniques for groundwater management. We hope that this reprint will improve the approach and understanding of techniques in groundwater science, especially in drought conditions.

Sang Yong Chung, Gyoo-Bum Kim, and Venkatramanan Senapathi
Editors

water

MDPI

Editorial

Drought and Groundwater Development

Sang Yong Chung [1], Gyoo-Bum Kim [2,*] and Venkatramanan Senapathi [3]

1 Department of Earth & Environmental Sciences, Institute of Environmental Geosciences, Pukyong National University, Busan 608737, Republic of Korea; chungsy@pknu.ac.kr
2 Department of Construction Safety and Disaster Prevention, Daejeon University, Daejeon 34520, Republic of Korea
3 Department of Disaster Management, Alagappa University, Karaikudi 630003, Tamil Nadu, India
* Correspondence: geowater@dju.kr

Groundwater is an important freshwater source that satisfies the needs of a significant portion of the world's population, industries, and ecosystems. It is estimated that the total amount of groundwater in the world is approximately four trillion cubic meters [1]. Particularly, in the Middle East Asian and African regions, surface water sources are scarce and polluted, and people rely heavily on groundwater [2] Due to climate change and increased evapotranspiration, the depth of groundwater production is becoming deeper. Recently, unusual droughts have occurred not only in these regions but also in mid-latitude regions, leading to water scarcity. Therefore, various technologies are being developed and applied to secure water resources.

Groundwater exists in aquifers, the world's largest water reservoir, and plays an important role in maintaining ecosystems. Particularly, in an era of climate change, groundwater plays a crucial role in helping humans adapt proactively to climate variability. The importance of groundwater has increased even more in recent years as it plays a crucial role in regulating the quantity of soil and surface water to cope with extreme climate events, such as droughts and floods, and it is a key factor affecting food productivity. Thus, deciding how to utilize groundwater resources as a means of coping with climate change on a global scale has become a critical issue [3].

Drought is a costly natural disaster that has a widespread impact on agriculture, ecosystems, water resources, social economy, and politics. Although drought is a common phenomenon that occurs in a hydrological cycle, its frequency and magnitude have been increasing in recent years because of climate change [4] When drought occurs, it has serious impacts on various sectors and sometimes leads to the over-exploitation of groundwater, which is an alternative water source. Moreover, the demand for agricultural and domestic water continues to increase worldwide as economic growth and living standards are improved. In water-scarce countries with inadequate water supply systems, surface water is vulnerable to weather changes, such as drought; therefore, alternative water sources, such as groundwater, are considered as a systematic means of water supply.

Artificial recharge is a technique that increases the total water amount by artificially injecting surface water, rainfall, reused water, or other water sources into underground aquifers. This involves the installation of artificial structures, such as artificial recharge basins, wetlands, canals, underground dams, and infiltration facilities, or changing the ground conditions to inject water artificially [5]. This technology is becoming more widely used in areas affected by climate change, such as major drought-prone regions in Africa and India. The biggest advantage of artificial recharge is that surplus water can be infiltrated into the ground for later usage during dry periods. It allows a larger amount of water to be stored in the subsurface compared to natural conditions and takes advantage of the natural purification capacity of aquifers during storage [6] The methods of artificial recharge vary greatly and include injection wells, bank filtration, ditches, recharge basins, in-channel modification, and others.

Citation: Chung, S.Y.; Kim, G.-B.; Senapathi, V. Drought and Groundwater Development. *Water* **2023**, *15*, 1908. https://doi.org/10.3390/w15101908

Received: 30 March 2023
Revised: 11 April 2023
Accepted: 12 April 2023
Published: 17 May 2023

In the United States, injection and infiltration technologies using wells have a long history, and aquifer storage and recovery (ASR) is one of the most widely used methods for storing and recovering groundwater. The main purpose of artificial recharge is to secure water for drinking water supply. It is stored underground to alleviate seasonal imbalances in water demand and is used when needed. In Europe, in particular Germany, the bank filtration technique, which is a form of artificial recharge, is widely used to secure water supply [7]. To design and investigate artificial recharge facilities for groundwater, various factors need to be analyzed, including terrain-related and meteorological characteristics, permeability and storage capacity of aquifers, clogging characteristics of soil, water supply for artificial recharge, and depth of groundwater level. In addition, the demand and supply of water and the specifications of the artificial recharge and extraction facilities used must be determined.

Groundwater helps alleviate droughts by providing underground flow to maintain stream flow during dry periods. The increased quantity of water resulting from artificial recharge not only contributes to increasing stream flow during drought periods, but it also slows down the rate of discharge in basins, thus contributing further to stream flow during such periods. Thus, artificial groundwater recharge can delay drought propagation and provide an efficient means of water supply during droughts. High technical challenges in the process of artificial recharge need to be addressed to increase the efficiency of artificial recharge facilities. The main cause of efficiency reduction is clogging, and various technologies have been developed over the past decades to evaluate and solve clogging, although current solutions are still inadequate. Land use and land cover (LU/LC) in a country are also important factors that influence groundwater recharge and surface runoff, and decision makers should consider LU/LC to increase natural groundwater recharge in the country's development [8].

This Special Issue focuses on various technologies to resolve water scarcity and contamination caused by abnormal droughts in Asia and Africa. It includes research on the utilization of groundwater level and precipitation data; drought prediction and diagnosis; and evaluation technologies for securing water resources [3,9,10]. In general, in an upstream watershed, water is quickly discharged, so drought damage appears there first. Then, the scope of the drought gradually widens, resulting in an increase in the scale of its damage. Generally, for widespread droughts, central governments show interest and quickly pursue various policies; however, for localized droughts that occur in upstream areas of a watershed, passive measures are mostly taken. In terms of water welfare, fundamental policies for drought relief and securing water resources should be pursued for these upstream areas that are marginalized [11]. Some of the technologies covered in this Special Issue are applicable to such water-welfare blind spots [12].

We conclude this Special Issue with the expectation that the research results presented here will contribute to solving water problems in various countries, particularly as solutions during periods of drought when surface water is limited.

Funding: This research was supported by the Basic Science Research Program through the National Research Foundation of Korea (NRF) funded by the Ministry of Education (2019R1D1A3A03103683).

Conflicts of Interest: The authors declare no conflict of interest.

References

1. Ritchie, H.; Roser, M. Water Use and Stress. Published Online at OurWorldInData.org. 2017. Available online: https://ourworldindata.org/water-use-stress (accessed on 30 March 2023).
2. Mengistu, T.D.; Chung, I.-M.; Kim, M.-G.; Chang, S.W.; Lee, J.E. Impacts and Implications of Land Use Land Cover Dynamics on Groundwater Recharge and Surface Runoff in East African Watershed. *Water* **2022**, *14*, 2068. [CrossRef]
3. Ani, M.; Jaunat, J.; Marin, B.; Barel, M.; Gnandi, K. Long-Term Evolution of Rainfall and Its Consequences on Water Resources: Application to theWatershed of the Kara River (Northern Togo). *Water* **2022**, *14*, 1976. [CrossRef]

4. Yun, S.-M.; Jeon, H.-T.; Cheong, J.-Y.; Kim, J.; Hamm, S.-Y. Combined Analysis of Net Groundwater Recharge Using Water Budget and Climate Change Scenarios. *Water* **2023**, *15*, 571. [CrossRef]
5. Choi, M.-R.; Kim, G.-B. Effects of Hybrid-Type Artificial Groundwater Recharge and Underground Barrier in a Small Basin. *Water* **2022**, *14*, 1849. [CrossRef]
6. Lee, J.-Y.; Woo, T.-Y. Performance Evaluation of Artificial Recharge–Water Intake System Using 3D Numerical Modeling. *Water* **2022**, *14*, 1974. [CrossRef]
7. Noh, J.-H.; So, S.-H.; Park, J.-I.; Kim, S.-Y.; Song, K.-G.; Choi, J.; Kim, G.-B.; Son, H.; Kim, H.; Maeng, S.-K. An Assessment of the Effectiveness of Riverbank Filtration in a Sewage Plant Effluent-Impacted River Using a Full-Scale Horizontal Well. *Water* **2022**, *14*, 1873. [CrossRef]
8. Mengistu, T.D.; Chang, S.W.; Kim, I.-H.; Kim, M.-G.; Chung, I.-M. Determination of Potential Aquifer Recharge Zones Using Geospatial Techniques for Proxy Data of Gilgel Gibe Catchment, Ethiopia. *Water* **2022**, *14*, 1362. [CrossRef]
9. Arumugam, M.; Kulandaisamy, P.; Karthikeyan, S.; Thangaraj, K.; Senapathi, V.; Chung, S.Y.; Muthuramalingam, S.; Rajendran, M.; Sugumaran, S.; Manimuthu, S. An Assessment of Geospatial Analysis Combined with AHP Techniques to Identify Groundwater Potential Zones in the Pudukkottai District, Tamil Nadu, India. *Water* **2023**, *15*, 1101. [CrossRef]
10. Sekar, S.; Kamaraj, J.; Poovalingam, S.; Raisamy, R.; Senapathi, V.; Chung, S.Y. Appraisal of Groundwater Vulnerability Pollution Mapping Using GIS Based GOD Index in Tiruchendur, Thoothukudi District, India. *Water* **2023**, *15*, 520. [CrossRef]
11. Chung, I.-M.; Lee, J.; Kim, M.G.; Kim, I.-H.; Yifru, B.A. Analysis of Water Supply Capacity of a Sand Dam. *Water* **2022**, *14*, 3039. [CrossRef]
12. Kim, B.R.; Lee, S.-I.; Yu, S.M. Conjunctive Operation of Sand Dam and Groundwater Well for Reliable Water Supply during Drought Conditions. *Water* **2022**, *14*, 2249. [CrossRef]

Short Biography of Author

Dr. Sang Yong Chung ompleted his Ph.D. program in the Department of Geological Sciences, University of Nevada, Reno, U.S.A., and his M.Sc. and B.Sc. degrees in the Department of Geological Sciences, Seoul National University, Republic of Korea. He is interested in hydrogeology, geostatistical and artificial intelligence applications to groundwater, and assessment of groundwater contamination vulnerability. He worked in the Department of Earth & Environmental Sciences, Pukyong National University, Republic of Korea, for 35 years, and he is now an Emeritus Professor. He served as the President of the Korean Society of Soil and Groundwater Environment (KOSSGE) 20 years ago and is currently an advisor of the KOSSGE. He has published over 110 articles, including SCI international and domestic publications, and has presented his research at more than 150 international and domestic academic conferences. His academic accomplishments are available for viewing at https://www.researchgate.net/profile/Sang Yong Chung.

Dr. Gyoo-Bum Kim completed his B.Sc., M.Sc., and Ph.D. in the Department of Earth and Environmental Sciences at Seoul National University in South Korea. From January 1991 to October 2015, he worked in various fields, including water resource policies, groundwater databases, groundwater investigations, and research at the Korea Water Resources Corporation. Since October 2015, he has been a professor of Disaster Safety Engineering at Daejeon University, where he is responsible for research and education in fields such as hydrogeology, geostatistics, artificial neural networks, artificial recharge, and environmental monitoring. He has served as the head of two large government-funded research projects, the "Riverside Groundwater Research Project" and the "Drought Response Groundwater Research Project (GW-SMART Project)", for 10 years and has published over 110 SCI international and domestic research papers to date.

Dr. Venkatramanan Senapathi completed his Ph.D. in the field of Environmental Geochemistry at Annamalai University, India, in 2013. He was a postdoctoral fellow (Brain Korea, BK21) at the School of Earth Environmental Hazard System, Pukyong National University, Busan, Republic of Korea, from 2013 to 2017. He also worked as a visiting research faculty member in the Department for Management of Science and Technology Development, Faculty of Applied Sciences, Ton Duc Thang University, Ho Chi Minh City, Vietnam, from 2016 to 2020. He is currently working in the Department of Disaster Management, Alagappa University, Karaikudi, Tamil Nadu, India. He has published more than 100 papers, which are indexed in Thomson Reuters and Scopus (https://www.researchgate.net/profile/Venkatramanan-Senapathi). He was also an alternate faculty member in the Department of Earth & Environmental Sciences at Pukyong National University during the 2015–2016 academic year. He has more than 10 years of experience in research and teaching. Recently, he edited two books on "GIS and Geo-statistical Techniques for Groundwater Science" and "Groundwater Contamination in Coastal Aquifers" published by Elsevier in 2019 and 2022. He is also currently editing books on "meso- and microplastic risk assessment in marine environment: New threats" with Elsevier. His research mainly focuses on the environmental geochemistry of water and sediments with respect to new contaminants and their sustainable management.

water

MDPI

Article

Combined Analysis of Net Groundwater Recharge Using Water Budget and Climate Change Scenarios

Sul-Min Yun [1], Hang-Tak Jeon [2], Jae-Yeol Cheong [3], Jinsoo Kim [4] and Se-Yeong Hamm [5,*]

1 Department of Geological Sciences, BK21 School of Earth and Environmental Systems, Pusan National University, Busan 46241, Republic of Korea
2 Geoscience Co., Busan 48058, Republic of Korea
3 Radwaste Technology & Research Institute, Korea Radioactive Waste Agency, Gyeongju 38062, Republic of Korea
4 Department of Spatial Information Engineering, Pukyong National University, Busan 48513, Republic of Korea
5 Institute of Environmental Studies, Pusan National University, Busan 46241, Republic of Korea
* Correspondence: hsy@pusan.ac.kr; Tel.: +82-51-510-2161

Abstract: Estimating the groundwater recharge rate is essential in all groundwater-related fields, including groundwater development, use, management, modeling, and contamination analysis. In this study, we proposed a combined method of water budget and climate change scenario for estimating the net groundwater recharge rate in the Nakdong River watershed (NRW), South Korea. For the climate change scenario method, the representative concentration pathway (RCP) 4.5 and 8.5 climate scenarios were adopted. First, using the water budget method from 2009 to 2018, the net groundwater recharge rate (NGRR) of 12.15–18.10% relative to annual precipitation (AP) was obtained, subtracting direct runoff (DR) of 21.18–25.32% relative to AP, evapotranspiration (EP) of 40.53–52.29% relative to AP, and baseflow of 12.42–17.84% relative to AP, from the AP (865–1494 mm). The average annual NGRR of the NRW was 200 mm (15.59%). Second, the mean NGRRs from 2009 to 2100 under the RCP 4.5 and RCP 8.5 scenarios were anticipated as 8.73% and 7.63%, respectively. The similarity between the water budget and climate change scenarios was confirmed using data from 2009 and 2018. According to the simple climate change scenario, it is predicted that annual precipitation will increase over the years while the groundwater level and net groundwater recharge rate will decrease. Nonetheless, the estimated NGRR by the water budget method in this study possesses uncertainty due to using potential ET instead of actual ET which should be estimated by considering soil water content.

Keywords: net groundwater recharge rate; water budget; simple climate change scenario; Nakdong river watershed; precipitation

Citation: Yun, S.-M.; Jeon, H.-T.; Cheong, J.-Y.; Kim, J.; Hamm, S.-Y. Combined Analysis of Net Groundwater Recharge Using Water Budget and Climate Change Scenarios. *Water* **2023**, *15*, 571. https://doi.org/10.3390/w15030571

Academic Editors: Bommanna Krishnappan and Chin H Wu

Received: 14 November 2022
Revised: 20 January 2023
Accepted: 28 January 2023
Published: 1 February 2023

1. Introduction

Groundwater originates mainly from rainfall and is recharged in aquifers. Watershed groundwater recharge is important for its development, use, conservation, and management. Moreover, it is critical for accurately estimating the groundwater recharge rate, which local and federal governments use for relevant policymaking. The groundwater recharge rate is one of the most challenging factors in assessing the quantitative evaluation of groundwater resources [1]. The groundwater recharge rate is calculated using methods, such as baseflow separation, water budget, water table fluctuation (WTF), groundwater modeling, climate change scenarios, and artificial intelligence. The net groundwater recharge rate (NGRR) using the water budget method is the amount of precipitation and groundwater inflow from other basins, excluding direct runoff, evapotranspiration, baseflow, pumping quantity, groundwater outflow into other basins, and coastal groundwater discharge [2,3]. The WTF method, a physical estimation method, estimates the water table rise in the phreatic aquifer and the groundwater recharge rate due to rainfall [4]. The hybrid water table fluctuation method (hybrid-WTF) combines the WTF method and the model of water content in unsaturated zones by considering the delay time of water reaching the water table from the land surface, as well as baseflow and groundwater level drops simultaneously in the condition of a permeable unsaturated zone and the underlying aquifer [5].

Moon et al. [6] estimated the groundwater recharge rate of four major river basins in South Korea using an improved WTF method based on groundwater level. In the rainy season (June–September) of 2005 in the uppermost watershed, Choi et al. [7] estimated average groundwater recharge rates of 24.75%, 13.4%, and 14.0%, respectively, using the WTF method, chloride mass balance method, and baseflow separation methods with chloride tracers. Chung et al. [8] calculated the groundwater recharge in the Mihocheon stream basin in South Korea using the SWAT-MODFLOW model [9], which is a surface water-groundwater integration model. In the Gangcheon Stream basin in Gyeonggi Province, South Korea, Noh et al. [10] calculated average groundwater recharge rates of 17.81% and 18.13%, respectively, using the hybrid-WTF method [5] and groundwater modeling. Bae [11] calculated the groundwater recharge rate using the natural resources conservation service curve number (NRCS-CN) and baseflow separation methods in five sub-watersheds and obtained a 30 mm/yr higher groundwater recharge rate using the baseflow separation and NRCS-CN methods. Lee and Bae [12] examined the variability in groundwater recharge rates due to urbanization and land cover changes. Viji et al. [13] calculated CN, including the antecedent moisture condition (AMC), for the Kundapallam watershed in India, considering land use and hydrological soil grouping (HSG). Ling et al. [14] proposed an improved soil conservation service (SCS)-CN method using nonparametric statistical analysis of rainfall and runoff data from the Wangjiaqiao watershed in China. Shi et al. [15] proposed the SCS-CN method by considering topographic slope, soil moisture, and storm duration factors, and applied the method to the Loess Plateau watershed in China.

Groundwater recharge rates are anticipated to change due to future climate change. Dripps and Bradbury [16] evaluated groundwater resources according to climate change by applying an improved soil water balance (SWB) model by considering soil moisture content, land cover, topography, climate data, and geographic information. Holman [17] analyzed the climate and socioeconomic impacts, uncertainty, and vulnerability of the groundwater recharge rate. Schindler et al. [18] investigated groundwater level variability according to soil type. They proposed a management plan for agricultural groundwater use in northeastern and central Germany, utilizing soil data at 368 monitoring points from 1951 to 2000 as well as using climate change scenarios from 2001 to 2055. Patricia et al. [19] proposed a standardized anomaly index for evaluating the impact of drought on a watershed scale in the Philippines, based on the spatiotemporal characteristics of groundwater level change. Jang et al. [20] estimated the groundwater level change from 2000 to 2012 in four major river basins in South Korea. Using climate change scenarios, they then predicted the groundwater level change from 2000 to 2100. Lee et al. [21,22] proposed coupled model development between groundwater recharge rate and climate change in river watersheds in Korea and calculated groundwater rate by using Visual HELP3. By using climate change scenarios for the Korean Peninsula and a geographic information system (GIS), Lee and Lee [23] estimated the groundwater recharge rates of 27.37% (1970 to 2000), 27.43% (2001 to 2030), 26.06% (2031 to 2050), and 27.88% (2051 to 2100) in the NRW. Lee and Lee [24] revealed the groundwater recharge mechanism in the Wonju and Uiwang areas through a time-series analysis of groundwater levels and rainfall. Kim et al. [25] conducted a time series analysis of the groundwater levels and precipitation on Geumjeongsan Mountain in Busan City. They suggested that the groundwater recharge rate of the bedrock aquifer is governed by fracture zones more than the direct rainfall effect.

The objective of this study is to estimate the net groundwater recharge rate (NGRR) from 2009 until 2100 by using a combined analysis of the water budget method and the climate change scenarios of representative concentration pathways (RCP) 4.5 and 8.5. Using data of 2009–2018, baseflow was calculated using streamflow data at the estuary of the

NRW, and the groundwater recharge rate was estimated based on the water balance. The long-term groundwater recharge until 2100 was predicted using precipitation from the RCP 4.5 and RCP 8.5.

2. Study Area

The Nakdong River watershed (NRW) consists of eight sub-basins in the upstream area (Andong Dam, Imha Dam, Mainstream I, Naeseongcheon, Yeonggang, Byeongseongcheon, Ssanggyecheon, and Wicheon Basin), seven sub-basins in the midstream region (Mainstream II, Gamcheon, Mainstream III, Geumho River upstream, Geumho River downstream, Main stream IV, Hoecheon Basin), six subbasins in the downstream western region (Hwanggang, Mainstream V, Namgang upstream, Yangcheon, Namgang Dam, and Namgang lower basin), six sub-basins in the downstream eastern region (Main stream VI, Miryanggang, Mulgeum, Yangsancheon, and Hagueon, Seobun) Nakdong River basin). The Nakdong River, which originates in Taebaek City, Gangwon Province, passes through Yeongnam Province and, flows into the South Sea through Busan Metropolitan City, with the longest length (510 km) and the second largest area (23,647 km^2) in South Korea.

A total of 122 wells under the National Groundwater Monitoring Network (NGMN) are distributed in the NRW (Figure 1). The surface elevation of the wells was the highest at 735.6 m (ND-3-SS well) and the lowest at 1.4 m (ND-3YN well). The highest groundwater level was 735.1 m, average mean sea level (amsl) (ND-3-SS well), and the lowest was −6.8 m amsl (ND-4-DC well), with an average groundwater level of 98.6 m amsl (Table 1). The highest groundwater temperature was 20.6 °C (ND-3-DGB well) and the lowest was 11.0 °C (ND-1-NA well), with the average groundwater temperature of all the wells being 15.0 °C. In addition, the average groundwater electrical conductivity (EC) of all the wells was 536 µS/cm, with the highest reading of 11,694.6 µS/cm (ND-4-DC well near the coast of the South Sea) and the lowest reading of 83.9 µS/cm (ND-4-USDC well).

Table 1. Well information statistics in the Nakdong River watershed for 2009–2018.

	Surface Elevation (m)	Well Depth (m)	Groundwater Level (m, amsl)	Temp. (°C)	EC (µS/cm)
Max	735.6	168.0	735.1	20.6	11,695
Min	1.4	40.0	−6.8	11.0	83.9
Arithmetic mean	105.6	79.7	98.6	15.0	536
Median	78.9	70.0	72.4	15.0	320
Standard deviation	105.4	19.8	104.6	1.2	1083
Range	734.2	128.0	741.9	9.6	11,611

Based on the precipitation distribution map of the NRW, prepared from precipitation data from 30 meteorological stations (Figure 2) from 1996 to 2019, the mean annual precipitation in the NRW was 1277 mm. The precipitation data of the 30 meteorological stations corresponded with the nearest monitoring well. The highest precipitation of 1989 mm/yr was observed in the ND-4-SH well, and the lowest precipitation of 917 mm/yr was observed in the ND-2-PC well. The precipitation in each well was used to calculate the net groundwater recharge.

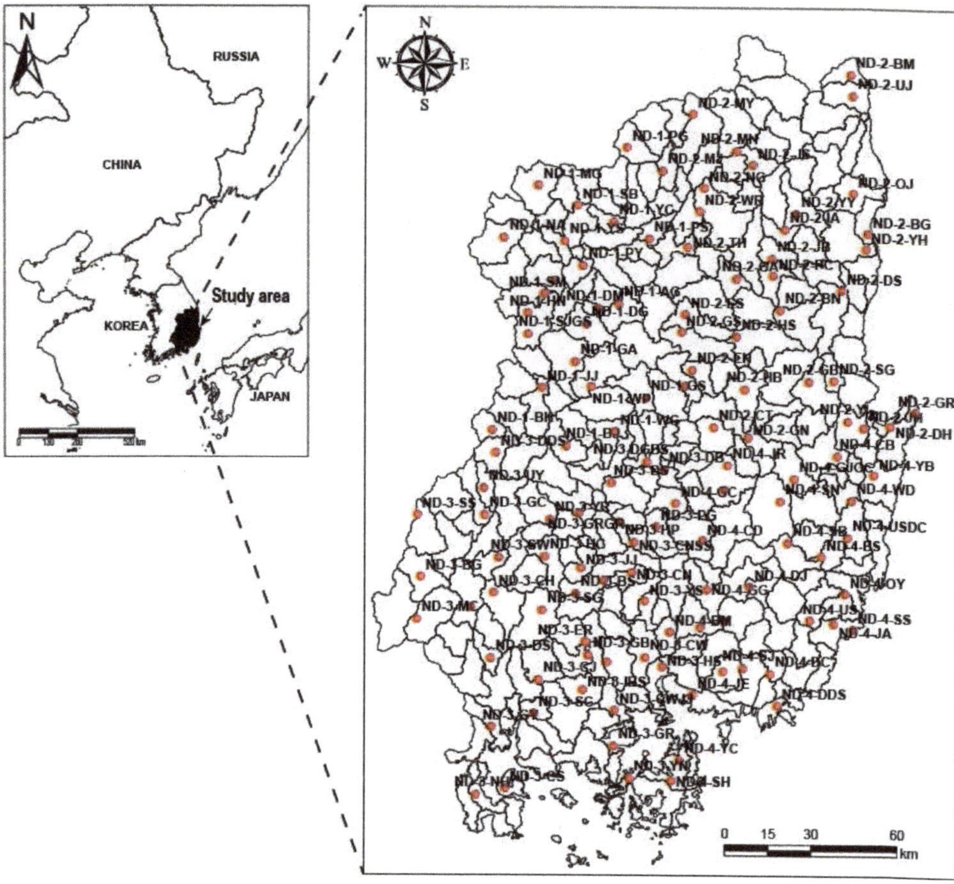

Figure 1. Location of the wells of the national groundwater monitoring network and the standard subwatersheds in the Nakdong River watershed.

Figure 2. Location of the weather monitoring stations in the Nakdong River watershed [26].

3. Methods

3.1. Method of Water Budget

In the hydrological cycle, the net groundwater recharge rate (*NGRR*) is the water that reaches the saturation zone through the unsaturated zone, and is calculated as the amount excluding direct runoff (*DR*), baseflow (*BF*), the outflow (*OU*) from the aquifer to the sea, the inflow (*RN*) of other geological layers, evapotranspiration (*ET*), as well as pumping (*DO*), if it occurs, from precipitation (*P*) [2,3]:

$$NGRR = P - DR - ET - BF - DO - OU + RN \quad (1)$$

Direct runoff (DR) is the amount that flows through the land surface and enters streams. If there is no RN and DO with no consideration of OU, then Equation (1) is transformed into:

$$NGRR = P - DR - ET - BF \tag{2}$$

Direct runoff (DR) is inversely proportional to the amount of infiltration and directly proportional to the water content in the soil layers. When there is no DR data, the DR is generally calculated using the SCS-CN method [27,28]. The SCS-CN method, DR should be greater than 0, and P should be greater than 0.2:

$$DR = \frac{(P - 0.2S)^2}{P + 0.8S} \tag{3}$$

where S is governed by the antecedent soil moisture condition (AMC) and depends on the runoff curve number (CN), which is inversely proportional to DR. The AMC values are determined by 5 or 3-day accumulated antecedent precipitation.

Evapotranspiration (ET) has been estimated in several ways [29–34]. Actual evapotranspiration (AET) is a fraction of potential evapotranspiration (PET) that is assumed as ET in the saturated condition of the soil. In this study, the PET (ET_0, mm/d) as an approximation of the AET was estimated by using the Penman–Monteith equation that is widely used in the world [35] and that considers atmospheric temperature, humidity, wind speed, and radiation. Additionally, Yang et al. [36] estimated the ratio of AET over PET to be 0.96 in the coastal area of Busan City belonging to the NRW.

Baseflow (BF), the water provided by riverside aquifers, is estimated by hydrograph separation [37,38] and the recession of BF is influenced by topography, watershed shape, soil, geological media, etc. The volume of BF discharged from riverside aquifers is given by Meyboom [37]:

$$V_{tp} = \frac{Q_0 t_1}{2.3026} \tag{4}$$

Finally, the baseflow (BF) during seasons adjacent to A and B is expressed by:

$$BF = \frac{2(Q_B - Q_A)t_1}{2.3026} \tag{5}$$

where Q_A is the baseflow at the critical time of recession after the hydrograph peak in season A and Q_B is the baseflow at the critical time of recession after the hydrograph peak in season B. Time t_1 is the time taken for the baseflow recession to decline by one log cycle.

3.2. Method of Climate Change Scenarios

Climate change scenarios calculate future climate factors, such as temperature, precipitation, wind, and humidity, by applying changes in radiative forcing caused by anthropogenic causes (greenhouse gases, aerosols, and land use changes) to the Earth system model. There are six climate change scenarios, A1 (A1FI, A1T, and A1BI), A2, B1, and B2 depending on the projected carbon dioxide emissions [39]. The IPCC (2007) determined the greenhouse gas concentration based on the amount of radiation exerted on the atmosphere by human activities according to the RCP. The Korea Meteorological Administration (KMA) predicted future climate change up to 2100 according to the RCP scenario after simulating past climates for natural and anthropogenic forcing from 1860 to 2005. The KMA developed a climate change scenario for the Korean Peninsula (www.climate.go.kr, accessed on 13 November 2022) using meteorological data (averaged from 2001 to 2010) obtained from the Automated Surface Observing System (ASOS) and the Automatic Weather System (AWS). The 1-km climate change scenario for South Korea was obtained by applying a 12.5-km resolution climate change scenario to the PRISM based downscaling estimation model (PRIDE). The calculated 1-km grid data is converted into administrative district-level data using GIS techniques. This 1-km climate change scenario can reflect the effects of complex topography in South Korea, which cannot be realized in the global climate model.

In this study, using the climate change scenarios of RCP 4.5 and RCP 8.5, $NGRR_i$ is calculated by Jang et al. [20]:

$$NGRR_i = P_i \cdot f \tag{6}$$

where recharge ratio f is determined by:

$$f = \Delta H_i \cdot S_y / P_i \tag{7}$$

where S_y is the specific yield. ΔH_i is the groundwater level difference between H_{imax} (the highest monthly groundwater level in the ith year) and H_{imin} (the lowest groundwater level in the ith year) for each year between 2009 and 2018. The values of f are computed by using Equation (7) with ΔH_i and P_i at each point in the historical period of 2009–2018. Finally, the mean f value of all the wells for the period of 2009–2018 is applied to each well with corresponding precipitation for the period of 2019–2100. P_i is annual precipitation of the ith year.

4. Results

4.1. Net Groundwater Recharge Rate by the Water Budget

The net groundwater recharge rate (*NGRR*) by the water budget was calculated for 264 standard sub-watersheds (Figure 1). For the determination of direct runoff (*DR*) by the SCS-CN method, the SCS-CN values were obtained using the land cover map and soil map of the NRW. The 21 land cover types were discovered in the study area, including forest 67.17% (broadleaf forest 30.58%, coniferous forest 25.39%, mixed forest 11.20%), residential areas 1.01%, industrial areas 1.32%, commercial areas 1.17%, cultural and sports areas 0.02%, transportation areas 0.95%, public facilities areas 0.01%, rice paddy fields 7.57%, farm fields 11.04%, greenhouse cultivation areas 2.39%, orchard 0.43%, other cultivation areas 0.21%, natural grasslands 1.51%, artificial grasslands 0.58%, inland wetlands 0.01%, coastal wetlands 0.01%, natural bare fields 1.05%, and artificial bare fields 0.44%. Drainage grades were divided into A (55.38%), B (20.58%), C (3.03%), and D (21.01%). The S values were calculated using the SCS-CN values, considering the area corresponding to each drainage class. Finally, annual *DR* values from 2009 to 2018 were calculated as 183.21 mm (21.18%) to 372.25 mm (25.32%) by using Equation (3). The average annual DR for the ten years is 308.28 mm, which is 24.14% of the average annual precipitation of 1277 mm (Table 2).

Table 2. Estimated *NGRR*, *DR*, ET_0, and *BF* of the Nakdong River watershed, with the ratios (%) to the annual average precipitation (*P*) during 2009–2018.

Year	*P*, mm	*DR*, mm (%)	ET_0, mm (%)	*BF*, mm (%)	*NGRR*, mm (%)
2009	1191	274.91 (23.08)	567.52 (47.65)	196.95 (16.54)	151.62 (12.73)
2010	1345	315.14 (23.43)	590.48 (43.90)	239.94 (17.84)	199.44 (14.83)
2011	1494	372.25 (24.92)	605.59 (40.53)	254.81 (17.06)	261.35 (17.49)
2012	1464	367.67 (25.11)	602.13 (41.13)	254.37 (17.38)	239.83 (16.38)
2013	1072	254.72 (23.76)	547.28 (51.05)	133.09 (12.42)	136.91 (12.77)
2014	1363	345.11 (25.32)	592.25 (43.45)	189.08 (13.87)	236.56 (17.36)
2015	1025	238.26 (23.24)	506.74 (49.44)	155.47 (15.17)	124.53 (12.15)
2016	1465	363.33 (24.80)	604.17 (41.24)	236.25 (16.13)	261.25 (17.83)
2017	865	183.21 (21.18)	452.27 (52.29)	122.64 (14.18)	106.88 (12.36)
2018	1482	368.18 (24.84)	604.89 (40.82)	240.64 (16.24)	268.29 (18.10)

Annual ET_0 was calculated using the Penman-Monteith method, by using the daily average temperature (14.5–15.7 °C), daily maximum temperature (32.5–37.3 °C), daily minimum temperature (−12.8–−6.0 °C), solar radiation (5.99–7.70 h), atmospheric pressure (100.7 to 100.9 kPa), daily average water vapor pressure (1.23 to 1.27 kPa), and daily average wind speed (3.1–3.4 m/s) from 2009 to 2018 (Table 3). Precipitation, temperature, and solar radiation amounts were obtained from 30 meteorological stations in the NRW. Due to the difficulty in directly measuring or calculating longwave radiation [40], 30% of the radiation energy was applied as the net radiation value according to the latitude effect provided by FAO. The annual *PET* from 2009 to 2018 ranged from 452.27 mm (40.53%) to 605.59 mm (52.29%), and the annual average ET for the ten years was 567.33 mm, which corresponded to 44.43% of the 10-year average annual precipitation (1277 mm).

Table 3. Factors for the estimation of ET_0.

	Average Daily Temp. (°C)	Daily Max. Temp. (°C)	Daily Min. Temp. (°C)	Solar Radiation (h)	Atmospheric Pressure (kPa)	Average Daily Water Vapor Pressure (kPa)	Average Daily Wind Speed (m/s)
2009	15.2	32.5	−7.6	5.99	100.7	1.25	3.4
2010	14.9	34.1	−8.1	6.25	100.8	1.26	3.3
2011	14.6	33	−12.8	6.37	100.9	1.23	3.3
2012	14.5	34.5	−9.9	7.12	100.7	1.27	3.3
2013	15.3	35	−10.7	7.7	100.7	1.24	3.4
2014	15.1	32.9	−6	6.89	100.8	1.25	3.2
2015	15.4	33.5	−7.8	7.04	100.8	1.24	3.1
2016	15.7	37.3	−10.2	6.99	100.8	1.26	3.1
2017	15.2	36.2	−7.7	7.55	100.8	1.25	3.2
2018	15.1	36.4	−9.9	7.41	100.8	1.24	3.2

The streamflow rates of peak and re-peak (2763–21,657 m^3/d), numbered from 1 to 10, with a time interval of 1.6–3.2 months between two peaks at the estuary of the NRW (Figure 3), were used for the calculation of the *BF* values. Using Equation (5), from 2009 to 2018, the *BF* values of the NRW were between 122.64 mm (12.42%) and 254.81 mm (17.84%), and the 10-year average *BF* value was 202.32 mm, which was 15.84% of the 10-year annual average precipitation (Table 2).

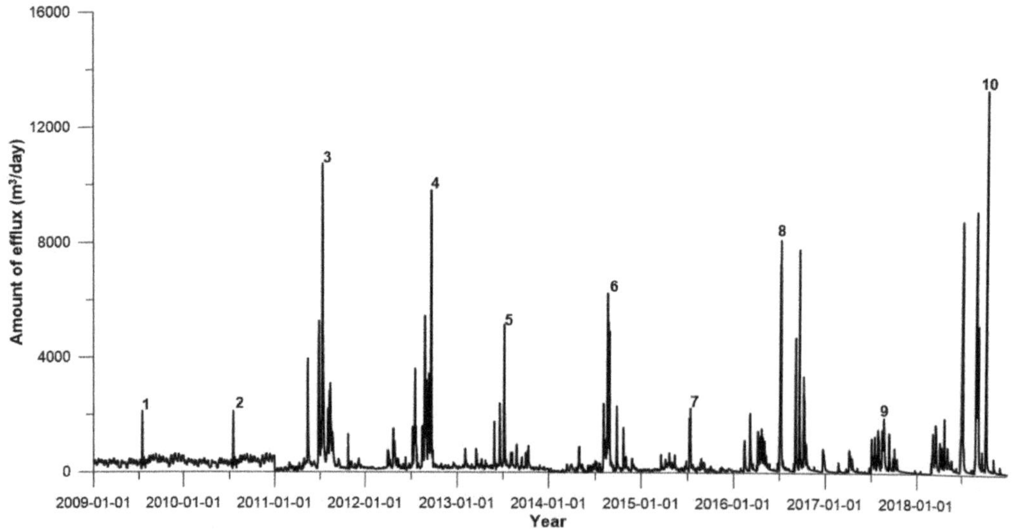

Figure 3. Streamflow in the Nakdong River watershed from 2009 until 2018. The numbers indicate the peaks for estimating baseflow.

From 2009 to 2018 in the NRW, the *NGRR* values were estimated to range from 12.15% to 18.10%. The average of the *NGRR* values in the NRW from 2009 to 2018 is 200 mm

(15.66%) when *DR* of 308 mm (24.12%), *PET* of 567 mm (44.40%), and *BF* of 202 mm (15.82%) are subtracted from the average *AP* (1277 mm).

4.2. Net Groundwater Recharge Rate under the Climate Change Scenarios

Before the RCP 4.5 and RCP 8.5 climate change scenarios, the recharge ratio (f), using the observed 2009–2018 data, was calculated by multiplying ΔH_i and specific yield (S_y) in Equation (7). The S_y value was determined to be 0.12 as a result of matching *NGRR* values with the *NGRR* values by the water budget method. In contrast, the mean S_y value of the NRW was reported to be 0.0134 by Moon et al. [41]. Next, the *NGRR* values from 2019 to 2100 were computed using the predicted precipitation data and f values in Equation (6), based on RCP 4.5 and RCP 8.5 climate change scenarios with a resolution of 1 km. Based on the RCP 4.5 and RCP 8.5 climate change scenarios, Figure 4 and Table 4 show mean AP (mm/yr) for the periods of 2009–2018, 2019–2039, 2040–2059, 2060–2079, and 2080–2100. The AP of the RCP 4.5 displays an increasing tendency from 2040 to 2079 and then a decreasing tendency from 2080 to 2100, while the AP of the RCP 8.5 displays an increasing tendency from 2019 until 2100.

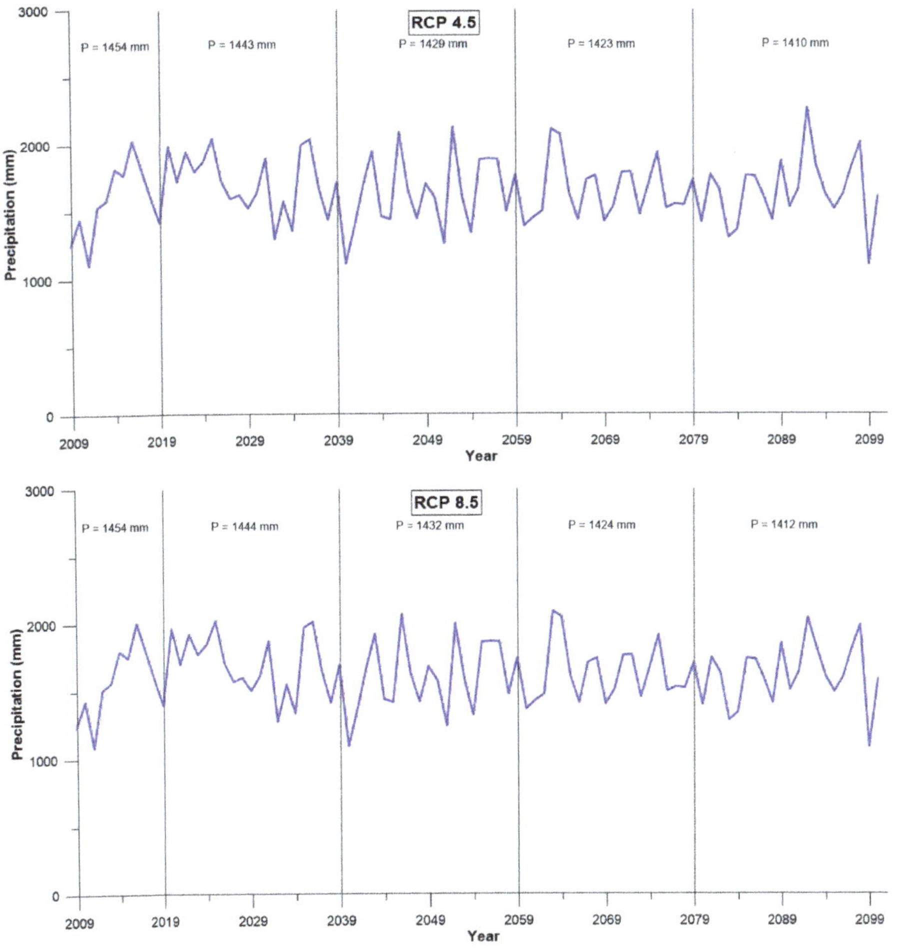

Figure 4. Predicted precipitation of 2009 to 2100 by the RCP 4.5 and 8.5 climate scenarios in the Nakdong River watershed.

Table 4. Estimated *NGRR* (%), annual precipitation (*P*, mm), recharge ratio (*f*), and specific yield (S_y) in the Nakdong River watershed for the periods of 2009–2018, 2019–2039, 2040–2059, 2060–2079, and 2080–2100 by the RCP 4.5 and RCP 8.5 climate change scenarios.

Period	RCP 4.5			RCP 8.5		
(Starting Year–Ending Year)	*P*	S_y	*NGRR*	*P*	S_y	*NGRR*
2009–2018	1455.1	0.12	13.73	1445.2	0.12	14.43
2019–2039	1312.6	0.12	7.74	1387.4	0.12	6.31
2040–2059	1493.1	0.12	7.81	1424.7	0.12	5.78
2060–2079	1538.5	0.12	7.30	1474.8	0.12	5.72
2080–2100	1414.4	0.12	7.05	1585.0	0.12	5.90

According to the climate change scenarios of RCP 4.5 and RCP 8.5, it is predicted that the decline rate of the groundwater level by RCP 4.5 and RCP 8.5 will be 0.023 m/yr and 0.024 m/yr, respectively, reaching a decline of 2.3 m and 2.4 m in 2100, respectively. In general, the decrease in groundwater level over a long period can be attributed to climate change and various man-made effects of groundwater pumping, irrigation, land-use change, etc. [4]. In this study, it is judged that the groundwater level decline with the annual precipitation decrease in the long term is caused by a decrease in infiltration rate into the subsurface as well as an increase in precipitation and an increase in direct runoff during the wet season relative to smaller precipitation in the dry season [20]. Moreover, a temperature increase of 3.5–4.0 °C until 2100 on the Korean peninsula [42] will increase the *ET* rate [43]. On the other side, urbanization can decrease *ET* rates by increasing runoff [44]. The *NGRR* values will be decreased to 7.05% based on the RCP 4.5 and to 5.90% based on the RCP 8.5 climate change scenarios in the period of 2080–2100 (Table 4), with a slightly decreasing tendency (Figure 5). The mean *NGRR* from 2009 to 2100 was estimated to be 8.73% in the RCP 4.5 scenario and 7.63% in the RCP 8.5 scenario, being similar to each other. Precipitation in summer for the periods of 2009–2019, 2020–2039, 2040–2059, 2060–2079, and 2080–2100 by using the RCP 4.5 and 8.5 climate change scenarios is intensified when going to 2100 (Table 5; Figure 6).

Table 5. Maximum, minimum, and mean precipitation (mm/mon) based on the RCP 4.5 and RCP 8.5 climate change scenarios for the periods of 2009–2018, 2019–2039, 2040–2059, 2060–2079, and 2080–2100.

Period	RCP 4.5					RCP 8.5				
(Starting Year–Ending Year)	Max		Min		Mean	Max		Min		Mean
2009–2018	520.2	August 2014	1.9	November 2017	107.6	520.2	August 2014	1.9	November 2017	107.6
2019–2039	502.3	July 2020	5.7	December 2020	109.9	580.8	July 2031	4.9	November 2035	115.9
2040–2059	702.7	July 2052	4.8	February 2042	124.4	657.0	July 2040	5.8	December 2058	118.7
2060–2079	824.1	July 2077	6.6	February 2079	128.2	549.1	July 2061	3.5	April 2072	122.9
2080–2100	980.3	July 2100	4.8	February 2098	117.9	790.2	July 2092	4.5	December 2089	132.1

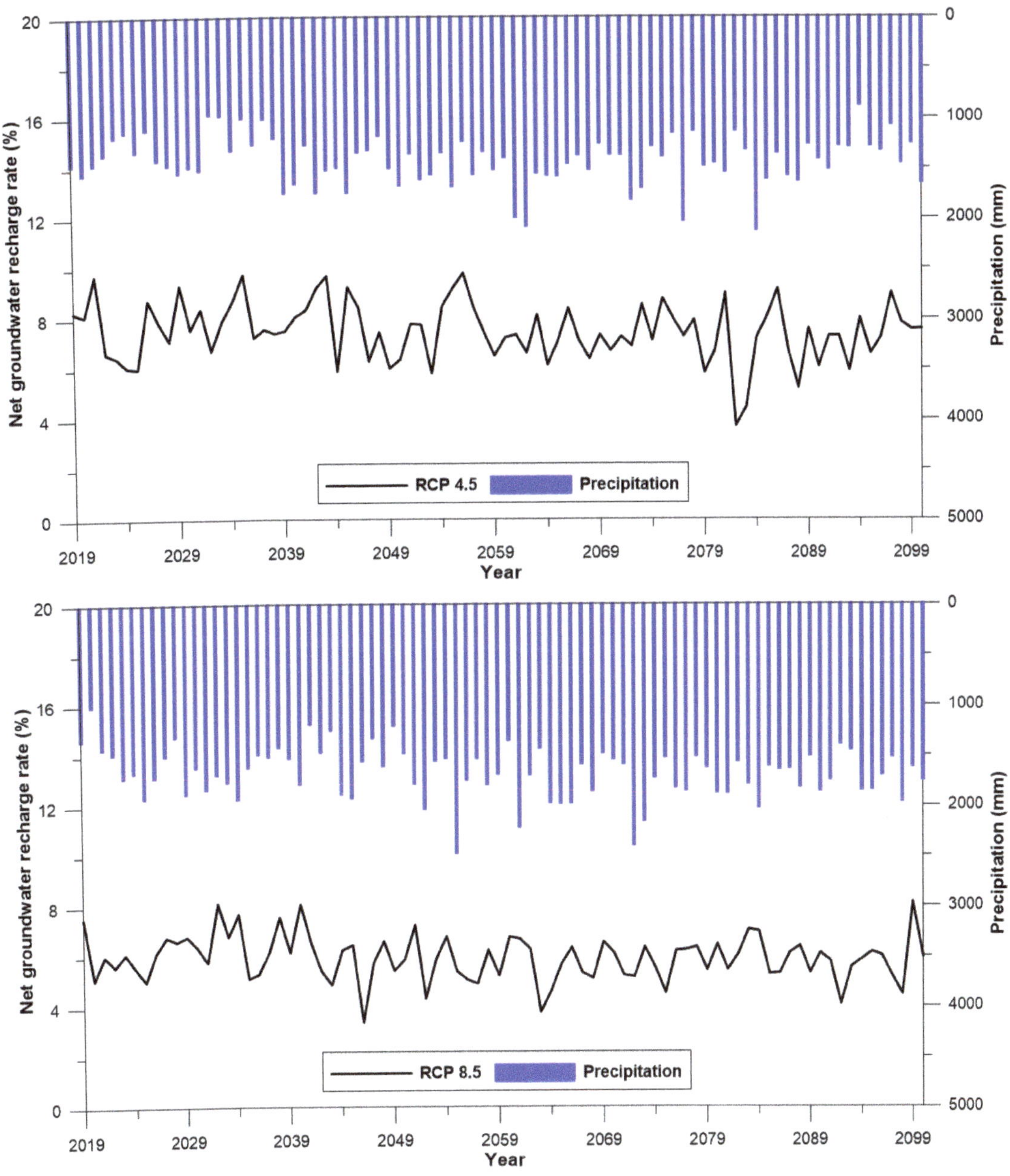

Figure 5. Predicted net groundwater recharge rate in the Nakdong River watershed for 2019–2100.

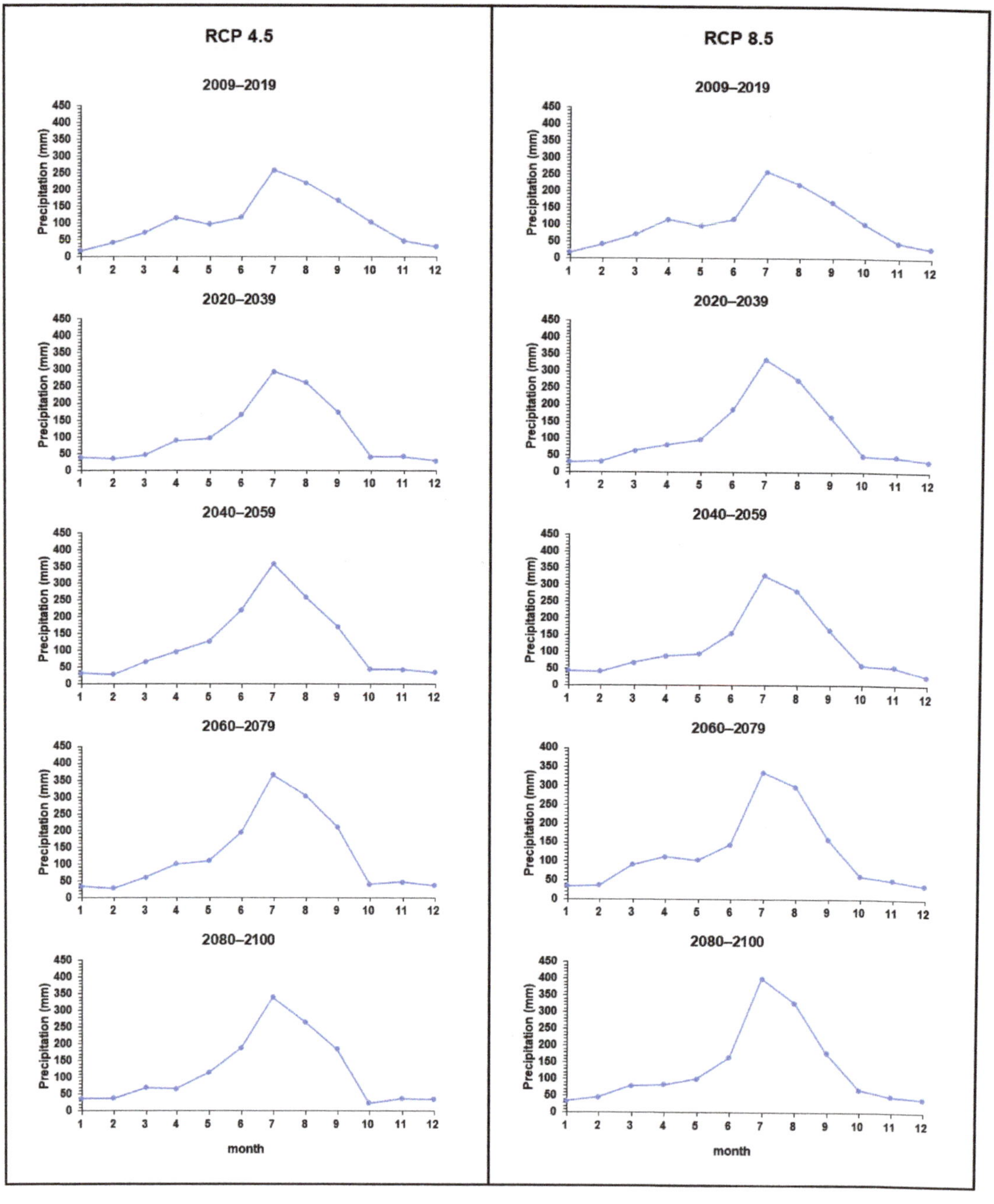

Figure 6. Seasonal trend of precipitation for the periods of 2009–2019, 2020–2039, 2040–2059, 2060–2079, and 2080–2100 using the RCP 4.5 and 8.5 climate change scenarios.

5. Discussion

Thus far, there has been no case of calculating the *NGRR* based on the long-term water budget method and simple climate change scenarios for the total area of the NRW, except in this study. Moon et al. [6] estimated the groundwater recharge rate in the NRW using the WTF method and obtained 6.1%, which is much smaller than the 15.54% of the *NGRR* according to the water budget method in this study. In this study, the observed streamflow rate was compared with the baseflow and calculated direct runoff by the SCS-CN method, showing good matching such that the observed streamflow rate of 323.59–622.42 mm/yr and the baseflow plus the calculated direct runoff of 321.89–617.99 mm/yr from 2009 to 2018 resulted in slightly greater values of the observed streamflow rate than those of the calculated direct runoff and baseflow (Figure 7).

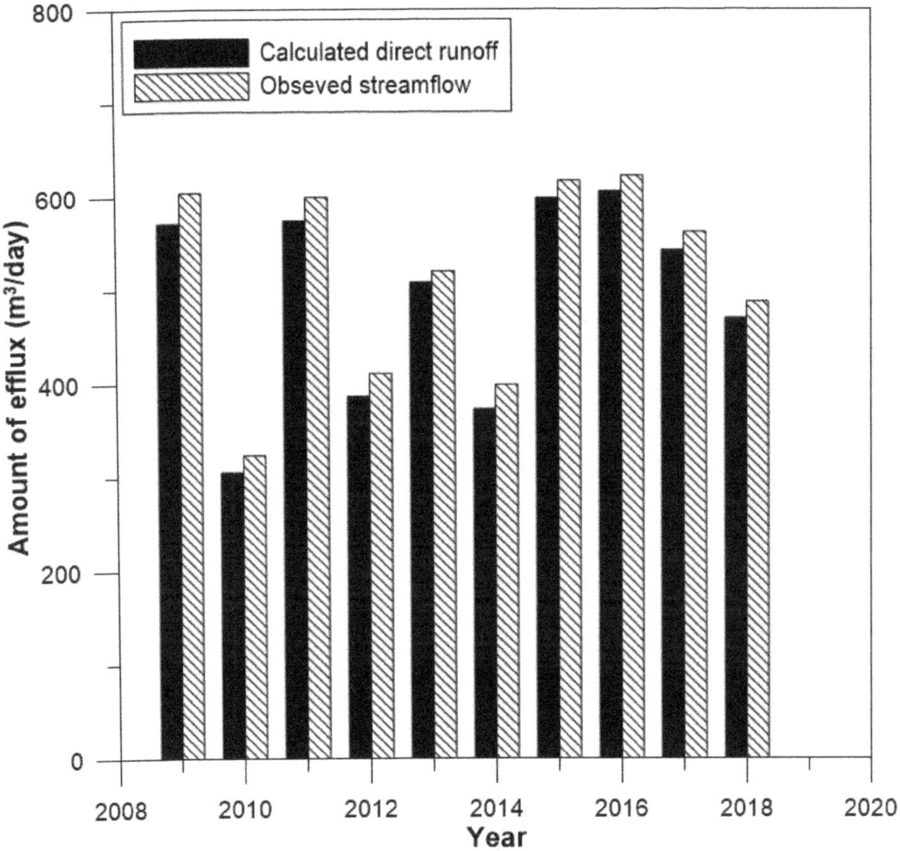

Figure 7. The observed streamflow and calculated direct runoff during 2009–2018.

The *NGRR* (200 mm, 15.66% of the AP) of the NRW in this study was compared with the average annual groundwater recharge rate (196 mm, 14.9% of the AP) in the Yanggok-ri subwatershed in the Geum River watershed from 2001 to 2018 [45]. Hence, the *NGRR* of the NRW is slightly greater than that of the Yanggok-ri subwatershed in the Geum River watershed, owing to the differences in the hydrogeological and climatic conditions. In addition, the NGRR of the NRW of 12.42–17.84% during 2009–2018 showed a much smaller variation than that of 3.6–28.2% in the Yanggok-ri subwatershed of the Geum River watershed during 2001–2018. In addition, the larger the watershed, the smaller the change in the *NGRR*. In this study, the mean of the *NGRR*s from 2009 to 2100 was 8.73% and 7.63%, respectively, using the RCP 4.5 and RCP 8.5 scenarios. In contrast, the *NGRR* of 26.7 mm to 432.5 mm (average 174.6 mm) in the Yanggok-ri subwatershed of the Geum

River watershed from 2019 to 2100 corresponds to 2.8–45.1% of the precipitation in 2100, based on the climate change scenario RCP 8.5 [38].

Looking at the change in the *NGRR* in the period of 2080–2100, the *NGRR* predicted by the climate change scenarios of the RCP 4.5 and RCP 8.5 is predicted to decrease from 13.73% to 7.05% and 14.43% to 5.90%, respectively. The *NGRRs* by the RCP 4.5 and RCP 8.5 scenarios are almost the same during 2009–2018, with a slight difference from the *NGRR* by the water budget analysis (Figure 8). Therefore, it is judged that the estimated values of *NGRR* in the climate change scenarios RCP 4.5 and RCP 8.5 are reasonable. Figure 9A shows the spatial distribution of the average *NGRRs* in the water budget analysis during 2009–2018. The *NGRR* showed high values in the eastern and southern regions but exhibited relatively low values in the inland areas, closely resembling precipitation distribution. In addition, the *NGRR* in the upstream area of the NRW was relatively smaller than that in the downstream area. On the other side, in Figure 9B, the *NGRR* distribution in 2080–2100 using the RCP 4.5 scenario shows overall higher values in the eastern region than those in the inland and southern areas, displaying a different spatial distribution of the *NGRR* from that using the water budget analysis in the period of 2009–2018 and showing smaller *NGRRs* than those by the water budget analysis.

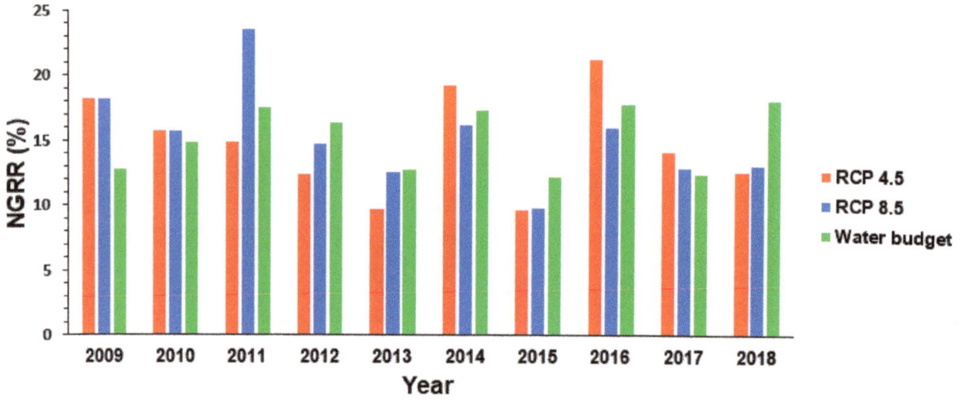

Figure 8. The NGRR (%) relative to the corresponding year from the water budget analysis and RCP 4.5 and 8.5 scenarios from 2009 until 2018.

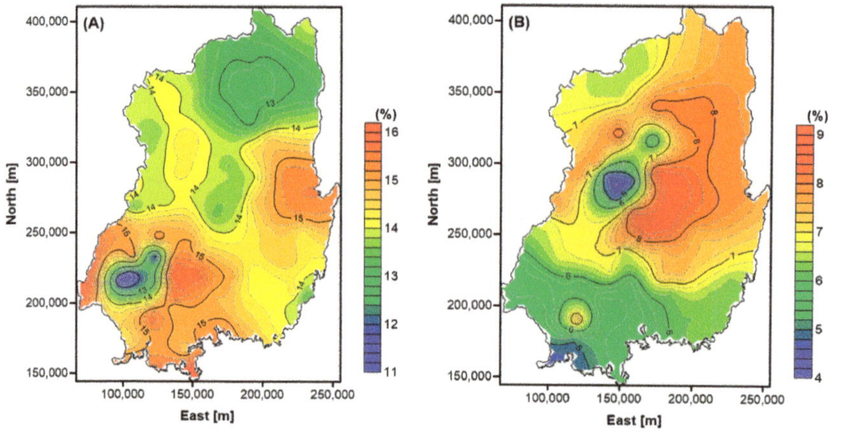

Figure 9. Distribution map of the average NGRRs in the Nakdong River watershed by (**A**) the water budget analysis of the period 2009–2018 and (**B**) the RCP 4.5 scenario in 2080–2100.

The estimated *NGRR* by the water budget analysis in this study possesses uncertainty due to using *PET* instead of *AET*, which can be estimated by considering soil water content [36,46,47]. A future study needs to reveal a more accurate *NGRR* in the NRW which was not possible due to insufficient data on soil water content. For the climate change scenarios, there is uncertainty about the specific yield (S_y), which is an average value (0.12) of the NRW since S_y is spatially different depending on geological formations in the NRW. Finally, the water budget analysis and the climate change scenarios also imply uncertainty due to not considering the effect of land cover change.

6. Conclusions

In this study, the change in the *NGRR* of the NRW was estimated using the water budget method and the climate change scenarios of RCP 4.5 and RCP 8.5. By using the water budget method and precipitation data from 2009 to 2018, direct runoff, evapotranspiration, and baseflow were estimated prior to determining the *NGRR*. The *DR* in the NRW by the SCS-CN method was calculated to be 183.21 mm (21.18%) to 372.25 mm (25.32%). The *PET* obtained by the Penman–Monteith method ranged from 452.27 mm (40.53%) to 605.59 mm (52.29%). The baseflow in the NRW using the baseflow separation method was 122.64 mm (12.42%) to 254.81 mm (17.84%).

The mean annual *NGRR* was determined to be 200 mm (15.66%) by subtracting the mean annual *DR* of 308 mm (24.12%), the mean annual *PET* of 567 mm (44.40%), and the mean annual *BF* of 202 mm (15.82%) from the mean annual precipitation of 1277 mm in the NRW for ten years (2009–2018), with an increasing trend in the *NGRR* over the ten years. The regional distribution of the *NGRR* showed high values in the eastern and southern regions but exhibited relatively low values in the inland areas, closely resembling precipitation distribution. In addition, the *NGRR* in the upstream area of the NRW was relatively smaller than that in the downstream area.

Under a decreased tendency of precipitation and the constant recharge ratio of the NRW, the mean *NGRR* for the entire NRW over the long term until 2100 is predicted to decrease from 13.73 to 7.05% based on the RCP 4.5 and from 14.43 to 5.90% based on the RCP 8.5. The *NGRR* distribution in 2080–2100 using the RCP 4.5 scenario shows overall high values in the eastern region and lower values in the inland and southern areas, similar to the spatial distribution of the *NGRR*, which displays a different spatial distribution of the *NGRR* from that using the water budget analysis in the period of 2009–2018, with smaller *NGRR* values than those by the water budget analysis.

The accuracy of the prediction of the *NGRR* according to the climate change scenarios RCP 4.5 and RCP 8.5 was verified by comparing the 2009 to 2018 water budget analysis and climate change scenarios. However, the robustness of the estimation of the *NGRR* is constrained by the uncertainty of the *PET*, the specific yield, and the land cover change. Therefore, it is judged that the estimated *NGRR* based on the climate change scenarios RCP 4.5 and RCP 8.5 was reasonable. The *NGRR* of the NRW presented in this study will be useful in the long-term development and management of groundwater resources in watersheds worldwide as well as in the NRW. Future research will focus on factors such as change in runoff, baseflow, and evapotranspiration over a long period.

Author Contributions: Conceptualization, S.-M.Y., H.-T.J. and S.-Y.H.; methodology, H.-T.J., J.-Y.C., S.-Y.H. and J.K.; software, S.-M.Y., H.-T.J. and J.-Y.C.; validation J.K. and S.-Y.H.; writing, original draft preparation, S.-M.Y., H.-T.J. and S.-Y.H.; writing, review and editing, J.-Y.C. and S.-Y.H.; visualization, S.-M.Y.; supervision, S.-Y.H.; project administration, S.-Y.H.; funding acquisition, S.-Y.H. All authors have read and agreed to the published version of the manuscript.

Funding: This study was financially supported by the National Research Foundation of Korea (NRF) grant funded by the Ministry of Science and ICT (2020R1A2B5B02002198).

Data Availability Statement: Data is unavailable due to privacy.

Conflicts of Interest: The authors declare no conflict of interest.

References

1. Sophocleous, M.A. Combining the soilwater balance and water-level fluctuation methods to estimate natural groundwater recharge: Practical aspects. *J. Hydrol.* **1991**, *124*, 229–241. [CrossRef]
2. Fetter, C.W. *Applied Hydrology*, 4th ed.; Prentice Hall: Englewood Cliffs, NJ, USA, 2003; 676p.
3. Schwartz, F.W.; Zhang, H. *Fundamentals of Groundwater*; John Wiley & Sons, Inc.: Hoboken, NJ, USA, 2003; 583p.
4. Healy, R.W.; Cook, P.G. Using groundwater levels to estimate recharge. *Hydrogeol. J.* **2002**, *10*, 91–109. [CrossRef]
5. Park, E.; Parker, J.C. A simple model for water table fluctuations in response to precipitation. *J. Hydrol.* **2008**, *356*, 344–349. [CrossRef]
6. Moon, S.-K.; Woo, N.C.; Lee, K.S. Statistical analysis of hydrographs and water-table fluctuation to estimate groundwater recharge. *J. Hydrol.* **2004**, *292*, 198–209. [CrossRef]
7. Choi, I.H.; Woo, N.C.; Kim, S.-J.; Moon, S.-K.; Kim, J. Estimation of the groundwater recharge rate during a rainy season at a headwater catchment in Gwangneung, Korea. *Korean J. Agric. For. Meteorol.* **2007**, *9*, 75–87. [CrossRef]
8. Chung, I.-M.; Kim, N.W.; Lee, J. Estimation of groundwater recharge by considering runoff process and groundwater level variation in watershed. *J. Soil Groundw. Environ.* **2007**, *12*, 19–32.
9. Kim, N.W.; Chung, I.-M.; Won, Y.S. An integrated surface water-groundwater modeling by using fully combined SWAT-MODFLOW model. *J. Korean Soc. Civil Eng.* **2006**, *26*, 481–488.
10. Noh, D.N.; Park, H.J.; Cheong, J.-Y.; Hamm, S.-Y. Groundwater recharge analysis and comparison using hybrid water-table fluctuation method and groundwater modeling: A case of Gangcheon basin in Yeoju City. *J. Geol. Soc. Korea* **2018**, *54*, 169–181. [CrossRef]
11. Bae, S.K. Applicability of NRCS-CN method for the estimation of groundwater recharge. *J. Korean Soc. Civ. Eng. B* **2005**, *25*, 425–430.
12. Lee, S.H.; Bae, S.K. Groundwater balance in urban area. *J. Environ. Sci.* **2011**, *20*, 1553–1560.
13. Viji, R.; Prasanna, P.R.; Ilangovan, R. Gis Based SCS-CN Method for estimating runoff In Kundahpalam watershed, Nilgries district, Tamilnadu. *Earth Sci. Res. J.* **2015**, *19*, 59–64.
14. Ling, L.; Yusop, Z.; Yap, W.-S.; Tan, W.L.; Chow, M.F.; Ling, J.L. A calibrated, watershed-specific SCS-CN Method: Application to Wangjiaqiao watershed in the three gorges area, China. *Water* **2019**, *12*, 60. [CrossRef]
15. Shi, W.; Wang, N. An improved SCS-CN method incorporating slope, soil moisture, and storm duration factors for runoff prediction. *Water* **2020**, *12*, 1335. [CrossRef]
16. Dripps, W.R.; Bradbury, K.R. A simple daily soil-water balance model for estimating the spatial and temporal distribution of groundwater recharge in temperate humid areas. *Hydrogeol. J.* **2007**, *15*, 433–444. [CrossRef]
17. Holman, I.P. Climate change impacts on groundwater recharge- uncertainty, shortcomings, and the way forward? *Hydrogeol. J.* **2005**, *14*, 637–647. [CrossRef]
18. Schindler, U.; Steidl, J.; Müller, L.; Eulenstein, F.; Thiere, J. Drought risk to agricultural land in Northeast and Central Germany. *J. Plant Nutr. Soil Sci.* **2007**, *170*, 357–362. [CrossRef]
19. Patricia, A.J.S.; Wang, L.; Koike, T. Modeling the hydrologic responses of the Pampanga River basin, Philippines: A quantitative approach for identifying droughts. *Water Resour. Res.* **2011**, *47*, 1–21.
20. Jang, S.; Hamm, S.-Y.; Yoon, H.S.; Kim, G.B.; Park, J.H.; Kim, M.S. Predicting long-term change of groundwater level with regional climate model in South Korea. *Geosci. J.* **2015**, *19*, 503–513. [CrossRef]
21. Lee, J.H.; Jeon, S.W.; Lee, M.J.; Hong, H.J. *Coupled Model Development between Groundwater Recharge Rate Quantity and Climate Change in River Watershed*; Korea Environment Institute: Sejong City, Republic of Korea, 2009; 09-06-52(4); 142p.
22. Lee, J.H.; Jeon, S.W.; Lee, M.J.; Hong, H.J. *Coupled Model Development between Groundwater Recharge Rate Quantity and Climate Change in River Watershed II*; Korea Environment Institute: Sejong City, Republic of Korea, 2010; 10-02-97(5); 149p.
23. Lee, M.J.; Lee, J.H. Coupled model development between groundwater recharge quantity and climate change using GIS, Green Growth Research Report. *J. Korean Assoc. Geogr. Inf. Stud.* **2011**, *14*, 36–51. [CrossRef]
24. Lee, J.Y.; Lee, K.K. A Comparative study on characteristics of waterlevel responses to rainfall in the two aquifer systems. *J. Soil Groundw. Environ.* **2002**, *7*, 3–14.
25. Kim, T.W.; Hamm, S.-Y.; Cheong, J.Y.; Ryu, S.M.; Lee, J.H.; Son, K.T.; Kim, N.H. Time series and groundwater recharge analyses using water fluctuation data in mountain Geumjeong area. *J. Environ. Sci.* **2008**, *17*, 257–267.
26. Jeon, H.-T. A Study of Long-Term Net Recharge Rate of Groundwater in Korea. Ph.D. Dissertation, Pusan National University, Busan, Republic of Korea, 2021; 167p.
27. Morel-Seytoux, H.J.; Verdin, J.P. *Extension of the SCS Rainfall-Runoff Methodology for Ungauged Watersheds*; Colorado State University: Fort Collins, CO, USA, 1981.
28. Aller, L.; Bennett, T.; Lehr, J.H.; Petty, R.H.; Hackett, G. *DRASTIC: A Standardized System for Evaluating Ground Water Pollution Potential Using Hydrogeologic Settings*; Prepared for U.S. EPA Office of Research and Development, Ada, OK; National Water Well Association: Dublin, OH, USA, 1987.
29. Thornthwaite, C.W. Report of the Committee on Transpiration and Evaporation, 1943–1944. *Eos Trans. Am. Geophys. Union* **1944**, *25*, 683–693.
30. Penman, H.L. Natural evapotranspiration from open water, bare soil, and grass. *Proc. R. Soc. Lond. Ser. A* **1948**, *193*, 120–145.
31. Blaney, H.F.; Criddle, W.D. *Determining Water Requirements in Irrigated Areas from Climatological and Irrigation Data*; USDA (SCS) TP-96; US Department of Agriculture: Washington, DC, USA, 1950.

32. Turc, L. Evaluation des besoins en eau d'irrigation. evapotranspiration potentielle, formulation simplifie et mise a jour. *Ann. Agron* **1963**, *12*, 13–49.
33. Doorenbos, J.; Pruitt, W.O.; Aboukhaled, A. Crop water requirements. In *Food and Agriculture Organization Irrigation and Drainage Paper 24, Technical Report*; FAO: Rome, Italy, 1977; 156p.
34. Allen, R.G.; Pereira, L.S.; Raes, D.; Smith, M. *Crop Evapo-Transpiration-Guidelines for Computing Crop Water Requirements*; FAO Irrigation and Drainage Paper 56; FAO: Rome, Italy, 1998; Volume 300, p. D05109.
35. Liu, Z. Estimating land evapotranspiration from potential evapotranspiration constrained by soil water at daily scale. *Sci. Total Environ.* **2022**, *834*, 155327. [CrossRef]
36. Yang, S.I.; Kang, D.H.; Kwon, B.H.; Kim, B.W. Influence of land use and meteorological factors for evapotranspiration estimation in the coastal urban area. *J. Environ. Sci. Int.* **2010**, *19*, 295–304. [CrossRef]
37. Meyboom, P. Estimating groundwater recharge from stream hydrographs. *J. Geophys. Res.* **1961**, *66*, 1203–1214. [CrossRef]
38. Rorabaugh, M.I. Estimating changes in bank storage and ground-water contribution to stream flow. *Inter. Assoc. Sci. Hydrol.* **1964**, *63*, 432–441.
39. IPCC. Climate Change 2007: The Physical Science Basis. In *The Fourth Assessment Report of the Intergovernmental Panel on Climate Change*; IPCC: Geneva, Switzerland, 2007.
40. Choi, D.-H.; Lee, B.-Y.; OH, H.-Y. Long and short wave radiation and correlation analysis between downtown and suburban area(II). *J. Korean Sol. Energy Soc.* **2013**, *33*, 101–110. [CrossRef]
41. Moon, S.-K.; Woo, N.C.; Lee, K.S. A study on the relation between types and recharges of groundwater: Analysis on national groundwater monitoring network data. *J. Soil Groundw. Environ.* **2002**, *7*, 45–59.
42. Keem, M.; Ko, I.; Kim, S. An analysis of the effect of climate change on Nakdong River flow condition using CGCM's future climate information. *J. Korean Soc. Water Qual.* **2009**, *25*, 863–871.
43. Snyder, R.; Moratiel, R.; Song, Z.; Swelam, A.; Jomaa, I.; Shapland, T. Evapotranspiration response to climate change. *Acta Hortic.* **2011**, *922*, 91–98. [CrossRef]
44. Mazrooei, A.; Reitz, M.; Wang, D.; Sankarasubramanian, A. Urbanization impacts on evapotranspiration across various spatio-temporal scales. *Earth's Future* **2021**, *9*, e2021EF002045. [CrossRef]
45. Ha, K.; Park, C.; Kim, S.; Shin, E.; Lee, E. Groundwater recharge evaluation on Yangok-ri area of Hongseong using a distributed hydrologic model (VELAS). *Econ. Environ. Geol.* **2021**, *54*, 161–176. [CrossRef]
46. Filgueiras, R.; Almeida, T.S.; Mantovani, E.C.; Dias, S.H.B.; Fernandes-Filho, E.I.; Cunha, F.F.; Venancio, L.P. Soil water content and actual evapotranspiration predictions using regression algorithms and remote sensing data. *Agric. Water Manag.* **2020**, *241*, 106346. [CrossRef]
47. Senay, G.B.; Kagone, S.; Velpuri, N.M. Operational global actual evapotranspiration: Development, evaluation, and dissemination. *Sensors* **2020**, *20*, 1915. [CrossRef] [PubMed]

water

MDPI

Article

Analysis of Water Supply Capacity of a Sand Dam

Il-Moon Chung, Jeongwoo Lee *, Min Gyu Kim, Il-Hwan Kim and Bisrat Ayalew Yifru

Department of Water Resources and River Research, Korea Institute of Civil Engineering and Building Technology, Goyang 10223, Korea
* Correspondence: ljw2961@kict.re.kr; Tel.: +82-31-910-0529

Abstract: A sand dam is an old technology used to trap sand materials and store water in dry riverbeds. Besides the volume of sand stored, the water storage and supply capacity of the sand dam depends principally on material properties including hydraulic conductivity (K) and the specific yield (S_y). In this study, the water supply capacity of a sand dam, applied in South Korea with a modified scheme, was analyzed. Imported sand materials were placed in three layers, and perforated drain pipes were installed to abstract the water to a collection tank. Using a Modular Finite-Difference Groundwater Flow (MODFLOW) model, the drain pipe discharge related to aquifer properties and supply capacity of a sand dam was evaluated. Using the model, the productivity of the sand storage with several cases of inflow conditions was assessed. For the period from 9 March 2020 to 16 May 2022, the results of calculation of water supply for the installation and operation of new sand dam are as follows. Regarding the actual inflow condition, compared to the average water supply of the existing water intake source, the sand dam installation increased water supply by 61.6%, and the sand dam installation showed an effect of improving the water supply by 36.0% for the minimum water supply. The calculated water supply for the condition where the inflow was reduced by 50% showed an effect of improving water supply by over 70% on average due to the installation of the sand dam. The findings would be a benchmark for future expansions of the scheme in related places, and management and maintenance of the system.

Keywords: water supply; sand dam; MODFLOW; drought; reliability; Korea

Citation: Chung, I.-M.; Lee, J.; Kim, M.G.; Kim, I.-H.; Yifru, B.A. Analysis of Water Supply Capacity of a Sand Dam. *Water* **2022**, *14*, 3039. https://doi.org/10.3390/w14193039

Academic Editors: Glen R. Walker and João Filipe Santos

Received: 31 August 2022
Accepted: 21 September 2022
Published: 27 September 2022

1. Introduction

In Korea, drought occurs every 2–3 years, depending on the region, and becomes extreme every seven years [1]. Major cities have reliable water sources and supply networks. However, surprisingly, the water supply network of South Korea has several blind spots, and, as a result, it is a prime concern for residents in the uppermost part of watersheds. Compared to national and regionals, small-scale supply sources have low water quality and are vulnerable to drought. For example, during a river water freezing-related drought in Gangwon Province in 2016, 3817 people (44 villages) from 1443 households (44 villages) received water supply (285 times, 1845 m^3) using a water truck. Such areas need locally adapted water supply sources and networks. In particular, in the upstream end of watersheds, usually mountainous areas, strategies and technologies need to be developed for alternative supply sources such as conjunctive use of groundwater and river water.

This study focuses on the application of sand dam technology in South Korea. A sand dam is usually built from concrete or beam on the impervious bedrock of a valley or river to accumulate sand materials and store water for later use [2–5]. Storing water in intermittent riverbeds has several advantages, including small-scale irrigation, water supply, and ecological use [6]. Even though sand dam technology has a long history, well-documented data on the worldwide application locations are unavailable [7–9]. Recently, the technology has attracted researchers' attention, and several publications have been produced in the last three years [7]. Besides the working principles and success stories, the factors affecting the sand dam performance, future directions, and many more aspects are reported [3,4,7,10]. However, most of the studies, particularly the success stories, are in arid and semi-arid regions, e.g., [7,11–13].

Evaporation losses and water pollution are less in sand storage compared to surface water storage dams, and the construction is also simple [6,14–16]. A sand dam is also considered as an artificial groundwater recharge method and has a groundwater level increasing effect [5]. In particular, when the connection between rivers and aquifers is strong, it can be used as an artificial recharge facility [17]. Although it is challenging to quantify inflow into the aquifer through the seepage flow from the riverbed, depending on the groundwater level in the area, continuous water exchange between the artificial aquifer formed by the sand storage and the natural aquifer in the region is an inevitable phenomenon [10,11,17,18]. This phenomenon will continue year-round, and, as a result, groundwater levels within the sand dam will rise [5,17]. Water stored in the sand increases resistance to drought even during dry seasons and is being promoted as a technology to cope with water shortages in long-term climate change [3,4,17,19,20].

Besides the general hydrological processes in sand dam areas [6,11,17] and failures related to sand material quality and water pollution [21], the detailed storage capacity based on layers of the sand is unexplored. The effect of depth-wise properties of accumulated material on the disposable amount of water is an untouched topic. Sand dam material selection lacks clear guidelines [7,22], even though both water quality and storage capacity rely on the sand particle property. In this study, the sensitivity of the principal sand material properties was evaluated using a Modular Finite-Difference Groundwater Flow (MODFLOW) model [23]. The principal goal of this study was to evaluate the water supply capacity of a sand dam constructed in Mullo-ri in Chuncheon, Gangwon-do, South Korea. The water supply in the area is from an intermittent river using a small reservoir to which the river water is supplied through seepage. The sand dam is built to enlarge the water supply capacity of this reservoir. Considering the existing water supply capacity, the amount of water that can be withdrawn from the sand dam was assessed. This technology is applied in Korea with a modified scheme for the first time. The results and conclusions of this study would be valuable inputs to adopt the technology in other parts of similar water supply blind spot mountainous areas of the country, and other parts of the world with related circumstances or climate conditions.

The rest of this paper is organized as follows: Section 2 outlines the study area, a brief description of sand dam construction, the model used in this study, and description and modeling processes. In Section 3, typical MODFLOW model simulation results are discussed. Section 4 summarizes the methods and main findings, including conclusions.

2. Materials and Methods

2.1. Description of the Study Area

The sand dam is built around Mulo-ri, Buksan-myeon, Chuncheon-si (Figure 1). The area is one of the water supply blind spots in the country. The regional supply network does not cover this area due to several factors, including the topography and population size. The area is small and could be supplied by water trucks during the 2016 drought. Currently, the local water supply capacity is 30 m^3/day. The water source is mainly from the intermittent river in the mountain, from which the water continuously flows through seepage. A small intake is used to divert water from the river. This bypass-type sand dam was built following the water intake at the side of the main waterway. The catchment area covers around 1.6 km^2. The topography of the catchment area varies from 298 to 950 m above sea level.

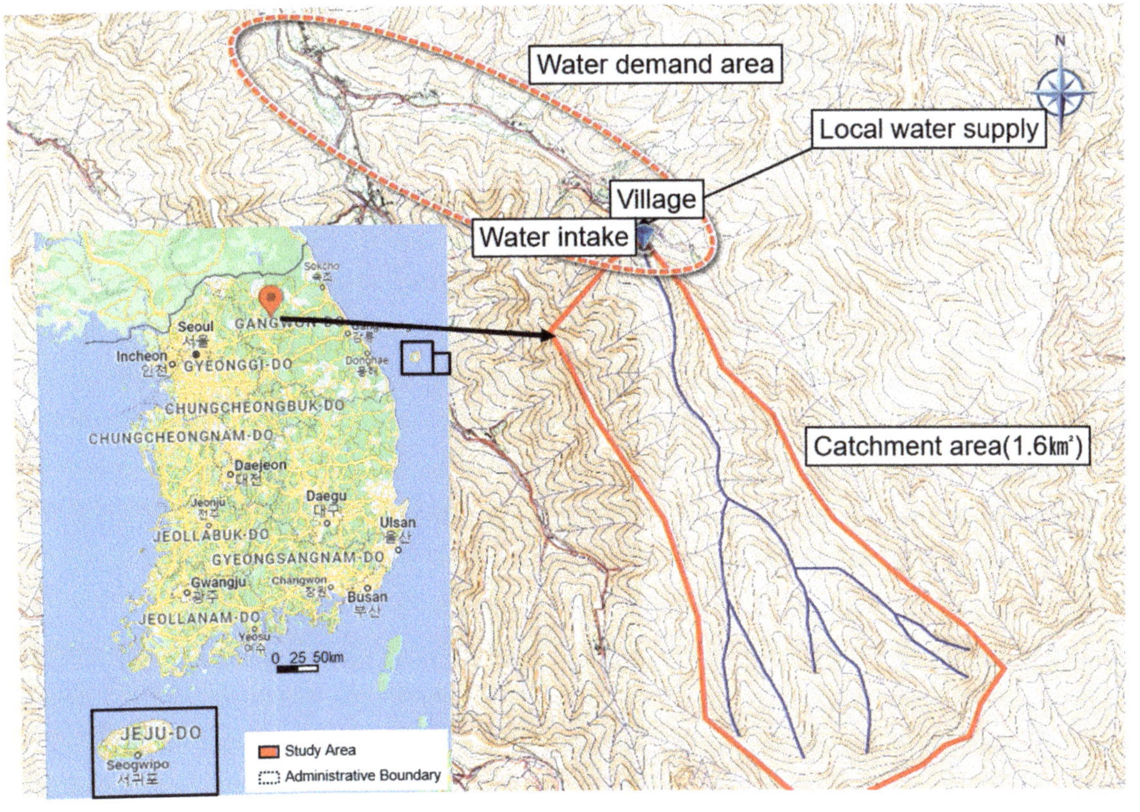

Figure 1. Sand dam location, the major streams, and the catchment area in the study area. The blue lines are the main waterways in the gorge, the bold red is the catchment area of the sand dam, and the dotted red circle shows the water demand area.

2.2. Sand Dam Construction

The dam is built from concrete, with a width of 5 m and a length of 10 m. Commonly, a sand dam is built on the main course of the river in stages and makes the water flow velocity decrease so that the transported sediment would have time to settle in the upstream area [6,24]. The accumulated sand materials during the first few flood events create an artificial aquifer that has the functionality to protect water from evaporation and pollution and stores for a longer time. Water abstraction methods also include scooping holes or infiltration galleries. Pumping wells are also used to withdraw water from the riverbed level in the sand dam area. Since the objective is to improve the existing small reservoir used to supply water for the residents in the region, the dam is built at the side of the main waterway connected to the existing small intake reservoir (pond), which is filled by seepage flow. Protection walls on both sides are installed and the storage area is filled with imported materials. The location of the sand dam is presented in Figure 2.

The right side of the sand dam is in contact with the valley wall. Large natural stone piling was carried out to secure the stability of the structure during flood periods. A maintenance pipe was installed in the downstream part of the sand dam so that portable cleaning technology, such as air surging, could be applied. The water entering the sand storage passes through a gabion to prevent the inflow of debris. The dam has a weir to discharge overflow and the flow rejoins the main downstream waterway.

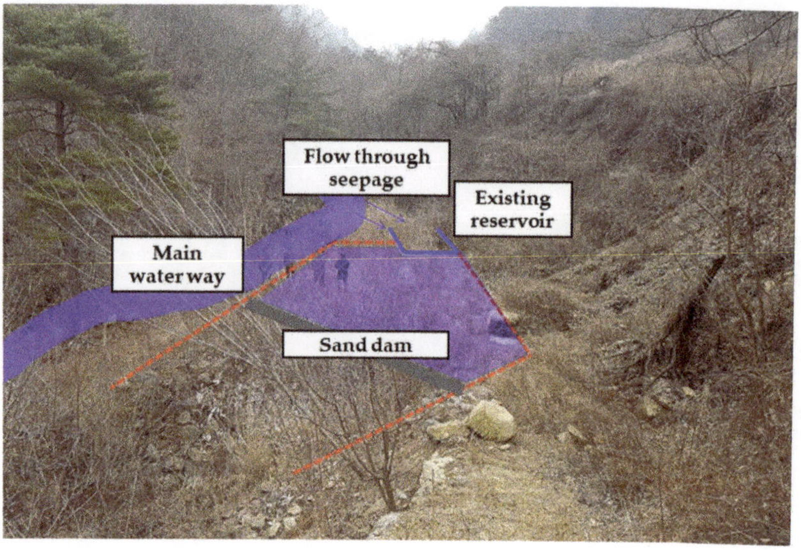

Figure 2. Photos showing the sand dam location.

As shown in Figure 3, before the sand dam construction, the foundation of the existing water intake was investigated. The geology shows a thick rock layer distributed evenly. Therefore, to build the foundation, a rock cutting was performed. In the dam body, a collection well and filter pack screen were installed. The sand storage area was filled with variable size layers of sand and rock. With the first layer, at the bottom, 2–5 mm size of sand about 0.8 m thick was laid, and at the top of this layer, a similar thickness of about 1–2 mm size filter sand was filled, then covered with 25 mm gravel on the top to prevent the loss of filler material.

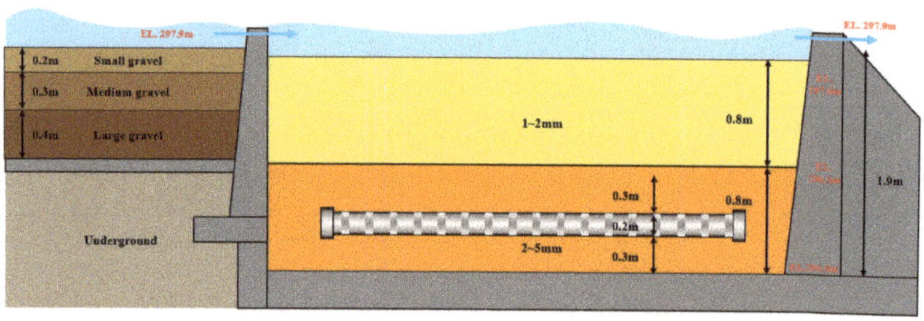

Figure 3. Cross section of existing intake reservoir (pond) and sand dam.

Figure 4 shows the cross-section of the sand dam environment. The water stored in the sand is first collected in an intake reservoir width of 1.5 m and a length of 3 m and then directly supplied to the water tank through a steel pipe. The water tank is on the downstream side of the sand dam, and the water could be piped with the help of gravity, but to make the water supply reliable, solar energy was also installed. Plan view of the sand dam and detail showing the connection between the old reservoir (pond) and the sand dam are shown in Figure 5. Figure 6 is schematic diagram of water supply system for the study area.

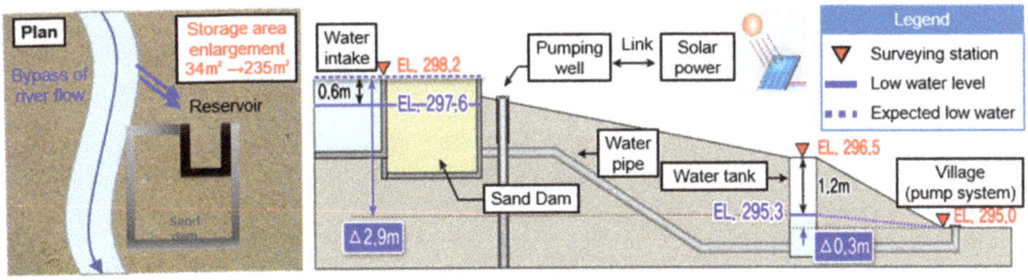

Figure 4. Schematics of the water supply from the sand dam.

Figure 5. Plan view of the sand dam (**a**) and detail showing the connection between the old reservoir (pond) and the sand dam (**b**).

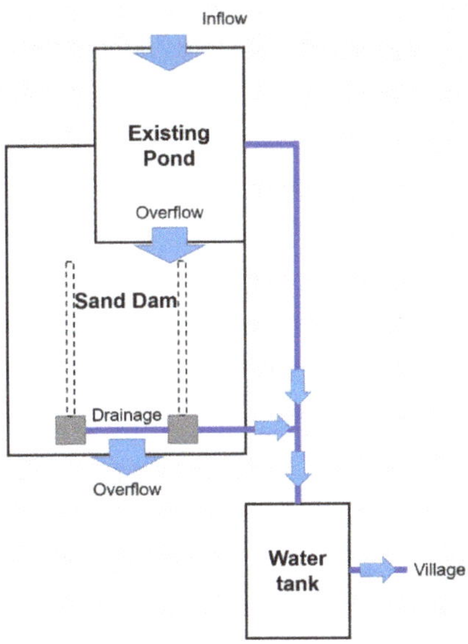

Figure 6. Schematic diagram of water supply system for the study area.

2.3. Groundwater Model

In this study, MODFLOW model [23] was used to analyze the water supply from the sand dam. Using the model, the amount of water that could be withdrawn from the sand dam via a perforated drain pipe was evaluated. MODFLOW is a three-dimensional groundwater flow model, a widely used groundwater analysis program, distributed by the United States Geological Survey (USGS). MODFLOW is a physically based finite-difference model that solves the three-dimensional groundwater flow through porous earth material. This model combines the continuity equation and Darcy's law to simulate the groundwater flow in a saturated zone and uses the three-dimensional groundwater equation, Equation (1), as the governing equation. It is widely used to compute the groundwater level (head), particularly in response to recharge and groundwater abstraction,

$$\frac{\partial}{\partial x}\left(K_{xx}\frac{\partial h}{\partial x}\right)+\frac{\partial}{\partial y}\left(K_{yy}\frac{\partial h}{\partial y}\right)+\frac{\partial}{\partial z}\left(K_{zz}\frac{\partial h}{\partial z}\right)\pm W=S_s\frac{\partial h}{\partial t} \quad (1)$$

where K_{xx}, K_{yy}, and K_{zz} are principal components of the hydraulic conductivity tensor in x, y, and z spatial directions, W is the source or sink, S_s is specific storage (1/L), h is the hydraulic head (L), and t is time [23]. MODFLOW has several packages and independent modules which make up the main program. Packages can be selected according to the user's needs. The basic (BAS) package forms the overall structure for model building, including the number of grids, the number of layers, boundary and initial conditions, simulation period, and time interval.

The BCF package (Block-Centered Flow package) takes the hydraulic parameters of the aquifer, such as the layer bottom elevation, hydraulic conductivity, and storage coefficient. In MODFLOW, groundwater abstraction is represented using the well package. The package enables inclusion of the well location, pumping rate and schedule, and other details. The natural groundwater recharge can be incorporated using the RCH package (ReCHarge package). Groundwater removal using drainage pipes is also commonly simulated using the DRN package (DRaiN package). The RIV package (RIVer package) simulates the phenomenon of supplying or draining to the surface water body based on the water head difference between rivers, lakes, etc., and groundwater.

2.4. Description of Modeling

In order to simulate the flow of groundwater in the sand dam and the amount of discharge through perforated drainage pipes in the 3rd layer and weir in 1st layer, it was divided into a grid of 1 m × 1 m in the horizontal and vertical directions as shown in Figure 7. Red dot means observation well.

The open channel flow region was discretized as a virtual medium layer. In addition, the virtual medium layer was set as an unconfined aquifer, and the lower two layers were set as a confined/unconfined convertible layer. The thickness of the 2nd and 3rd layers, which are the sand filling layers, is 0.8 m, respectively, and for the hydraulic conductivity values of each layer, 172 m/d and 280 m/d, the values calculated through the indoor hydraulic test, are inputted, respectively. As shown in Figure 3, the specific yield of the 2nd (1–2 mm) and 3rd (2–5 mm) layers were entered as 0.45 and 0.39, respectively, considering the porosity, and the storage coefficient was entered as 1/100 of this value considering the sand layer. For the first floor, which is a hypothetical layer with 100% porosity, a storage coefficient value of 1.0 was input, and a hydraulic conductivity value of 50,000 m/d, which was calibrated through observed and simulated groundwater levels, was input. Only the inside of the sand dam was set as an active cell, and the outer part and the bottom of the 3rd floor were given a no-flux boundary condition. As the upper boundary condition of the simulation area, the amount overflowing from the existing pond and flowing into the sand dam was applied only to the active cell according to the location of the calculated water level. As internal boundary conditions, the weir of the 1st layer and the perforated drain pipe of the 3rd layer were simulated using a drain package. The drainage conductance was calculated through calibration of the model. The initial water level and head were entered as the first-floor bottom level for all floors.

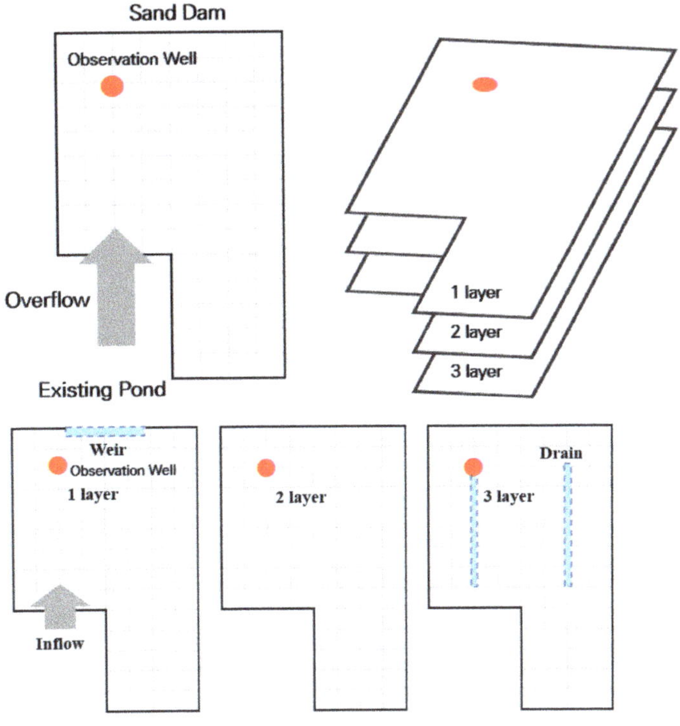

Figure 7. Discretization of modeling domain.

MODFLOW's drain package was used to simulate the two perforated drain pipes (drain pipes) installed on the third layer of the sand dam. In the drain package, it is assumed that the amount of discharge from the aquifer (Q_d) is proportional to the difference between the head of the aquifer (h) and the elevation of the drainage pipe (d). For the elevation of the drain pipes, the central elevation of the drain pipe is usually used, and if the calculated head is smaller than this, it is treated as no drainage. The equivalent conductivity coefficient C_d (drainage conductivity coefficient) of a drain pipe varies depending on the hydraulic conductivity and head distribution around the pipe, the thickness and hydraulic conductivity of the filter media, and the filtration capacity of the drain pipe, such as the number of screen openings. In this study, it is assumed that the value of the drain conductivity per unit length is the same as the value of the hydraulic conductivity of the lowest model layer.

Excess water from the sand dam flows downstream through the installed weir. This natural weir discharge was conceptualized in MODFLOW using a drain package. Weir discharge was made in proportion to the difference between the calculated water level and weir height. The equivalent conductivity coefficient C_W (weir conductivity coefficient) of the weir was set as the same as the hydraulic conductivity value of the uppermost layer.

3. Results and Discussion

3.1. Model Calibration and Validation

After the sand dam was completed in October 2021, filled with sand material, a water quality check, structural stability check such as for leaks, water level observation hole installation, valve and flow meter installation, a drainage performance check of the perforated drain pipe and maintenance pipe, etc., were carried out. The actual operation started from May 2022. Thus, calibration of the model was performed for the operation period, during which the water level measurement was made in the sand dam.

Figure 8 shows the comparison between the observed data converted to the water level by the water pressure sensor installed in the filler material on the upstream side of the sand dam, about 2 m from the downstream end of the sand dam, and the calculated water level in the cell. Until 15 May 2022, the valve of the drain pipe was shut off and discharge was allowed only through the weir. Therefore, for the valve shut-off period, repeated simulations were performed until the observed and simulated water levels were well-matched while changing the hydraulic conductivity of the first floor, which is the virtual layer, and the drainage coefficient of the weir. The value of the drainage conductance of the drain pipe was also calculated through repeated observation and simulated water level fitting process from 16 May to 7 July.

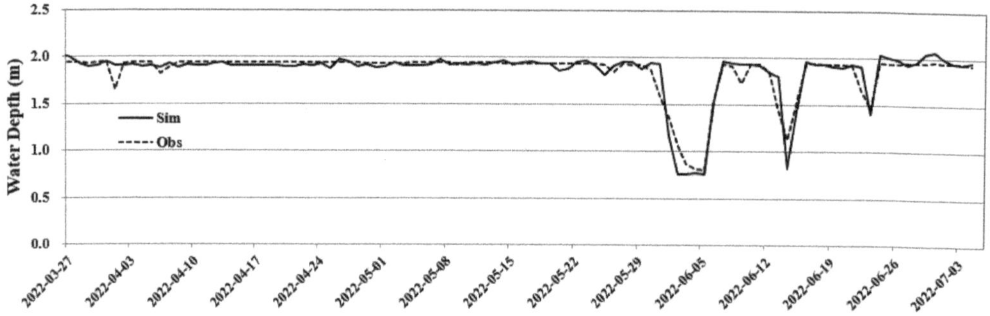

Figure 8. Comparison of simulated and observed water depth in sand dam.

For the hydraulic conductivity values of the second and third layer fillers, the values calculated through the hydraulic tests were used as they were without calibration. As a result of calibration, the hydraulic conductivity of the first layer was 50,000 m/d, and the drainage conductance per unit length of the perforated drain pipe on the third floor was calculated to be 7 m/d. If the hydraulic conductivity or drainage conductance is smaller than this, the simulated water level is formed above the observed water level.

As shown in Figure 8, it can be seen that the simulated water level fits the observed level well. During the valve shut-off period, the water level is maintained at a value slightly above about 1.9 m, and the water level fluctuates according to the amount of inflow into the sand dam and the amount of discharge from the drain pipe after the valve was opened on 16 May 2022.

The discharge measured by the flow meter at the time of the weir discharge of the sand dam was about 200 m^3/day, and the simulated discharge by formula was also calculated as an amount close to this, indicating that the calibration of the model was well-performed. In the future, we plan to measure the flow rate continuously and accurately by installing an automatic flow meter in the pipeline connected to the water tank.

3.2. The Effect of Water Supply Enhancement

The flow rate was measured several times in the pipeline connected to the drain pipe of the existing water intake source using an ultrasonic flow meter, and the pipe drainage conductance value of the drain pipe was calculated assuming that the flow rate and water level have a linear relationship when the pressure is assumed zero inside the perforated pipe. The value of the drainage conductance per unit length of the existing water intake was calculated to be about 12 m/d, which is slightly larger than that of the sand dam (7 m/d). The perforated drain pipe of the sand dam is a steel screen and has a large opening rate, but it has a filter pack screen, so it is small compared to the existing water intake source.

For the existing water intake source, the discharge amount of the continuous drain pipe was calculated by multiplying the water level automatically observed every hour from 9 March 2020 by the drainage conductance. Figure 9 shows the discharge amount of the existing water intake for the period immediately before the operation of the new sand dam from the time when the existing water intake was observed. It can be seen that the discharge amount decreased due to the decrease in water level in the winter in January and February, and it ranges from a minimum of 117.5 m^3/day to a maximum of 210.4 m^3/day. During the analysis period, an average of 179.9 m^3/day of water is supplied through the existing water intake source.

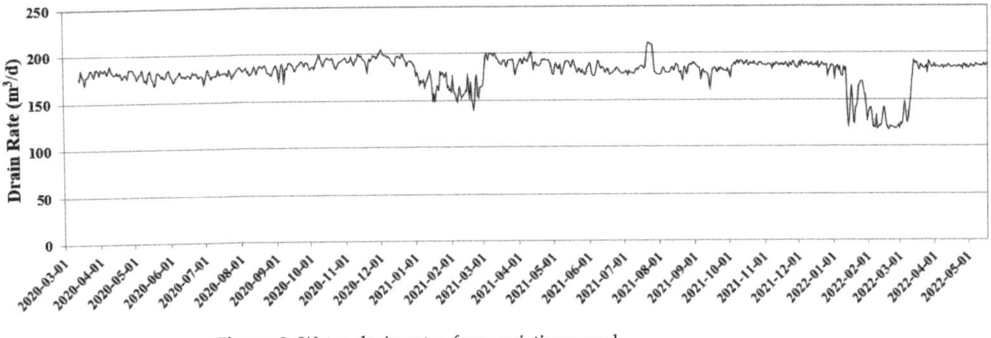

Figure 9. Water drain rates from existing pond.

For the same period, numerical modeling was performed on the conditions assuming that the sand dam was installed and operated to simulate the amount of discharge through the perforated drain pipe in the sand dam. Figure 10 shows the simulated results of sand dam discharge as a unit, and indicates the amount of water supplied in addition to the discharge from the existing water intake source. It overflows from the existing water intake and there is no inflow into the sand dam, so there are cases where the discharge is 0 m^3/day, and it occurs up to 251 m^3/day during the flood season. The average discharge amount during the simulation period was calculated to be 110.8 m^3/day, and it was analyzed that the water supply could be increased by 61.6% of the average discharge amount of the existing water intake source. Therefore, it is possible to bring about the effect of improving the water supply in the research area by artificially incubating the surplus water overflowing from the existing water intake source as a sand dam and using it again.

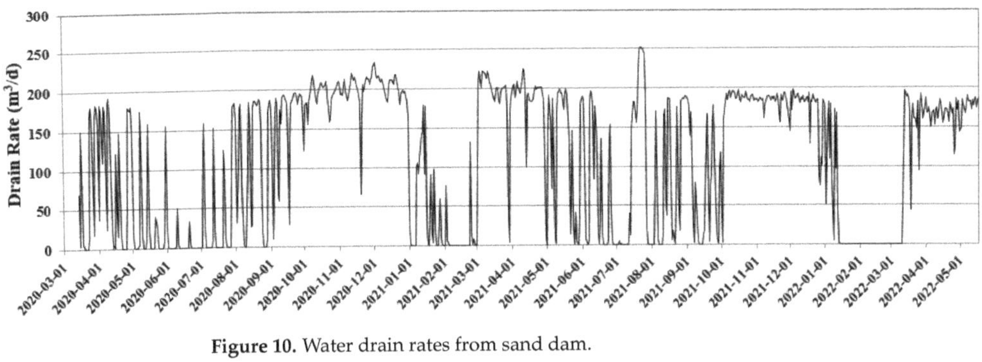

Figure 10. Water drain rates from sand dam.

Figure 11 shows the amount of discharge through the drain pipe when the existing water intake source and the new sand dam are operated together. In other words, the result in Figure 10 is the same as the sum of results from Figures 8 and 9. It shows a minimum discharge of 117.5 m^3/day and a maximum of 461.6 m^3/day. The increased availability of water can be used for purposes other than drinking water and can also contribute to meeting the growing demand in the region.

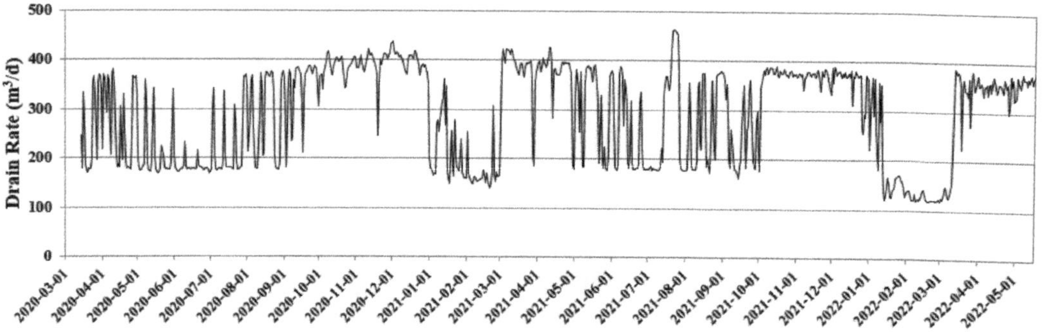

Figure 11. Water drain rates from existing pond and sand dam.

The study area suffered from water shortages such as reliance on water trucks due to limited water supply from existing water sources due to an extreme drought in 2016. Therefore, it is necessary to evaluate the water supply capacity when the inflow to the water intake source is very small. For this, the existing water intake source and sand flow were assumed to be reduced by 50% (Case1) and 90% (Case2) compared to the actual inflow during the simulation period from 9 March 2020 to 16 May 2022. The amount of discharge through the drain pipe of the dam was simulated.

Figure 12 shows the simulated drain pipe discharge when the inflow is reduced by 50%. On average, 139.0 m^3/day from the existing water intake source, 97.8 m^3/day from the new sand dam, and 236.8 m^3/day of water discharged from the two facilities, were analyzed to be possible. This is 80% of the average water supply of 290.7 m^3/day for the actual inflow condition.

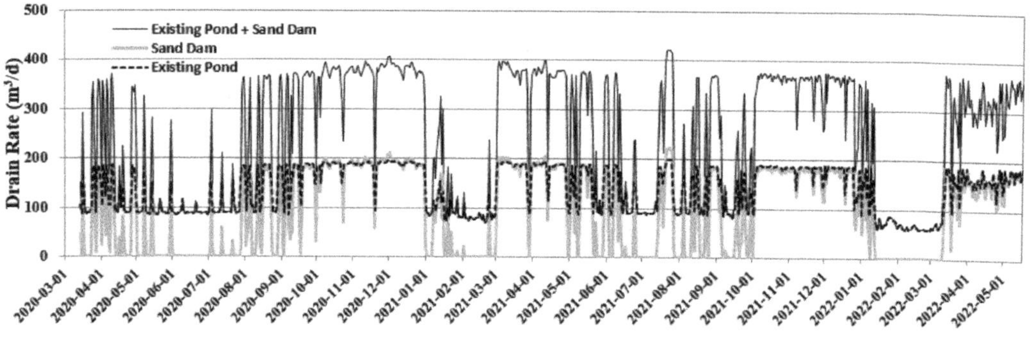

Figure 12. Water drain rates for 50% reduction in inflow.

Figure 13 shows the simulated results when the inflow is greatly reduced, by 90% compared to the actual. On average, it was analyzed that 87.2 m^3/day from the existing water intake source, 42.8 m^3/day from the new sand dam, and 133.0 m^3/day from the two facilities, would be possible. This is about 50% of the actual inflow condition. However, the minimum supply amount was 11.8 m^3/day in February 2022, which was evaluated as not meeting the minimum domestic water demand of 15.0 m^3/day in this area.

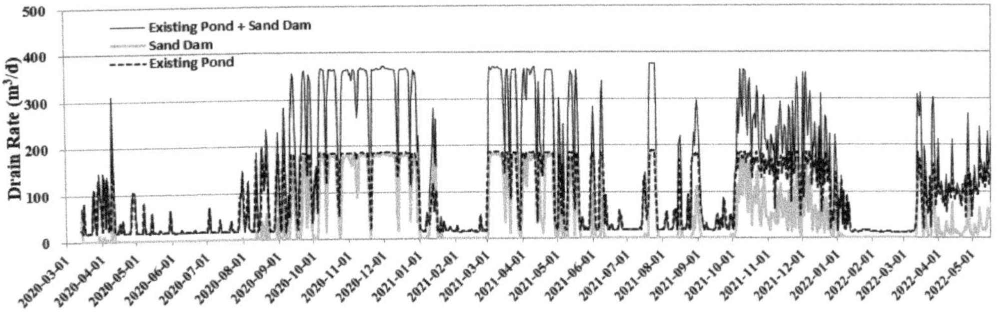

Figure 13. Water drain rates for 90% reduction in inflow.

4. Summary and Conclusions

In this study, the water supply capacity of the dam in comparison with the current demand in the study area was assessed. The sand dam was built on an intermittent mountainous stream in South Korea to improve the water supply of an existing small reservoir. Even though the technology has been used widely to harvest water from dry riverbeds in arid and semi-arid regions, the application in South Korea is new. In the study area, the sand dam and the water supply scheme were modified from the usual scheme used in the rural part of arid and semi-arid countries. Commonly, a sand dam is built in stages so that the natural river flow in the rainy season slows and lets the water drop the transported sand. After a few flood events, the upstream area fills with sand materials and creates an artificial aquifer. The water supply and storage capacity of the sand dam depends on the volume and property of the accumulated sand. Thus, understanding of hydraulic properties of the sand material is vital. However, most studies on the characterization of supply and storage capacity of sand storage rely on simplified calculations using dimensions of the artificial aquifer and a few numbers on material properties, i.e., hydraulic conductivity and specific yield. In this study, the sand dam site is mountainous, and the gorge is big. Moreover, the plan was to improve the existing small reservoir, which takes and stores water from the main gorge. Due to these site-specific circumstances, the sand dam was built following the existing reservoir. After the construction of the dam wall, three layers of different sizes and property sand materials were filled. To abstract water from the sand storage, perforated drain pipes were installed in the bottommost layer.

Using the MODFLOW model, the flow of groundwater in the sand dam and the amount of discharge through perforated drainage pipes in the third layer and weir in first layer has been analyzed. The open channel flow region was discretized as a virtual medium layer. In addition, the virtual medium layer was set as an unconfined aquifer, and the lower two layers were set as a confined/unconfined convertible layer. For the first floor, which is a hypothetical layer with 100% porosity, a storage coefficient value of 1.0 was input, and a hydraulic conductivity value of 50,000 m/d, which was calibrated through observed and simulated groundwater levels, was input. Only the inside of the sand dam was set as an active cell, and the outer part and the bottom of the third floor were given a no-flux boundary condition. As the upper boundary condition of the simulation area, the amount overflowing from the existing pond and flowing into the sand dam was applied only to the active cell according to the location of the calculated water level. As internal boundary conditions, the weir of the first layer and the perforated drain pipe of the third layer were simulated using a drain package. The drainage conductance was calculated through calibration of the model. The initial water level and head were entered as the first-floor bottom level for all floors.

For the period from 9 March 2020 to 16 May 2022, the results of calculation of water supply for the installation and operation of existing water sources, new sand dams, and underground water collection tanks are as follows. Regarding the actual inflow condition, compared to the average water supply of the existing water intake source, the sand dam installation increased water supply by 61.6%, and the sand dam installation showed improvement by 36% for the minimum water supply. The calculated water supply for the condition where the inflow was reduced by 50% showed an effect of improving water supply by over 70% on average, due to the installation of the sand dam.

Author Contributions: Conceptualization, J.L. and I.-M.C.; methodology, J.L.; validation, J.L.; investigation, M.G.K. and I.-H.K.; resources, M.G.K. and I.-H.K.; data curation, J.L.; writing—original draft preparation, J.L. and I.-M.C.; writing—review and editing, B.A.Y. and I.-M.C.; visualization, J.L.; supervision, I.-M.C.; project administration, I.-M.C.; funding acquisition, I.-M.C. All authors have read and agreed to the published version of the manuscript.

Funding: This study was funded by the Korea Ministry of Environment (MOE) as a Demand Responsive Water Supply Service Program (Grant No. 146515).

Conflicts of Interest: The authors declare no conflict of interest.

References

1. Min, S.K.; Kwon, W.T.; Park, E.H.; Choi, Y. Spatial and Temporal Comparisons of Droughts over Korea with East Asia. *Int. J. Climatol.* **2003**, *23*, 223–233. [CrossRef]
2. Kamel, A.H.; Almawla, A.S.; Hatem, A. Experimental Investigation about the Effect of Sand Storage Dams on Water Quality. *ZANCO J. Pure Appl. Sci.* **2016**, *28*, 485–491.
3. Lasage, R.; Aerts, J.C.J.H.; Verburg, P.H.; Sileshi, A.S. The Role of Small Scale Sand Dams in Securing Water Supply under Climate Change in Ethiopia. *Mitig. Adapt. Strateg. Glob. Chang.* **2015**, *20*, 317–339. [CrossRef]
4. Aerts, J.; Lasage, R.; Beets, W.; de Moel, H.; Mutiso, G.; Mutiso, S.; de Vries, A. Robustness of Sand Storage Dams under Climate Change. *Vadose Zone J.* **2007**, *6*, 572–580. [CrossRef]
5. Yifru, B.; Kim, M.-G.; Woo Chang, S.; Lee, J.; Chung, I.-M. Numerical Modeling of the Effect of Sand Dam on Groundwater Flow. *J. Eng. Geol.* **2018**, *28*, 529–540. [CrossRef]
6. Ertsen, M.; Hut, R. Two Waterfalls Do Not Hear Each Other. Sand-Storage Dams, Science and Sustainable Development in Kenya. *Phys. Chem. Earth Parts A/B/C* **2009**, *34*, 14–22. [CrossRef]
7. Yifru, B.A.; Kim, M.-G.; Lee, J.-W.; Kim, I.-H.; Chang, S.-W.; Chung, I.-M. Water Storage in Dry Riverbeds of Arid and Semi-Arid Regions: Overview, Challenges, and Prospects of Sand Dam Technology. *Sustainability* **2021**, *13*, 5905. [CrossRef]
8. Standen, K.; Costa, L.R.D.; Monteiro, J.-P. In-Channel Managed Aquifer Recharge: A Review of Current Development Worldwide and Future Potential in Europe. *Water* **2020**, *12*, 3099. [CrossRef]
9. Ritchie, H.; Eisma, J.A.; Parker, A. Sand Dams as a Potential Solution to Rural Water Security in Drylands: Existing Research and Future Opportunities. *Front. Water* **2021**, *3*, 31. [CrossRef]
10. Hut, R.; Ertsen, M.; Joeman, N.; Vergeer, N.; Winsemius, H.; van de Giesen, N. Effects of Sand Storage Dams on Groundwater Levels with Examples from Kenya. *Phys. Chem. Earth Parts A/B/C* **2008**, *33*, 56–66. [CrossRef]
11. Quilis, R.O.; Hoogmoed, M.; Ertsen, M.; Foppen, J.W.; Hut, R.; Vries, A. de Measuring and Modeling Hydrological Processes of Sand-Storage Dams on Different Spatial Scales. *Phys. Chem. Earth Parts A/B/C* **2009**, *34*, 289–298. [CrossRef]
12. Jadhav, M.V.; Shaikh, E.; Gite, E.; Yadav, E. Sand Dam Reservoir—Need of Semi Arid Areas. *Int. J. Eng. Res. Appl.* **2012**, *2*, 1690–1694.
13. de Trincheria, J.; Wibbing, J.; Leal Filho, W.; Otterpohl, R. Practical Recommendations to Prevent, Restore and Rehabilitate Silted-up Sand Storage Dams in Arid and Semi-Arid Areas. In Proceedings of the 7th RWSN Forum "Water for Everyone" 2016, Abidjan, Côte d'Ivoire, 29 November–2 December 2016.
14. Baurne, G. "Trap-Dams": Artificial Subsurface Storage of Water. *Water Int.* **1984**, *9*, 2–9. [CrossRef]
15. Stern, J.H.; Stern, A. Water Harvesting through Sand Dams. Technical Note. 2011. Available online: https://www.ircwash.org/sites/default/files/Stern-2011-Water-Techn.note.pdf (accessed on 2 May 2022).
16. Villani, L.; Castelli, G.; Hagos, E.Y.; Bresci, E. Water Productivity Analysis of Sand Dams Irrigation Farming in Northern Ethiopia. *J. Agric. Environ. Int. Dev.* **2018**, *112*, 139–160. [CrossRef]
17. Quinn, R.; Rushton, K.; Parker, A. An Examination of the Hydrological System of a Sand Dam during the Dry Season Leading to Water Balances. *J. Hydrol. X* **2019**, *4*, 100035. [CrossRef]
18. Hoogmoed, M. Analyses of Impacts of a Sand Storage Dam on Groundwater Flow and Storage: Groundwater Flow Modeling in Kitui District, Kenya. Master's Thesis, VU University Amsterdam, Amsterdam, The Netherlands, 2007.
19. Lasage, R.; Aerts, J.; Mutiso, G.-C.M.; de Vries, A. Potential for Community Based Adaptation to Droughts: Sand Dams in Kitui, Kenya. *Phys. Chem. Earth Parts A/B/C* **2008**, *33*, 67–73. [CrossRef]
20. Ryan, C.; Elsner, P. The Potential for Sand Dams to Increase the Adaptive Capacity of East African Drylands to Climate Change. *Reg. Environ. Chang.* **2016**, *16*, 2087–2096. [CrossRef]

21. Graber Neufeld, D.; Muendo, B.; Muli, J.; Kanyari, J. Coliform Bacteria and Salt Content as Drinking Water Challenges at Sand Dams in Kenya. *J. Water Health* **2020**, *18*, 602–612. [CrossRef] [PubMed]
22. Ngugi, K.N.K.; Gichaba, C.M.M.; Kathumo, V.M.V.; Ertsen, M.W.M. Back to the Drawing Board: Assessing Siting Guidelines for Sand Dams in Kenya. *Sustain. Water Resour. Manag.* **2020**, *6*, 58. [CrossRef]
23. McDonald, M.G.; Harbaugh, A.W. *A Modular Three-Dimensional Finite-Difference Ground-Water Flow Model*; U.S. Geological Survey: Reston, VI, USA, 1988.
24. Sivils, B.E.; Brock, J.H. Sand Dams as a Feasible Water Development for Arid Regions. *J. Range Manag.* **1981**, *34*, 238. [CrossRef]

water

MDPI

Article

Conjunctive Operation of Sand Dam and Groundwater Well for Reliable Water Supply during Drought Conditions

Bo Ram Kim, **Sang-Il Lee** * and **Su Min Yu**

Department of Civil and Environmental Engineering, Dongguk University, Seoul 04620, Korea; skysee0325@naver.com (B.R.K.); gpm0691@naver.com (S.M.Y.)
* Correspondence: islee@dongguk.edu

Abstract: Some mountainous regions without water service facilities are among the areas most vulnerable to drought. In these locations, it is particularly essential to establish practical alternatives to cope with the increase in the intensity and duration of droughts caused by climate change. This study proposes a methodology for the conjunctive use of a sand dam and groundwater well under various drought conditions. The method has been applied to a small mountainous area in South Korea. Owing to the scarcity of observational data, it is crucial to properly estimate the hydrological components necessary for judging the feasibility and reliability of conjunctive operations. The step-by-step procedures for performing the tasks are presented in this study. For the inflow of the sand dam, which is a portion of the basin runoff, two different approaches were employed and compared: the Kajiyama formula and a simple two-parameter monthly water balance model (TPM). Water budget analysis allowed for the determination of whether the current and increased water demand could be met under various drought conditions. Preliminary analysis revealed that a sand dam alone could not reliably meet the demand for 10-year or more severe drought conditions. Various water allocation scenarios between surface water (i.e., sand dam) and groundwater were tested. Conjunctive use of a sand dam and groundwater well turned out to increase the reliability of the water supply. As water demand increases and droughts become more severe, the role of groundwater increases. With appropriate resource allocation, 100% water supply reliability could be achieved, even for one year-lasting 50-year drought. We demonstrated how a system could be flexibly operated to meet the target demands monthly, given the system reliability level.

Keywords: conjunctive use; sand dam; groundwater well; water budget analysis; runoff estimation; water balance model; Kajiyama formula

Citation: Kim, B.R.; Lee, S.-I.; Yu, S.M. Conjunctive Operation of Sand Dam and Groundwater Well for Reliable Water Supply during Drought Conditions. *Water* **2022**, *14*, 2249. https://doi.org/10.3390/w14142249

Academic Editors: Sang Yong Chung, Gyoo-Bum Kim and Venkatramanan Senapathi

Received: 13 June 2022
Accepted: 15 July 2022
Published: 17 July 2022

1. Introduction

Climate change is projected to increase the intensity and frequency of droughts by increasing temperature and evapotranspiration [1–4]. Moreover, water shortages are expected to intensify in drought-prone areas, where water service facilities are not provided. Although the water supply rate in Korea reached 97.4% in 2020, some mountainous areas remain vulnerable to drought, and people living in these areas chronically suffer from water shortages. For areas with difficulties in developing water service facilities due to economic or environmental limitations, it is essential to develop alternatives for a stable water supply with maximum utilization of available water resources.

Sand dam installation can increase the water supply capability without significantly damaging the natural environment. As shown in Figure 1, a sand dam can be constructed on impervious bedrock, storing water in the pores of sand depositions. Various types can be designed and built, depending on the field conditions, as illustrated in Figure 2. A standalone sand dam can serve as an intake source with an erosion control function. However, it may require periodic maintenance, incurring high costs. A multistage sand dam can steadily supply water by separating the erosion control function. A by-pass sand dam induces flow into a specially designed structure, causing less disturbance to the streamflow and making the system less susceptible to contamination.

Sand dams are widely used in Africa, as well as in North/South America, Asia, and the Middle East [5–13]. They are mainly used for supplying domestic or agricultural water and their ability to reduce evaporation and pollution makes them favorable, especially in arid and semi-arid regions.

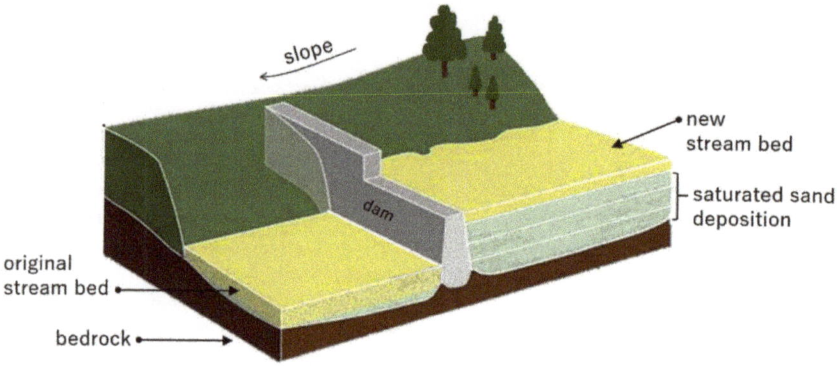

Figure 1. Configuration of sand dam (modified from [14]).

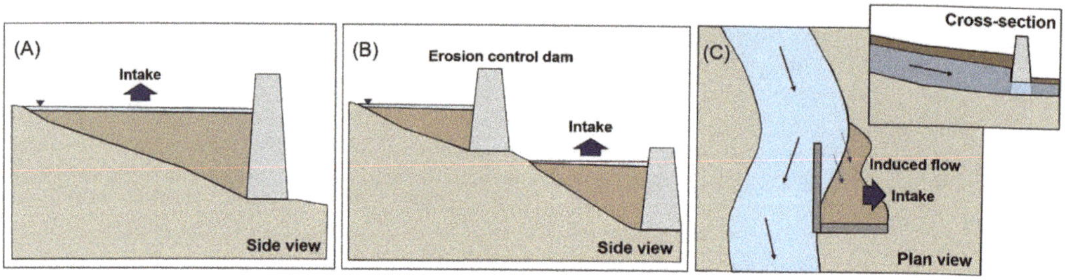

Figure 2. Types of sand dams: (**A**) Stand-alone, (**B**) Multi-stage, (**C**) By-pass (modified from Kim et al. [15]).

The conjunctive use of available water resources can contribute to the alleviation of dependence on a single source and provide resilience to the water supply system from various adverse situations, such as temporary interruption of water intake for maintenance or system failure. Conjunctive use does not imply the simple addition of water amount. Rather, it pursues efficient allocation of water resources. Conjunctive use has been widely employed to solve water resource management issues such as securing domestic/agricultural water, managing water quality, and preventing saltwater intrusion. Rafipour-Langeroudi et al. [16] proposed an operating rule for the conjunctive use of rivers and canals in the Tehran Plain, India. Tabari and Yazdi [17] developed an optimal water supply allocation ratio between the surface water and groundwater of the Urmia Lake Basin, Iran. Dai et al. [18] suggested a cost-effective surface and subsurface water allocation scheme for the Yunnan Province, China. Barlow et al. [19] simulated a management model for the stream/aquifer system of Rhode Island, USA. Khare et al. [20] evaluated the possibility of optimal conjunctive use of surface water and groundwater for canal operations in Andra Pradesh, India. Kim and Lee [21] provided a more extensive review of this subject.

The main objective of this study is to propose a methodology for the conjunctive operation of a sand dam and a nearby groundwater well, which has rarely been investigated elsewhere. We aim to provide useful guidelines for effective water supply under drought conditions by demonstrating how the methodology works. Information on the study site is presented in Section 2. Section 3 presents the step-by-step methodology. The results of the methodology applied to the study area are presented in Section 4. In Section 5, we conclude our study and provide suggestions for future work.

2. Study Site

The study area, Mullo-ri, is in the mountainous region of Chuncheon-si, Gangwon-do, South Korea (Figure 3). The catchment basin has a 2.1 km^2 area and 2.2 km stream length. Topographical characteristics of the 1708 m high Taebaek Mountains make the slope of the basin quite steep, causing rapid surface outflow. According to the 1992–2021 30-year rainfall data from the Chuncheon meteorological station (37.90262° N and 127.7357° E), July and August account for 53% of the total annual rainfall, June to September for 72%, and May to October for 83% [22]. With no modern facilities, seasonal fluctuations of streamflow and low yields of groundwater well make the water supply problem of this small village difficult. Overall, the precipitation pattern and the characteristics of the location lead to the low water supply potential especially during the dry season.

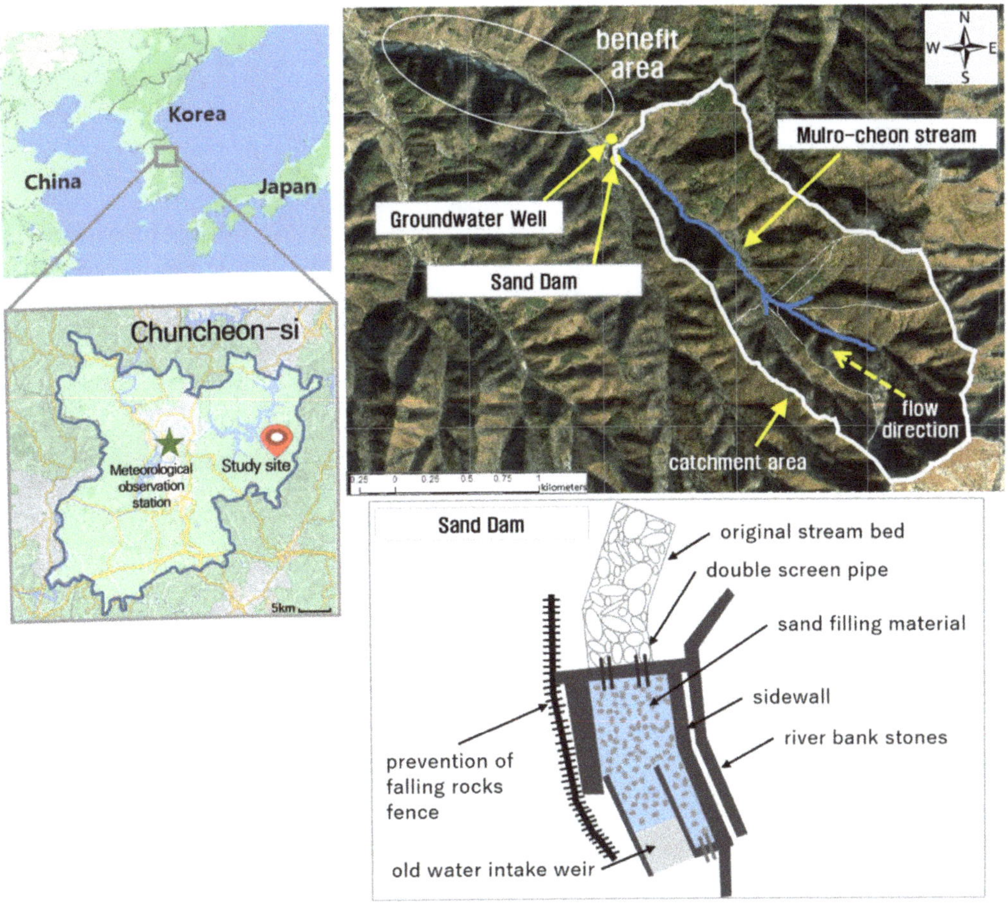

Figure 3. Study site and sand dam schematics (modified from [23]).

Figure 4 shows the standard precipitation index (SPI) plots for the data observed at the Chuncheon meteorological station. The SPI is a drought index based on the probability of precipitation at any time scale [24]. For example, a 3-month SPI at the end of March compares the January–February–March precipitation total in that particular year with the same period precipitation totals of all years. SPI plots with time scales of 3–12 months revealed frequent episodes of precipitation lower than in other years. The blue dotted lines indicate that major droughts caused significant damage to people living in the region, creating social issues.

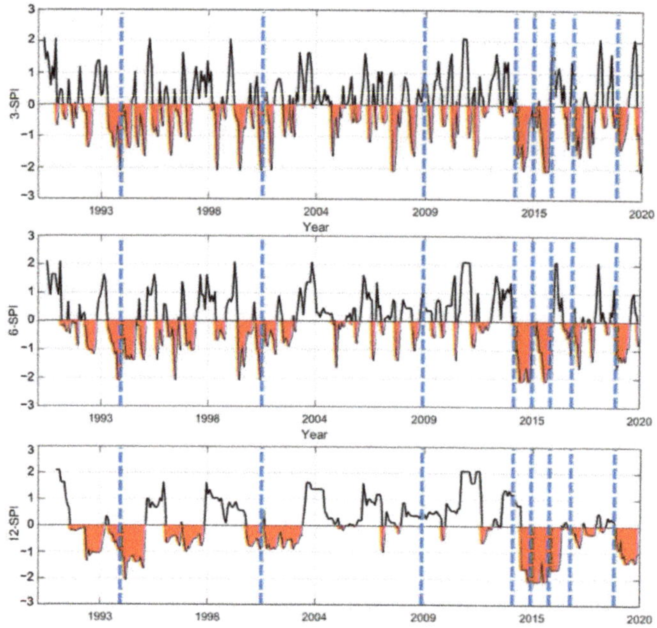

Figure 4. Standard precipitation index (SPI) for study site. The blue dotted lines indicate major droughts caused significant damage.

Originally, Mullo-ri utilized an old water intake weir (5 × 10 m) for water supply. As the entire village depended on a small facility, residents (population size of 50) frequently suffered from water shortages. In 2021, a by-pass type sand dam was built downstream of the old water-intake weir. In addition, a groundwater well was installed 470 m downstream from the sand dam. Table 1 lists the characteristics of the two water resources in this region [23]. Figure 5 shows the configuration of the well.

Table 1. Characteristics of the sand dam and groundwater well.

Sand Dam		Groundwater Well	
Top width (m)	13	Type of rock	Sedimentary rock
Bottom width (m)	13	Excavation depth (m)	120
Height (m)	1.9	Discharge pipe diameter (mm)	20
Design rainfall (mm/h)	109	Pumping capacity (m^3/day)	23
Runoff coefficient	0.8	Pump power (hp)	1
Maximum discharge (m^3/s)	3.18	Depth to pump (m)	100

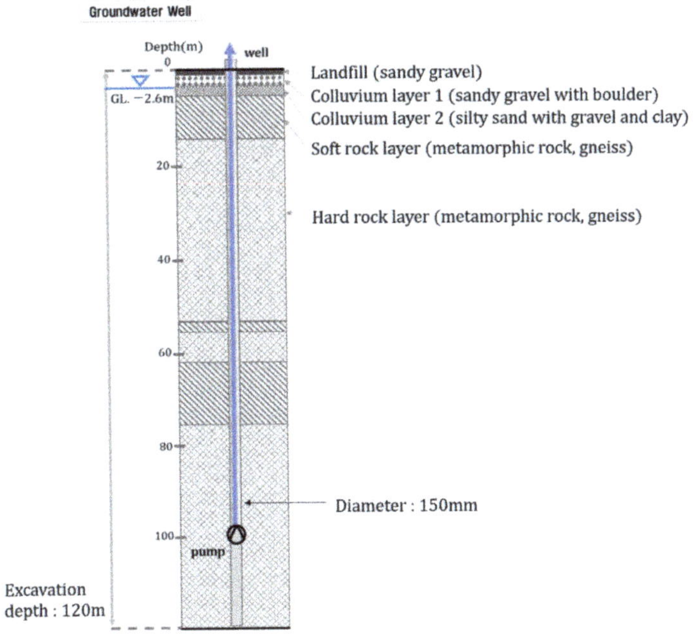

Figure 5. Cross-sectional view of the well configuration.

3. Methodology

Conjunctive use of the effective water supply from multiple sources requires a systematic approach. We proposed a six-step procedure, as indicated in Figure 6. The details are explained in the following sections.

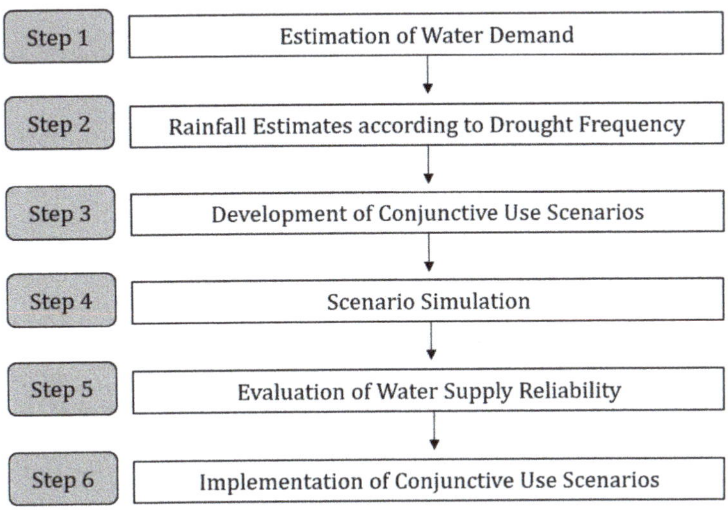

Figure 6. Procedure for conjunctive use of multiple water sources.

3.1. Estimation of Water Demand

The first step included the prediction of water demand. The water demand can be defined as the total amount of domestic water supplied through the conjunctive use of water sources. When there are no officially recognized data sources on daily water consumption in areas such as the study site, water demand can be calculated as the product of daily water consumption per person (lpcd) and population. The target water demand can be estimated by considering the following characteristics: location (a mountainous village in this study), age structure, and water usage. Future demand, along with the current amount of water use, can be obtained and used for scenario simulation (Step 4).

3.2. Rainfall Estimates According to Drought Frequency

Drought is a natural stochastic phenomenon caused by a severe shortage of precipitation. Although drought gradually progresses, it can exhibit longer-lasting effects than other meteorological hazards. To properly understand the impact of drought on the conjunctively operated system, rainfall data is a prerequisite. Probable rainfalls according to drought frequency are estimated in the second step. They were used as the input data for the water budget analysis (Step 4). Typical rainfall-frequency analysis methods can also be used. We employed the NDIC-FAT software program for this purpose [25]. The NDIC-FAT was developed by the K-water Institute and can facilitate drought frequency analysis. In this study, probable rainfalls for various drought frequencies (normal, 5-year, 10-year, 20-year, and 50-year) were estimated.

3.3. Development of Conjunctive Use Scenarios

The third step included the development of water supply scenarios that allocate water supply amounts to each water source under drought conditions. Furthermore, water supply scenarios for a sand dam and groundwater well are developed according to the frequency of drought. As the sand dam can be more vulnerable to drought than groundwater well, this requires consideration. In a normal year, the sand dam solely covers the demands without placing a burden on groundwater. As the drought intensifies, the role of groundwater in each scenario increases. However, the maximum groundwater allocation must be less than the pumping capacity, as excessive pumping can increase the risk of groundwater depletion or land subsidence [26].

3.4. Scenario Simulation

In the next step, the feasibility of the water supply was investigated through a simulation based on the developed scenarios. With the target water demand preset, the monthly water budget was analyzed according to a scenario (i.e., the ratio allocated to each water source).

Even for a scenario in which groundwater allocation is at its maximum, supplying groundwater will have no problem because the allocated water is less than the pumping capacity. This means that the feasibility of the water supply depends entirely on the success of sand dam operation.

The sand dam can be considered a reservoir, and the water budget equation (Δ Storage = inputs − outputs) is applied to it. To simulate time-varying storage levels, the water budget equation is formulated as follows:

$$\begin{array}{ll} RS_{i,j} = RS_{i-1,12} + I_{i,j} - WD_{i,j} - EV_{i,j} - OF_{i,j} & \text{when } j = 1 \\ RS_{i,j} = RS_{i,j-1} + I_{i,j} - WD_{i,j} - EV_{i,j} - OF_{i,j} & \text{otherwise} \end{array} \quad (1)$$

where $RS_{i,j}$ is the reservoir storage in year i and month j, $I_{i,j}$ is the inflow to the reservoir, $WD_{i,j}$ is the water demand, $EV_{i,j}$ is evaporation, and $OF_{i,j}$ is overflow. The water level was estimated and the water supply was considered feasible when the level was higher than the dead storage level.

It is not uncommon for inflow data ($I_{i,j}$) to be unavailable in rural areas. Depending on the type of sand dam, the entire amount or some portion of the runoff from the watershed would become inflow. Among the many ways to estimate watershed runoff, we utilized two approaches: the Kajiyama formula and a simple two-parameter monthly water balance model (TPM).

The Kajiyama formula has been widely used in practice in Korea [27,28]. It was developed based on relatively old data (1916–1927) but has been considered useful, especially when there is not much available data. The mathematical expression is as follows:

$$Q(t) = \sqrt{P(t)^2 + (138.6f + 10.2)^2} - 138.6f + \epsilon \quad (2)$$

where $Q(t)$ is the monthly runoff depth (mm), $P(t)$ is the monthly rainfall (mm), f is the runoff characteristic coefficient, and ϵ is the monthly correction factor (mm). Details of the runoff characteristic coefficient (f) and monthly correction factor (ϵ) can be found in the literature [27].

A simple two-parameter monthly water balance model (TPM) was proposed by Xiong and Guo [29]. It has been widely used because of its high accuracy and applicability. Given the observed monthly precipitation $P(t)$ and pan evaporation $EP(t)$, the actual monthly evapotranspiration $E(t)$ can be determined by the following:

$$E(t) = c \times EP(t) \times \tanh\left[\frac{P(t)}{EP(t)}\right] \quad (3)$$

where c is the first model parameter to consider the effect of the change in time scale, that is, from year to month. The monthly runoff $Q(t)$ is also assumed to be a hyperbolic tangent function of the soil water content $S(t)$, which is given by the following:

$$Q(t) = [S(t-1) + P(t) - E(t)] \times \tan h[\{S(t-1) + P(t) - E(t)\}/SC] \quad (4)$$

Here, the second parameter SC is the field capacity of the catchment (mm). The water content at the end of the t-th month can be calculated according to the water conservation law as follows:

$$S(t) = S(t-1) + P(t) - E(t) - Q(t) \quad (5)$$

3.5. Evaluation of Water Supply Reliability

Once the water supply scenarios are simulated, the reliability of the water supply system can be quantified [30]. This can be defined as the ratio of the period of continuous water supply to the total simulation period.

$$\text{water supply reliability} = \left(1 - \frac{\text{water shortage period}}{\text{total simulation period}}\right) \times 100 \quad (6)$$

A 100% water supply reliability means that a continuous water supply is possible throughout the year under certain drought conditions. The reliabilities of 91.7% and 83.3% corresponded to 11 and 10 months of continuous water supply, respectively. If a sand dam is designed to achieve a water supply reliability of, for example, 95% for a 20-year drought, one can expect higher reliability when groundwater well is added for conjunctive use.

3.6. Implementation of Conjunctive Use Scenarios

The last step was to implement conjunctive use scenarios for monthly operations based on real precipitation data. The drought conditions (i.e., frequency and duration) can be determined regularly at a certain time of a month by comparing the observed rainfall with the probable drought rainfall. Subsequently, an appropriate scenario was selected and applied to meet the target demand and reliability. To avoid excessive dependence on a certain water source and its resultant depletion, changing the scenarios is recommended if other options are available.

4. Results and Discussion

4.1. Water Demand

Two water demands were considered as the target supply amounts: the current amount of water use (Demand 1) and 150% of the current amount of water use (Demand 2). Demand 1 (9.16 m^3/day) corresponds to 187 L per capita per day (lpcd), which is the water consumption of typical small villages in South Korea. Demand 2 (13.7 m^3/day) accounts for 280.5 lpcd, which assumes future lifestyle changes, including water consumption patterns. It is the median value between 187 lpcd for small villages and 362 lpcd for the nearby city of Chuncheon [31].

There are various population projection methods [21] and the estimation of water demand based on such projection is feasible. Here, we did not follow the procedure due to the small size of the population (<50). Two water demands were devised by taking into account the local circumstances.

4.2. Probable Rainfall for Drought

Probable rainfall for drought was calculated according to the procedure described in Section 3.2. Rainfall data from the last 30 years were used, and the Gumble type was chosen for the probability distribution [23]. The calculated annual rainfall for the 5-year, 10-year, 20-year, and 50-year droughts was 1074.2, 986.2, 921.5, and 855.9 mm, respectively. These annual rainfall events were disaggregated into months according to the characteristics of past data. The estimated monthly rainfall according to drought frequencies is shown in Table 2. Considering the precipitation pattern of the monsoon watershed, four typical drought periods were assumed for the scenario simulation. They were (i) two months (July–August), (ii) four months (June–September), (iii) six months (May–October), and (iv) twelve months (January–December).

Table 2. Estimated monthly rainfall according to the drought frequency (unit: mm).

Drought Frequency	Jan.	Feb.	Mar.	Apr.	May	June	July	Aug.	Sept.	Oct.	Nov.	Dec.
Normal	18.6	26.5	35.9	73.9	102.2	122.6	383.0	322.3	125.3	50.6	49.3	22.8
5-year	15.0	21.4	28.9	59.6	82.4	98.8	308.6	259.7	101.0	40.8	39.8	18.3
10-year	13.8	19.6	26.6	54.7	75.6	90.7	283.3	238.4	92.7	37.4	36.5	16.8
20-year	12.9	18.3	24.8	51.1	70.7	84.7	264.7	222.8	86.6	35.0	34.1	15.7
50-year	11.9	17.0	23.1	47.4	65.6	78.7	245.9	206.9	80.5	32.5	31.7	14.6

4.3. Conjunctive Use Scenarios

Scenarios were developed to allocate the water supply to each water source under drought conditions (Table 3). In each scenario, the amounts of water supplied to the sand dam and groundwater well were assigned to meet the target demands. Each scenario with a different allocation ratio was tested through a simulation to determine whether the designated supplies would meet the target demand. For all scenarios, the sand dam supplies the entire demand with no help from the groundwater well during normal years. As drought becomes more severe, groundwater supply increases. The scenarios were designed to examine the effect of the increased role of groundwater in the context of conjunctive use. The maximum groundwater allocation can reach 40% of the total supply (S4, 50-year drought) but remains under the pumping capacity.

According to the Korean groundwater regulations, if the pumping rate exceeds 100 m^3/day, a groundwater impact assessment must be conducted before the development of a groundwater well. If it is below 100 m^3/day, it is regarded that no environmental problems are expected and the impact assessment is exempted. In our study, the maximum groundwater allocation (S4 with 50-year drought frequency) is 5.5 m^3/day, which is lower than 100 m^3/day and way below the pump capacity of 23 m^3/day. Pumping of such amount for several months would create little environmental problems. Of course, rigorous simulations may be needed for accurate environmental impact assessment.

Table 3. Scenarios for water allocation ratio (%) and supply (m^3/day) under various drought conditions.

Scenario	Frequency	Sand Dam %, Supply 1, Supply 2	Groundwater Well %, Supply 1, Supply 2
S0 (Base Scenario)	Normal	100, 9.16, 13.74	0, 0.00, 0.00
	5-year	100, 9.16, 13.74	0, 0.00, 0.00
	10-year	90, 8.25, 12.37	10, 0.92, 1.37
	20-year	90, 8.25, 12.37	10, 0.92, 1.37
	50-year	85, 7.79, 11.68	15, 1.37, 2.06
S1	Normal	100, 9.16, 13.74	0, 0.00, 0.00
	5-year	95, 8.70, 13.06	5, 0.46, 0.69
	10-year	90, 8.25, 12.37	10, 0.92, 1.37
	20-year	90, 8.25, 12.37	10, 0.92, 1.37
	50-year	85, 7.79, 11.68	15, 1.37, 2.06
S2	Normal	100, 9.16, 13.74	0, 0.00, 0.00
	5-year	95, 8.70, 13.06	5, 0.46, 0.69
	10-year	90, 8.25, 12.37	10, 0.92, 1.37
	20-year	85, 7.79, 11.68	15, 1.37, 2.06
	50-year	80, 7.33, 11.0	20, 1.83, 2.75
S3	Normal	100, 9.16, 13.74	0, 0.00, 0.00
	5-year	95, 8.70, 13.06	5, 0.46, 0.69
	10-year	90, 8.25, 12.37	10, 0.92, 1.37
	20-year	85, 7.79, 11.68	15, 1.37, 2.06
	50-year	70, 6.41, 9.62	30, 2.75, 4.12
S4	Normal	100, 9.16, 13.74	0, 0.00, 0.00
	5-year	95, 8.70, 13.06	5, 0.46, 0.69
	10-year	90, 8.25, 12.37	10, 0.92, 1.37
	20-year	80, 7.33, 11.0	20, 1.37, 2.06
	50-year	60, 5.50, 8.25	40, 3.67, 5.50

4.4. Simulation of Conjunctive Use Scenarios

The scenarios in Table 3 were applied to simulate the sand dam/well operation. The purpose was to determine whether the target water demand could be met under drought conditions. As the scenarios were designed such that even the maximum groundwater allocation (S4 with 50-year drought frequency) would not cause any environmental problems, the pumping operation according to all scenarios would always be possible. This implies that the judgment depends only on the simulation results of the sand dam.

The sand dam can be considered as a rectangular box with a dimension of 7.5 m × 13 m × 2.0 m with three layers of filling material (Figure 7). The porosities of the coarse, medium, and fine sands were 45, 35, and 25%, respectively, resulting in a total pore volume of 82.5 m^3. The water surface area was 97.5 m^2, the high water level was 1.9 m from the bottom, and the dead water level was 0.2 m due to the diameter of the drainage pipe. The surface evaporation was calculated using the daily pan evaporation dataset from 2017 to 2020. Evaporation from the sand-fill volume was not considered.

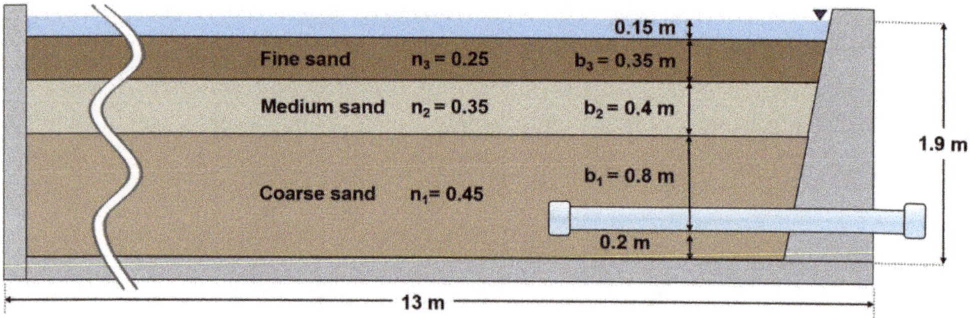

Figure 7. Sand dam with three layers of sand-fill material.

4.4.1. Inflow Estimation

When the water budget was analyzed using Equation (1), the inflows to the sand dam were used from the two runoff models described in Section 3.4. For the Kajiyama formula, Equation (2), $P(t)$ corresponds to the monthly rainfall for different drought frequencies in Table 2. The value of f was 0.8, owing to the steep slope of the study area, and ϵ was determined according to monthly rainfall [27].

Regarding TPM, Equation (3), locally observed data were used for the soil water content and evaporation [32]. Parameters from a study of the Hongchun sub-basin, 4 km from the study site, were adopted [33].

References [33–35] compared the Kajiyama formula and TPM in terms of the accuracy of runoff estimation. They analyzed the performance of the models against the observed data. However, the study area was ungagged and lacked runoff data. Therefore, we intend to compare the two models from the aspect of conjunctive use, not to determine which model is more accurate.

In 2021, observations were made to determine the amount of water directed through the by-pass structure. The ratio between the stream and inflow to the sand dam was 9:1, indicating that 10% of the watershed runoff was the inflow to the sand dam. The estimated runoff depth (Q) multiplied by the watershed area provided the total runoff and 10% was used as the inflow to the sand dam.

Figure 8 shows the inflow to the sand dam estimated using the Kajiyama formula and TPM. The first subplot compares the estimated values with observed ones made in 2020. Frequency analysis indicated that the precipitation corresponded to a normal year. Overall, the TPM resulted in a lower inflow than that of the Kajiyama formula, apart from the fall/winter season when precipitation was not high. The Kajiyama formula, an empirical equation originally developed to relate the rainfall and runoff for relatively larger rivers than ours, may have difficulties in reflecting characteristics such as slope or vegetation of the mountainous watershed.

In contrast, TPM considers various hydrologic components, such as monthly precipitation, evapotranspiration, and soil water content. As a result, the peak flows were much lower than those of the Kajiyama formula and shifted accordingly. On a more scientific basis, TPM was considered suitable for scenario analysis.

As it is common for observed runoff data to be scarce for small mountainous watersheds, the estimation of runoff is crucial to water budget analysis. The Kajiyama formula, which is commonly used in practice, is easy to use but tends to result in higher runoff than the TPM. A large amount of inflow to the sand dam overwhelmed the water demand and did not make any difference between the scenarios. TPM is recommended based on its scientific background and credibility.

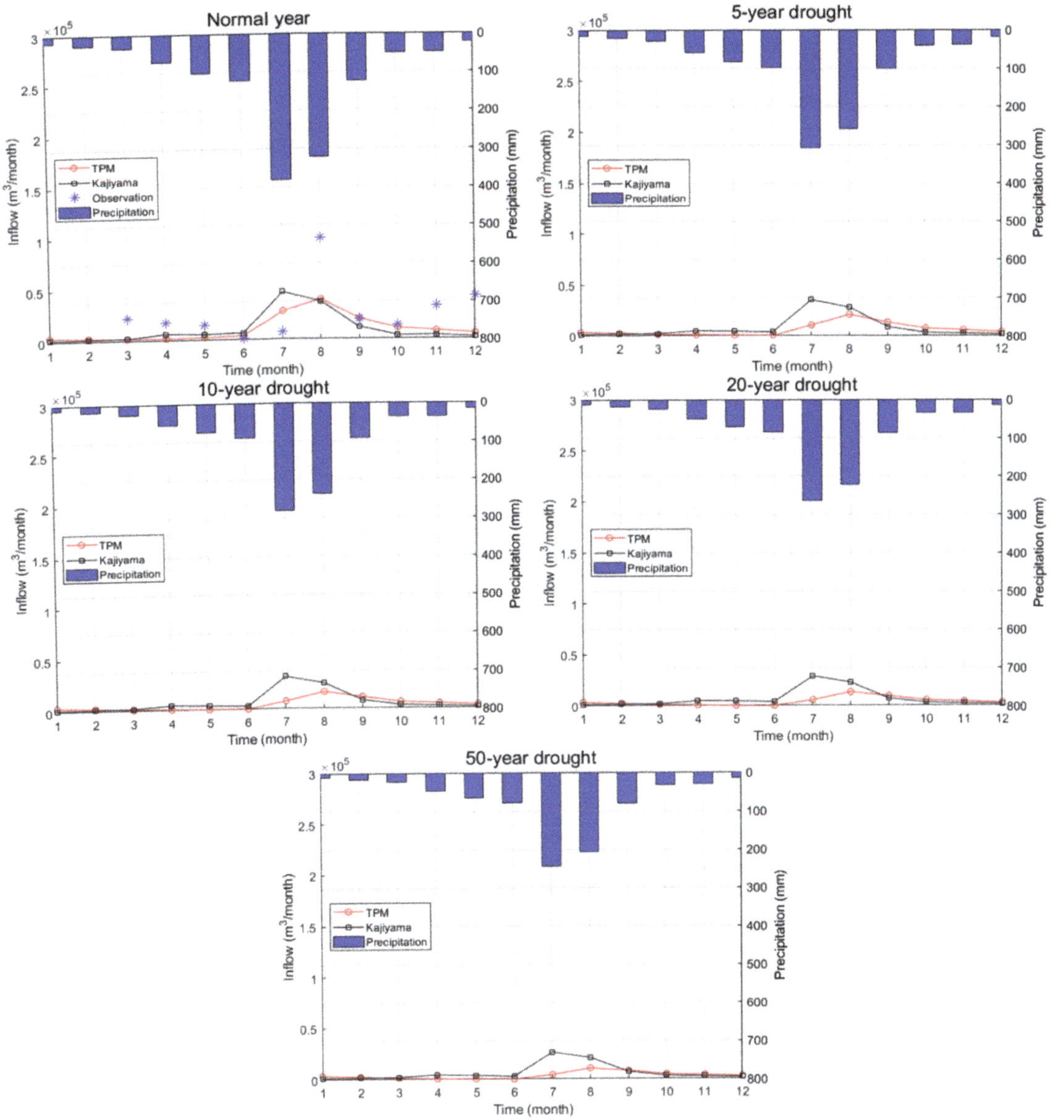

Figure 8. Precipitation and inflow to sand dam for various drought conditions (normal, 5-, 10-, 20-, and 50-year frequency). Data observed in a normal year are shown along with model estimates.

4.4.2. Water Supply Reliability

Before simulating the conjunctive use scenarios, one must test whether meeting the two target demands would be feasible with only the sand dam. If it fails, its conjunctive use with the aid of groundwater would be justified. As shown in Table 4, the water demand was satisfied for all cases when the inflow was estimated using the Kajiyama formula. As for the TPM, the water supply would cause no problems for normal years, and it would be the same for drought situations of 5-year frequency. The water supply started to fail in more severe droughts, showing the reliability of 91.7% for 10-year droughts and 83.3% for worse cases. These results call for measures to ensure a reliable water supply.

Table 4. Water supply reliability when only the sand dam is operated. A year-round drought is assumed.

Inflow Estimation	Water Demand (m^3/day)	Drought Frequency				
		Normal	5-Year	10-Year	20-Year	50-Year
Kajiyama formula	Demand 1 (9.16)	100	100	100	100	100
	Demand 2 (13.7)					
TPM	Demand 1 (9.16)	100	100	91.7	83.3	83.3
	Demand 2 (13.7)	100	100	83.3	83.3	83.3

Table 5 presents the comprehensive simulation results for the conjunctive operation. Scenario-wise outcomes of the possibility and reliability of a continuous water supply are tabulated. Black circles indicate that the conjunctive operation succeeds in meeting the demand, whereas the white circles correspond to the water shortage. Outcomes for both Demand 1 (current amount) and Demand 2 (increased amount) are shown. When a water shortage occurs, system reliability is given in parentheses.

When the Kajiyama formula was used for the inflow estimation, continuous water supply was possible for all cases, regardless of water demand and drought conditions. Sufficient inflow was believed to make the water supply system robust, even for a 50-year drought lasting 12 months. However, when the TPM was used for inflow estimation, this may not be the case. Water could be supplied to meet the demands of drought episodes of any return period lasting up to four months. Yet, when a drought lasted for six months, the water supply failed (S0, S1, S2). Failure could occur under the most severe conditions, that is a 50-year drought frequency with increased demand (Demand 2). The reliability dropped to 91.7%, implying that water shortages may have occurred for a month. Even though groundwater covered 15–20% of the target demand (see Table 3), the water storage in the sand dam could fall below the dead water level. Such water shortage problems can be solved by increasing the groundwater supply to 30% or more (S3 and S4).

When the drought period was extended to 12 months (last column in Table 5), the water supply capability weakened. Unlike the six-month drought, in which no problems occurred for droughts of the 20-year return period, the system could not meet Demand 2 even for the precipitation of the 10-year return period (S0). Not to mention, more severe drought. The water supply became infeasible for both demands, and the reliability could fall to 83.3% depending on the cases (S1 and S2). The input from the groundwater well, accounting for 30% of the total, could not completely solve the water shortage problem (S3). The water supply was made possible throughout the year only when the ratio between the sand dam and groundwater well became 6:4 (S4).

In this study, we presented the reliability of a water supply system. At a later stage, it would be meaningful to consider similar but different features as well: resiliency and vulnerability may be the candidates. The former refers to how rapidly a system will likely return from a failure state, and the latter describes the degree to which a system is susceptible to adverse effects of failure.

Table 5. Possibility and reliability of continuous water supply with conjunctive operation.

Inflow Estimation	Scenario	Drought Frequency	Duration			
			2 Months (July–Aug.)	4 Months (June–Sept.)	6 Months (May–Oct.)	12 Months (Jan.–Dec.)
Kajiyama formula	All	All	•/•	•/•	•/•	•/•
TPM	S0	Normal	•/•	•/•	•/•	•/•
		5-year	•/•	•/•	•/•	•/•
		10-year	•/•	•/•	•/•	•/○ (91.7)
		20-year	•/•	•/•	•/•	○/○ (91.7/83.3)
		50-year	•/•	•/•	•/○ (91.7)	○/○ (83.3/83.3)
	S1	Normal	•/•	•/•	•/•	•/•
		5-year	•/•	•/•	•/•	•/•
		10-year	•/•	•/•	•/•	•/○ (91.7)
		20-year	•/•	•/•	•/•	○/○ (91.7/83.3)
		50-year	•/•	•/•	•/○ (91.7)	○/○ (83.3/83.3)
	S2	Normal	•/•	•/•	•/•	•/•
		5-year	•/•	•/•	•/•	•/•
		10-year	•/•	•/•	•/•	•/○ (91.7)
		20-year	•/•	•/•	•/•	○/○ (91.7/83.3)
		50-year	•/•	•/•	•/○ (91.7)	○/○ (83.3/83.3)
	S3	Normal	•/•	•/•	•/•	•/•
		5-year	•/•	•/•	•/•	•/•
		10-year	•/•	•/•	•/•	•/○ (91.7)
		20-year	•/•	•/•	•/•	○/○ (91.7/83.3)
		50-year	•/•	•/•	•/•	○/○ (83.3/83.3)
	S4	Normal	•/•	•/•	•/•	•/•
		5-year	•/•	•/•	•/•	•/•
		10-year	•/•	•/•	•/•	•/•
		20-year	•/•	•/•	•/•	•/•
		50-year	•/•	•/•	•/•	•/•

Note: •/•: Possible for Demand 1 and 2. •/○: Possible for Demand 1, but a water shortage for Demand 2. The number (%) in parentheses indicates reliability. ○/○: Water shortages for Demands 1 and 2.

4.5. Operation with Conjunctive Use Scenarios

In the previous section, we saw that droughts of six to twelve months could result in water shortages and that such a situation could be resolved by increasing the allocation of water supply from groundwater. Given the simulation results, system operators or managers must make decisions to supply water in a sustainable manner. It is common for such a decision to be made monthly, even in a multi-purpose dam. In this section, we demonstrate the practical implementation of conjunctive use scenarios.

For instance, if a drought is at the end of a month and the target amount of water supply and reliability provided, the first step in determining the scenario for the next month's operation is to compare the precipitation of that month with the probable precipitation. With the frequency of drought known, the duration of drought can be determined by referring to the records of the previous months. The applicable scenarios were identified based on the simulation results shown in Table 5. Special attention must be paid when sudden changes in precipitation occur. A 'critical month' concept may be useful for developing some operational guidelines. A month when a severe drought starts abruptly is an example of a critical month. In such a case, scenarios with high groundwater allocation could be considered. Another example of a critical month is a transitional case, in which a long and severe drought is interrupted by normal precipitation. Since there is always a possibility that drought conditions will resume again, a sudden change to a surface water source might not be ideal. Concurrently, as there is a high chance of using groundwater for a long time, a policy change to a surface water source (i.e., sand dam) might be desirable. The final decision must be made considering various aspects such as water availability and environmental impacts.

Figure 9 illustrates an example of the operation of the sand dam and groundwater well in the study area, with conjunctive use scenarios. Precipitation sequences (hypothetical for demonstration purposes) are displayed, along with probable rainfall with various drought frequencies. The target was to meet Demand 2, and the design reliability of the system was set to 100%. Each month, the drought frequency is decided (third row in the table), and inapplicable/applicable scenarios are classified using simulation results, as shown in Table 5 (fourth and fifth rows). Critical months are identified (sixth row), and the final determination of the applicable scenario is made based on the guidelines mentioned above (seventh row).

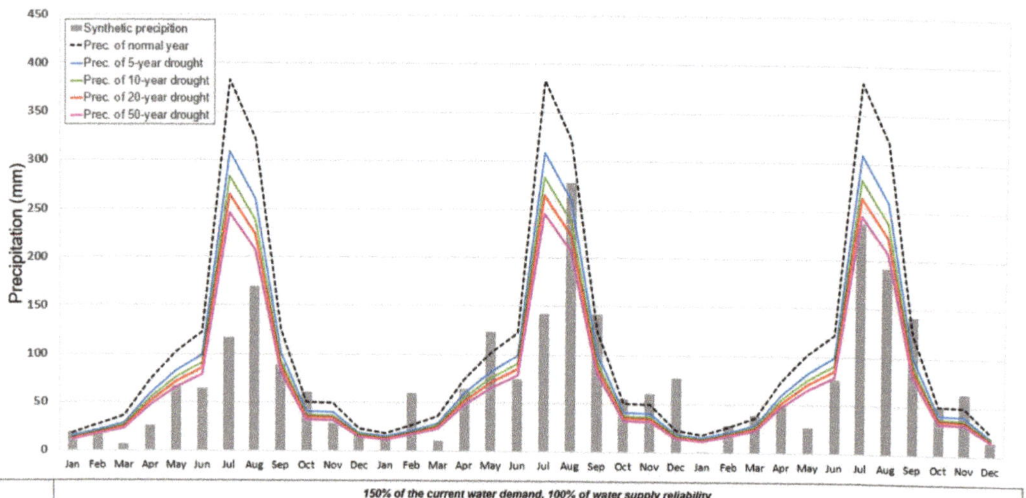

Target	150% of the current water demand, 100% of water supply reliability																																		
Month	1	2	3	4	5	6	7	8	9	10	11	12	13	14	15	16	17	18	19	20	21	22	23	24	25	26	27	28	29	30	31	32	33	34	35	36
Drought frequency [year]	Normal	5	50	50	50	50	50	50	10	Normal	50	10	10	Normal	50	5	Normal	50	50	5	Normal	Normal	Normal	Normal	50	Normal	Normal	50	50	50	50	50	Normal	Normal	Normal	50
Inapplicable scenario	S0 S1 S2	-	-	-	-	-	-	S0 S1 S2	-	-	-	-	-	-	-	-	-	-	-	-	-	-	-	-	-	-	-	-	-	-	-	-	-	-	-	-
Operational scenario	S3 S4	All	All	All	All	All	All	S3 S4	All	All	All	All	All	All	All	All	All	All	All	All	All	All	All	All	All	All	All	All	All	All	All	All	All	All	All	All
Critical month	ⓐ						ⓑ		ⓒ						ⓓ																					
Application	S3 S4	S0	S0	S0	S0	S0	S4	S4	S0	S0	S0	S0	S0	S0	S3	S0	S0	S0	S0	S0	S0	S0	S0	S0	S0	S0	S0	S0	S0	S0	S0	S3	S0	S0	S0	S0

Figure 9. Illustrative operation with conjunctive use scenarios.

An additional explanation may be helpful for a better understanding of the final decision-making process. In this illustration, it is assumed that a 50-year drought lasted for six months prior to month 1. Since it is the first month experiencing an abrupt change, it is transitional (critical month ⓐ). Normal precipitation might be temporary, and drought can resume. Therefore, scenarios with high dependency on surface water (S0, S1, S2) are not recommended, and S3 or S4 can be applied. As for month 7 (ⓑ), it is the fifth month of a 50-year drought and one more month of drought would see a water shortage. The simulation results proved that S0 could not be maintained (Table 5). The adoption of S4 is in order. At ⓒ, a 50-year drought that lasted for a relatively short period (two months) ended. A return to the base scenario (S0) appears to be a reasonable decision. In month 15 (ⓓ), S3 might be reasonable for the preparation of a possible drought owing to the significant reduction in rainfall. The rationale for the change to S3 in month 32 (ⓔ) is similar to ⓑ.

As operational records accumulate, the methodology presented here can be adjusted and modified. Such issues will be dealt with elsewhere. Last but not least, it is worth mentioning that our approach is suitable for application to data-scarce regions, since sand dams are typically built in rural or ungaged areas.

5. Conclusions

In the era of climate change, drought-prone areas, such as mountain regions where modern water facilities are not available, could suffer from chronic water shortages. To establish practical alternatives to overcome water shortage problems, this study proposed a methodology for the conjunctive use of a sand dam and groundwater well under various drought conditions. It employed a scenario-based approach in which the water allocation ratios between sources varied with the drought frequency. Simulations were carried out to determine whether the target water demand could be met under various drought conditions. The results revealed that the effect of the conjunctive operation began to appear in droughts for more than 5-year frequency in terms of water supply reliability. With detailed simulation results for various scenarios, decisions can be made for the monthly operation of the combined surface/subsurface water supply system.

According to our analysis, the system could achieve 83.3–100% water supply reliability through conjunctive use, even in the extreme case of a 50-year drought lasting an entire year. A higher water demand generally led to lower reliability, and the increased supply from the groundwater made the system more robust. Through an illustrative operation, it was shown that flexible application of scenarios can be achieved based on simulation results and practical guidelines concerning environmental impacts.

The methodology proposed here is not limited to a combination of a sand dam and a groundwater well. Other types of water resources can be conjunctively used by adopting this drought scenario-based method. As hydrologic and operational data accumulate, the system can evolve towards real-time operation. Fluctuations in demand, inflow, or groundwater level can be considered, which will make the use of valuable water resources more flexible.

Author Contributions: Conceptualization, B.R.K. and S.-I.L.; methodology, B.R.K. and S.-I.L.; software, B.R.K.; validation, B.R.K. and S.-I.L.; formal analysis, B.R.K. and S.M.Y.; investigation, B.R.K.; resources, B.R.K.; data curation, B.R.K.; writing—original draft preparation, B.R.K.; writing—review and editing, B.R.K., S.-I.L., and S.M.Y.; visualization, B.R.K., S.-I.L., and S.M.Y.; supervision, S.-I.L.; project administration, S.-I.L.; funding acquisition, S.-I.L. All authors have read and agreed to the published version of the manuscript.

Funding: This research was supported by the Korea Environment Industry & Technology Institute (KEITI) through the Demand Responsive Water Supply Service Program funded by the Korea Ministry of Environment (MOE) (146515).

Institutional Review Board Statement: Not applicable.

Informed Consent Statement: Not applicable.

Data Availability Statement: Not applicable.

Acknowledgments: The authors would like to thank the reviewers for their valuable comments and suggestions.

Conflicts of Interest: The authors declare no conflict of interest.

References

1. Mukherjee, S.; Mishra, A.; Trenberth, K.E. Climate change and drought: A perspective on drought indices. *Curr. Clim. Chang. Rep.* **2018**, *4*, 145–163. [CrossRef]
2. Trenberth, K.E.; Dai, A.; Schrier, G.V.D.; Jones, P.D.; Barichivich, J.; Briffa, K.R.; Sheffield, J. Global warming and changes in drought. *Nat. Clim. Chang.* **2014**, *4*, 17–22. [CrossRef]
3. Briffa, K.R.; van der Schrier, G.; Jones, P.D. Wet and dry summers in Europe since 1750: Evidence of increasing drought. *Int. J. Climatol.* **2009**, *29*, 1894–1905. [CrossRef]
4. Dai, A.; Trenberth, K.E.; Qian, T. A global dataset of Palmer drought severity index for 1870–2002: Relationship with soil moisture and effects of surface warming. *J. Hydrometeorol.* **2004**, *5*, 1117–1130. [CrossRef]
5. Lasage, R.; Aerts, J.C.C.; Verburg, P.H.H.; Sileshi, A.S. The role of small scale sand dams in securing water supply under climate change in Ethiopia. *Mitig. Adapt. Strateg. Glob. Chang.* **2015**, *20*, 317–339. [CrossRef]
6. Lasage, R.; Aerts, J.; Mutiso, G.C.M.; de Vries, A. Potential for community based adaptation to droughts: Sand dams in Kitui, Kenya. *Phys. Chem. Earth Parts A/B/C* **2008**, *33*, 67–73. [CrossRef]
7. Eisma, J.A.; Merwade, V.M. Investigating the environmental response to water harvesting structures: A field study in Tanzania. *Hydrol. Earth Syst. Sci.* **2020**, *24*, 1891–1906. [CrossRef]
8. Ertsen, M.; Hut, R. Two waterfalls do not hear each other. Sand-storage dams, science and sustainable development in Kenya. *Phys. Chem. Earth* **2009**, *34*, 14–22. [CrossRef]
9. Villani, L.; Castelli, G.; Hagos, E.Y.; Bresci, E. Water productivity analysis of sand dams irrigation farming in northern Ethiopia. *J. Agric. Environ. Int. Dev.* **2018**, *112*, 139–160.
10. Kamel, A.H.; Almawla, A.S. Experimental investigation about the effect of sand storage dams on water quality. *Zanco J. Pure Appl. Sci.* **2016**, *28*, 485–491.
11. Rao, S.V.R.; Rasmussen, J.A. Technology of small community water supply systems in developing countries. *J. Water Resour. Plan. Manag.* **1987**, *113*, 485–497. [CrossRef]
12. Nishigaki, M.; Kankam-Yeboah, K.; Komatsu, M. Underground dam technology in some parts of the world. *J. Groundw. Hydrol.* **2004**, *46*, 113–130. [CrossRef]
13. Baurne, G. "Trap-Dams": Artificial subsurface storage of water. *Water Int.* **1984**, *9*, 2–9. [CrossRef]
14. Rotary Beeston. Available online: https://www.rotary-ribi.org/clubs/page.php?PgID=573352&ClubID=1288 (accessed on 6 July 2022).
15. Kim, M.G.; Chung, I.-M.; Chang, S.W.; Lee, J. Method of installation of By-pass type sand dam in Mullori, Chuncheon. In Proceedings of the 2020 Korean Society of Civil Engineers Convention (KSCE 2020), Jeju, Korea, 21–23 October 2020; pp. 262–263. (In Korean).
16. Rafipour-Langeroudi, M.; Kerachian, R.; Bazargan-Lari, M.R. Developing operating rules for conjunctive use of surface and groundwater considering the water quality issues. *KSCE J. Civ. Eng.* **2014**, *18*, 454–461. [CrossRef]
17. Tabari, M.M.R.; Yazdi, A. Conjunctive use of surface and groundwater with inter-basin transfer approach: Case study Piranshahr. *Water Resour. Manag.* **2014**, *28*, 1887–1906. [CrossRef]
18. Dai, C.; Cai, Y.P.; Lu, W.T.; Liu, H.; Guo, H.C. Conjunctive water use optimization for watershed-lake water distribution system under uncertainty: A case study. *Water Resour. Manag.* **2016**, *30*, 4429–4449. [CrossRef]
19. Barlow, P.M.; Ahlfeld, D.P.; Dickerman, D.C. Conjunctive-management models for sustained yield of stream-aquifer systems. *J. Water Resour. Plan. Manag.* **2003**, *129*, 35–48. [CrossRef]
20. Khare, D.; Jat, M.; Sunder, J.D. Assessment of water resources allocation options: Conjunctive use planning in a link canal command. *Resour. Conserv. Recycl.* **2007**, *51*, 487–506. [CrossRef]
21. Kim, B.R.; Lee, S.-I. Conjunctive Operation of Surface and Subsurface Dams based on Drought Severity. *Water* **2021**, *13*, 847. [CrossRef]
22. KMA: Korea Meteorological Administration. Available online: http://www.kma.go.kr/ (accessed on 22 March 2022).
23. Korea Institute of Civil Engineering and Building Technology. *Development of Advanced Technology for Conjunctive Use of Sand Dam and Water Intake: Annual Report*; Korea Institute of Civil Engineering and Building Technology: Goyang-si, Korea, 2021. (In Korean)
24. Zargar, A.; Sadiq, R.; Naser, B.; Khan, F.I. A review of drought indices. *Environ. Rev.* **2011**, *19*, 333–349. [CrossRef]
25. National Drought Information-Analysis Center. NDIC-FAT Drought Frequency Analysis Program User's Manual. 2022. Available online: http://www.drought.go.kr/menu/m50/m50.do (accessed on 31 March 2022).
26. Foster, S.S.D.; Chilton, P.J. Groundwater: The processes and global significance of aquifer degradation. *Philos. Trans. R. Soc. B* **2003**, *358*, 1957–1972. [CrossRef] [PubMed]
27. Seo, J.; Lee, D.; Lee, G.; Kim, J.; Kim, K.-S.; Lim, K.J. Assessment and improvement of monthly coefficients of Kajiyama formular on climate change. *J. Korean Soc. Agric. Eng.* **2018**, *60*, 81–93. (In Korean)
28. Kim, S.M.; Lee, S.I.; Kim, B.C. Effective use of water resources through conjunctive use-(2) application. *Korea Water Resour. Assoc.* **2004**, *37*, 799–812. (In Korean) [CrossRef]
29. Xiong, L.; Guo, S. A two-parameter monthly water balance model and its application. *J. Hydrol.* **1999**, *216*, 111–123. [CrossRef]
30. Shamir, U.; Howard, C.D.D. Water supply reliability theory. *J. Am. Water Works Assoc.* **1981**, *73*, 379–384. [CrossRef]
31. Ministry of Environment (MOE). *2019 Statistics of Waterworks*; Ministry of Environment: Sejong City, Korea, 2018. (In Korean)

32. WAMIS: Water Resources Management Information System. Available online: http://www.wamis.go.kr/ (accessed on 31 May 2022).
33. Kim, S.; Hong, S.J.; Kang, N.; Noh, H.S.; Kim, H.S. A comparative study on a simple two-parameter monthly water balance model and the Kajiyama formula for monthly runoff estimation. *Hydrol. Sci. J.* **2016**, *61*, 1244–1252. [CrossRef]
34. Lim, D.S.; Kim, H.S.; Seo, B.H. A study on computation methods of monthly runoff by water balance method. *J. Korea Water Resour Assco.* **2001**, *34*, 713–724. (In Korean)
35. Jung, S.; Bae, Y.; Kim, J.; Joo, H.; Kim, H.S.; Jung, J. Analysis of small hydropower generation potential: (1) Estimation of the potential in ungaged basins. *Energies* **2021**, *14*, 2977. [CrossRef]

water

MDPI

Article

An Assessment of Geospatial Analysis Combined with AHP Techniques to Identify Groundwater Potential Zones in the Pudukkottai District, Tamil Nadu, India

Muruganantham Arumugam [1], Prabakaran Kulandaisamy [1], Sivakumar Karthikeyan [1], Kongeswaran Thangaraj [1], Venkatramanan Senapathi [2,*], Sang Yong Chung [3], Subagunasekar Muthuramalingam [4], Muthuramalingam Rajendran [1], Sathish Sugumaran [1] and Siva Manimuthu [1]

1 Department of Geology, Faculty of Science, Alagappa University, Karaikudi 630003, Tamil Nadu, India
2 Department of Disaster Management, Alagappa University, Karaikudi 630003, Tamil Nadu, India
3 Department of Earth and Environmental Sciences, Pukyong National University, Busan 608737, Republic of Korea
4 Department of Rural Development, Gandhigram Rural Institute, Chinnalapatti 624302, Tamil Nadu, India
* Correspondence: venkatramanansenapathi@gmail.com

Abstract: Groundwater is critical to the socioeconomic development of any region. Infiltration of surface water into the ground is influenced by a variety of factors such as soil pores, folds, fractures, faults, and joints, all of which contribute to groundwater recharge. Groundwater is an important source of freshwater in the drought-prone Pudukkottai district of Tamil Nadu, India. Therefore, the search for groundwater potential zones (GWPZs) is critical. The present study focuses on the investigation of potential groundwater zones using geospatial techniques. Geology, land use and land cover, geomorphology, soil, drainage density, lineament, and groundwater levels were obtained from state and non-state associations. ArcGIS version 10.8 was used to create all thematic layers and classified grids. The intensive use of groundwater in arid and semiarid regions is becoming a problem for the public to meet their freshwater needs. The condition of arid and semi-arid regions due to intensive groundwater extraction has become one of the most important environmental problems for the public. In this study, a powerful groundwater potential mapping technique was developed using integrated remote sensing data from GIS-AHP. Using AHP techniques, thematic layers for geology, geomorphology, and soil followed by drainage, drainage density and lineament, lineament density, slope, water level, and lithological parameters were created, classified, weighted, and integrated into a GIS environment. According to the results of the study, it is estimated that 14% of the groundwater potential in the study area is good, 49% is moderate and 36% is poor. A groundwater level map was used to verify the groundwater potential. In addition, the model was validated with a single-layer sensitivity analysis, which showed that geology was the most influential layer and water level was the least influential thematic layer. The low-potential areas identified on the groundwater potential map can be used for further study to identify ideal locations for artificial recharge. In low potential areas, the groundwater potential map can be used to find ideal locations for artificial recharge. The water table in the area must be raised by artificial recharge structures such as infiltration basins, recharge pits, and agricultural ponds. Artificial recharge structures such as infiltration basins, recharge pits, and agricultural ponds can be used for groundwater development in the low potential zones. The GWPZ map was successfully validated with three proxy data, such as the number of wells, groundwater level, and well density, obtained from well inventory information. The results of this study will improve our understanding of the geographic analysis of groundwater potential and help policy makers in this drought-prone area to create more sustainable water supply systems.

Keywords: remote sensing; GIS; thematic layers; AHP; weighted overlay; groundwater potential zone

Citation: Arumugam, M.; Kulandaisamy, P.; Karthikeyan, S.; Thangaraj, K.; Senapathi, V.; Chung, S.Y.; Muthuramalingam, S.; Rajendran, M.; Sugumaran, S.; Manimuthu, S. An Assessment of Geospatial Analysis Combined with AHP Techniques to Identify Groundwater Potential Zones in the Pudukkottai District, Tamil Nadu, India. *Water* **2023**, *15*, 1101. https://doi.org/10.3390/w15061101

Academic Editor: Jianhua Xu

Received: 9 November 2022
Revised: 25 February 2023
Accepted: 8 March 2023
Published: 13 March 2023

1. Introduction

Since the turn of the millennium, surface water accessibility has been compromised by insufficient precipitation due to global climate change, increased urbanization, and industrial development in rain-fed areas. As a result, potential groundwater areas have been designated [1]. On a global scale, the quantity and accessibility of drinking water are two of the most important environmental, social, and political issues. It is challenging for environmental engineers and hydrologists to monitor water quantity and draw conclusions from the data, as every step from sampling to analysis is subject to uncertainty [2]. Predicting groundwater potential zones is difficult in rare areas, so standard methods for managing groundwater resources are needed. In the absence of understanding regarding groundwater accessibility, the main issues worldwide are the depletion of the water table and loss in well yield caused by excessive pumping. Below the surface of the earth, in the fissures of geologic strata, is where we can find groundwater, a precious natural resource [3–6]. Planning strategies to stop groundwater depletion and preserve the wellbeing of the groundwater ecosystem can benefit from assessments of the volume and spatial distribution of groundwater storage [7].

Hydrogeologists use a variety of techniques to determine the groundwater potential zone. In recent years, airborne electromagnetic (AEM) surveys have become the accepted method for determining groundwater pathways in fractured crystalline hard rock [8]. Traditional scientific methods for locating groundwater include the electrical resistivity technique and the magnetic method. The modern approach includes integrated remote sensing and GIS. Various thematic maps derived from satellite imagery and field surveys were overlaid to identify the potential groundwater zones GIS. The AHP technique was created in 1977 by Professor Thomas Saaty [9]. In addition, the Analytic Hierarchy Process (AHP) [10–13] is internationally recognized and employed a quantifiable technique. The evaluation of possible groundwater resource zones in rapidly urbanizing areas looks to be a flexible decision-making tool for multi-criteria problems. It enables problem hierarchy and guarantees that during the evaluation process, both qualitative and quantitative components of a problem are taken into account. Multi criteria Decision Making (MCDM), which gives judgments structure, verifiability, transparency, and correctness, has been shown in numerous studies to be a useful technique for managing water resources [14]. Even if these low-cost techniques are restricted to small-scale explorations, the variability of the Earth's subsurface makes it more difficult to locate groundwater potential zones. Using remotely sensed satellite photography, several surface characteristics can be utilized to determine the presence or absence of groundwater. In AHP inference, the mapping from a given input to an output is formulated using AHP logic [15–18]. The model that transforms input data into input membership categories is present in each AHP inference system—rules into a collection of output features, output features into output membership functions, and output membership functions into a single-valued output or a decision related to the output.

Geographic information system (GIS) based studies evaluated static groundwater storage volumes but did not offer information on how widely applicable the results are in identifying potential groundwater development areas [19,20]. On the other hand, current delineation methods are based on either a single indicator that may be insufficient to reflect numerous elements of groundwater development or an excessive number of indicators for which data are not readily available for a target area. For example, existing methods rely on the length of screened units of the aquifer. Volume of aquifer Satellite imagery has been used extensively and successfully to map regional groundwater potential zones, which is a cost-effective method. Identification of groundwater potential zones is improved when remote sensing and GIS are used together [21–25]. Pudukottai is a drought-prone, arid district in southern India. It is frequently affected by cyclic droughts due to failures of the monsoon in the last 100 years. The objective of the present study is to delineate groundwater potential zones to meet the freshwater needs of the region during droughts. It also aimed at producing a map using remote sensing and GIS to help reduce uncertainty in the identification of groundwater potential zones. By integrating GIS and AHP, a model was created to produce the map of groundwater potential zones. In addition, the model was validated with the single-layer sensitivity analysis.

2. Study Area

In the Pudukkottai district of Tamil Nadu, droughts are common. It lies between 9°50′ and 10°40′ north latitude N and 78°25′ to 79°15′ east longitude. The toposheets prepared by the Survey of India (SOI) are 58 J/9, 10, 11, 14, 15, 16, 58 N/2, 3, 4 and 58 O/1&2 at a scale of 1:50,000. It is a large festival with an area of about 4663 km^2. Pudukkottai is divided into 11 taluks, 13 blocks and 16 towns with 750 villages. Thanjavur is located in the northeast and east of the Pudukkottai district, Palk Road in the southeast, Ramanathapuram and Sivagangai in the southwest and Tiruchirappalli in the northwest. (Figure 1). The Pudukkottai district receives an average of 821 mm of rainfall annually. Most of the rainfall in this district occurs during the northeast monsoon (397 mm), followed by the southwest monsoon (303 mm). The main aquifers of the district consist of fractured and weathered crystalline rocks, mainly hornblende gneisses, granitic gneisses and pink granites. Sedimentary formations, ranging in age from Cretaceous to recent, include sandstones, limestones, shales, and unconsolidated alluvium. Precipitation patterns, surface conditions, land use, soils, and geology all affect water quality. Summer and winter precipitation amounts are 81 and 40 mm, respectively. Precipitation increases from east to southwest in the district. The main occupations of the population are agriculture and tourism. Of the total agricultural land, 1420.24 km^2 (31.97%) is arable land, 18.12 km^2 (0.38%) is fallow land, and 110.87 km^2 (2.37%) is plantation land. About 1188 km^2 of the land was fallow. Distribution of the study area: 24.32% of the total land use and land cover of the district. Shrubland and other fallow land are classified as wasteland. A total of 241.07 km^2 is covered by forests. The study region consists of Cauvery basin and sub-basins of Vellar, Agniyar, Ambuliyar, Koraiyar, Gundar and Pambar. At Manamelkudi, the Vellar is the main river that flows into the Bay of Bengal after flowing in an east-southeast direction. In this study, cropland, agricultural fallow land, and plantations constitute the agricultural land use. Built-up areas include both rural and urban settlements, industries and mines. Tanks, canals, rivers, coastal wetlands, backwaters, and impounded wetlands are examples of wetlands. Scrublands, alkaline soils, sandy areas, and salt pans are examples of bars. Finally, there is the category of forest use/cover. Maximum land use Thirumayam > Viralimalai > Manalmelkudi > Gandharvakottai. This has developed especially in the study area; it could be very useful for studying changes in land use/cover to predict weather extremes. Storativity in sedimentary formations ranges from 4.9×10^{-6} to 4.4×10^{-4}, while in hard rocks it ranges from 3.26×10^{-5} to 5.2×10^{-5}. In sedimentary rocks, the specific yield is 23% and in hard rocks it is 2.1%.

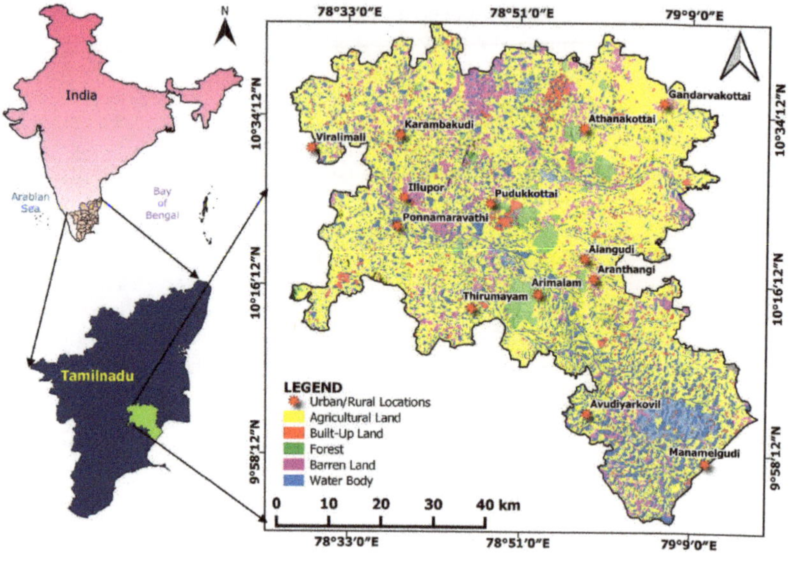

Figure 1. Pudukkottai district map.

3. Materials and Methods

Groundwater potential (GWP) mapping requires the following thematic layers: geology, geomorphology, soil, drainage density, line density, and slope. A base layer of the study area was created using the toposheet, and all other thematic layers were created using remote sensing data. The approach developed for this study is shown as a flowchart in Figure 2, and the thematic layers were also categorized. A 1:50,000 scale toposheet was used to create the basemap of the study area. IRS -P6 LISS-IV MX Satellite imagery with SRTM (Shuttle Radar Topographic Mission) DEM data with a resolution of 50 m was used to create the different types of thematic layers such as geomorphology (GM) and geology (GG), drainage density (DD), line density (LD), and slope gradient (SL) and soil type (ST) related to the occurrence of groundwater. These thematic layers are more commonly used for mapping groundwater potential as they control recharge, infiltration, runoff and groundwater movement. The lithologic nature of a rock outcrop is critical to groundwater recharge because weathering and fracturing create secondary porosity as an aquifer. Therefore, the geology of the study area was considered along with other strata because of its influence on water percolation and groundwater availability. The subsurface lithology and structural features of a study area influence the geomorphology. Visual interpretation of processed satellite imagery for geomorphologic mapping can identify and delineate structural features and different landforms. Soil infiltration conducts precipitation water from the surface to groundwater by gravity and capillary forces and is therefore an important criterion for identifying potential zones. Soil texture and structure, soil moisture and density, biological shells and vegetation controls the infiltration and are influenced by soil temperature as well as human activities at the soil surface. Drainage represents the lithology of the surface and subsurface. The distance between channels is represented by the unit of drainage density, km/km^2. Drainage density, expressed as the number of km/km^2, can be used to calculate the distance between channels. The drainage, catchment area and its specifications were determined using SRTM Dem data. The following formula was used to determine the drainage densities (DD) for the drainage map:

$$DD = \Sigma TLWS/TAWS$$

where TLWS is the total stream length in the watershed and TAWS is the total area of the watershed.

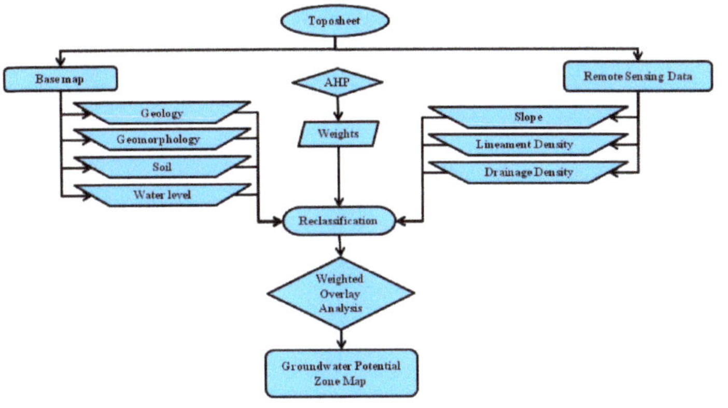

Figure 2. Flowchart of methodology Adopted.

Lineaments indicate groundwater recharge zones, and the density of lineaments indicates high groundwater potential [26–28]. The lineaments of the study area were identified by edge formation techniques using high-pass and low-pass filtering, and lineament density was calculated using information on the sources of lineament density. The degree of slope affects the infiltration of groundwater into the subsurface, providing information about the potential groundwater zone. On steep slopes, stormwater drains more quickly, while on gentle slopes it drains more slowly, which increases infiltration because stormwater remains on the surface of the slope longer. GIS Software such as ERDAS IMAGINE: 8.7 and ArcGIS 10.8 were used to allow processing of the digital images using statistics and geographic analysis. The base map was used to draw ground control points (GCP) for geometric rectification of the satellite data [29–31]. AHP logic is a type of site selection that is commonly used [32,33]. It assigns membership values between 0 and 1 to sites (ESRI). Based on their relative position in the groundwater determination, the weights for each layer were controlled using analytical hierarchical processing (AHP). The variables controlling surface runoff, infiltration, and groundwater flow are used in the pairwise comparison method to determine the more favorable layer. The pairwise comparison matrix developed by Saaty (1980) was checked for accuracy using the formulas below (Equations (1)–(5)). The standardized principal eigenvector is given in Table 1. The pairwise comparison matrix table (1) was developed based on the relative importance of the thematic strata in determining groundwater potential. Here, geology is the most fundamental layer that forms the aquifer of the water-bearing formation in a study area. Geomorphology is the second most important layer, followed by soil, drainage density, lineament density, slope and water table. In general, the eigenvalue (λ) describes the scalar factor change of a vector as a result of a linear transformation. Using equation 4, the maximum eigenvalue estimate was calculated for all layers (Table 2).

$$Wi = \frac{\sum_{i=1}^{n}\left[\frac{Ci}{\sum_{i=1}^{n} Ci}\right]}{n} \quad (1)$$

$$Cj = \frac{\sum_{i=1}^{n}[Ci \times Wi]}{Wi} \quad (2)$$

$$\lambda max = \frac{\sum Cj}{n} \quad (3)$$

$$CI = \frac{\lambda max - n}{n-1} \quad (4)$$

$$CR = \frac{CI}{RI} \quad (5)$$

where Ci is the indicator value consigned. Each measure from the pairwise comparison matrix has a weight called Wi. Cj stands for consistency judgement factor, the calculation of the greatest Eigen value is λmax, and the number of criteria is n.

Table 1. Pairwise comparison matrix.

Layers	Geology	Geomorphology	Soil	Drainage Density	Lineament Density	Slope	Water Level
Geology	1.000	3.000	4.000	5.000	6.000	7.000	8.000
Geomorphology	0.333	1.000	3.000	4.000	5.000	6.000	7.000
Soil	0.250	0.333	1.000	3.000	4.000	5.000	6.000
Drainage density	0.200	0.250	0.333	1.000	3.000	4.000	5.000
Lineament density	0.167	0.200	0.250	0.333	1.000	3.000	4.000
Slope	0.143	0.167	0.200	0.250	0.333	1.000	3.000
Water level	0.125	0.143	0.167	0.200	0.250	0.333	1.000
SUM	2.218	5.093	8.950	13.78	19.58	26.33	34.00

Table 2. Normalized principal eigenvector for comparison matrix.

Layers	Geology	Geomorphology	Soil	Drainage Density	Lineament Density	Slope	Water Level	Weights
Geology	0.451	0.589	0.447	0.363	0.306	0.266	0.235	0.404
Geomorphology	0.150	0.196	0.335	0.290	0.255	0.228	0.206	0.243
Soil	0.113	0.065	0.112	0.218	0.204	0.190	0.176	0.150
Drainage density	0.090	0.049	0.037	0.073	0.153	0.152	0.147	0.092
Lineament density	0.075	0.039	0.028	0.024	0.051	0.114	0.118	0.055
Slope	0.064	0.033	0.022	0.018	0.017	0.038	0.088	0.032
Water level	0.056	0.028	0.019	0.015	0.013	0.013	0.029	0.024
Eigen vector	1.000	1.000	1.000	1.000	1.000	1.000	1.000	1.000

CI, *RI*, and *CR* are denoted for consistency index, random index and consistency ratio of derived weights. The presence of lineaments indicates rechargeable zones in the aquifer, and the density of lineaments indicates excellent groundwater potential. Edge development with low- and high-pass filters was utilized to discover the lineaments in the study region using the source of lineament closeness [34]. The weights are consistent under all conditions and can be used in a weighted linear combination. Using Equation (6) and the weighted linear combination approach, the groundwater potential index was obtained. The groundwater potential index was computed using the weighted linear combination method.

$$\text{Groundwater Potential Index (GWPI)} = \sum_{i=1}^{n}(\text{GGw.wi, GMw.wi, STw.wi, DDw.wi, LDw.wi, SLw.wi}) \quad (6)$$

Here, GWPI—groundwater potential index, GG—geology, GM (geomorphology), ST (soil type), DD (drainage density), LD (line density), and SL (slope) are used. The index "w" represents the standardized weights of each layer and the index "wi" represents the normalized weights of each thematic layer [35]. The groundwater potential area is divided into five categories based on the GWPI values: poor potential, moderate potential, and good potential-excellent, good, moderate, low, and poor. The groundwater potential index is also converted to a GIS database file, which is then used to create the groundwater potential zone map (GPZM). The groundwater potential zone map (GPZM) is created in the recommendation by first converting the groundwater potential index to a GIS database file. A random survey of the region was conducted to confirm the groundwater benchmarks.

3.1. Weighted Overlay Analysis

Each thematic map received an AHP comparative weight assignment to create a collective weight of the individual thematic maps. The weight value of each map was then determined based on the actual field conditions. The consistency ratio of their thematic maps and the instantaneous weights of the features of the various thematic layers were calculated and assigned for each thematic map. All thematic layers were integrated with ArcGIS 10.8 to produce the GWPZ map by applying the below Equation (7):

$$\text{GPZM} = ((\text{TM1w} \times \text{SC1r}) + (\text{TM2w} \times \text{SC2r}) + (\text{TM3w} \times \text{SC3r}) + (\text{TM4w} \times \text{SC4r}) + (\text{TM5w} \times \text{SC5r}) + (\text{TM6w} \times \text{SC6r}) + (\text{TM7w} \times \text{SC7r}) + (\text{TM8w} \times \text{SC8r})) \quad (7)$$

where GPZM stands for the groundwater potential zone map, TM1–TM8 stands for the major criterion (1 to 8 thematic layer maps), w stands for the thematic map's weighting, SC1–SC8 stands for the sub-conditions of each thematic layer map, and r stands for the sub-criteria class rating [36].

3.2. Sensitivity Analysis

The relative importance of each thematic layer can be evaluated by sensitivity analysis. It also helps to evaluate the feasibility of the chosen method by calculating the uncertainty of the results. In the present study, a single-layer sensitivity analysis was performed. This method was used to evaluate the influence of each thematic layer on GWPI. It defines the effective weight of each thematic layer. Equation (8) gives the effective weight of the selected thematic layer, which can be compared with the AHP weight assigned to the same layer:

$$W = \left(\frac{TLi.Sj}{GWPI}\right) \times 100 \quad (8)$$

where W is the effective weight of the selected layer, TL_i is the assigned ranking, and S_j is the AHP-derived weight for the selected layer. The overall index of groundwater potential was given by *GWPI*. The results of the above equation help to identify the thematic layer that has a greater effective weight than the AHP-derived weight and thus has a greater impact on the overall GWPI [37].

4. Discussion

4.1. Geology

Hard rocks from the Archean era and Quaternary sedimentary deposits make the study area's geological structure (Figure 3). The entire research region falls within the hard rock and sedimentary rock categories geologically. While sedimentary formations developed on the east side of the study area, hard rocks developed on the west. The research region is made up of around 45% Archean hard rocks and 55% sedimentary rocks that range in age from the Precambrian to the Quaternary. Along with the primary rock types, the Pudukkottai block's central and southern regions are mostly home to charnockites, hornblende gneiss, biotite gneiss, granite, and quartzite charnockites and granitic rocks. In the western portion of the study region, gneissic rocks of various sorts can be discovered. The Annavasal and Thirumayam blocks have modest amounts of quartzite deposits. In this area, mining is taking place. Shaley sandstone, sand, clay and gravel make up the sedimentary deposits of this area. Sandstone, clay and mudstone make up the tertiary sedimentary deposits. The Arantangi, Gandharvakottai, Alangudi, and Thiruvarankulam blocks include these deposits, which create a healthy groundwater zone. Sand, gravel, and silt comprise the unconsolidated coastal alluvium that lines the riverbank. Silt and clay deposits from the Quaternary can be found in the Avudaiyarkoil and Manamelkudi blocks [38]. Near the Pudukkottai district's coastline limit, the beach has sand deposits with ridges and dunes. The geology of an area forms the aquifers of the region. When a geologic formation has the properties to hold water, it is called an aquifer. Often, aquifers are distinguished based on their location and the arrangement of the overlying formations. Therefore, geology plays an important role in identifying groundwater potential zones. In this study, geological formations were classified into three groups, namely good groundwater potential, moderate groundwater potential, and poor groundwater potential. This classification is based on the evaluation of the water storage capacity of each formation. The reclassification of the region based on the geological formation shows that 49% have poor potential, 40% have moderate potential, and 11% have good potential (Figure 4). However, this classification refers only to the primary relative capacity of the formations and does not take into account the capacity created by secondary processes. To overcome these limitations, other thematic strata were used to determine groundwater potential.

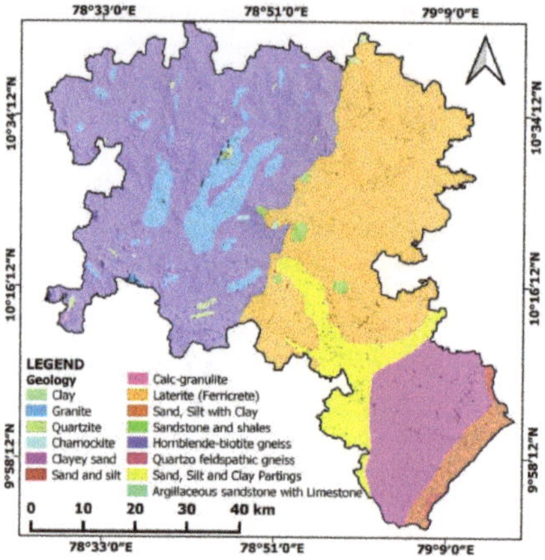

Figure 3. The study area geology map.

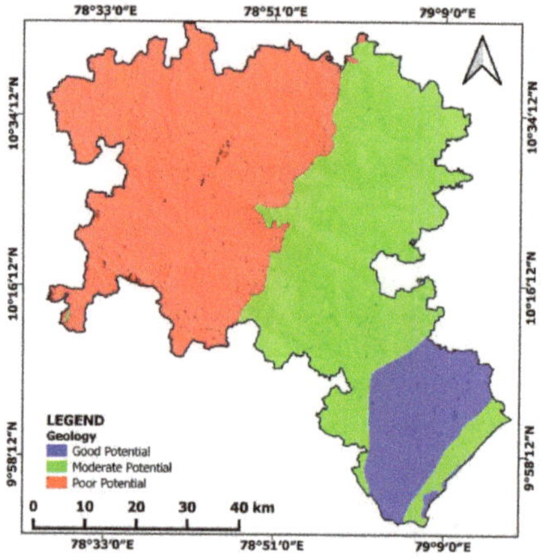

Figure 4. Ground water Potential map based on Geology.

4.2. Geomorphology

Seven zones are identified on the geomorphological map (Figure 5) as majorly plains of lateritic, alluvial, pediment and floods, structural hills, uplands, and denudation hills. In the Pudukkottai region, the alluvial plain makes up the majority (49.1%), followed by the alluvial plain (35.6%), upland (11.5%) and flood plain (1.7%) mixed with lateritic plain cover (1.2%), structural hills around 0.7%, and denudation hills (0.4%). The district's north, west, and southern regions have residual hills that create topographical undulations and flat terrain with alluvial plains in the eastern portion of the district. While it is 1.5 m above sea level close to the coast, the terrain in the western section of the district is roughly 125 m above sea level. Denudation, structural, and fluvial processes dominate the geomorphological evolution of the area. The development of the different landforms has been substantially influenced by the geologic formations' resistance to these processes. In the region, there are pediments, buried pediments, erosional plains, residual mounds, and deltaic plains. Lateritic Plains > Alluvial Plain > Pediplain > Structural Mounds > Upland > Buried Mounds. With thin soil cover, the shallow pediments produce low to moderate yields. Groundwater is abundant in the buried summits and delta plain [39]. In general, geomorphology represents the features of the earth's surface created by various geological impacts. Based on the nature of these surface features, an area can be differentiated by groundwater potential zones. Geomorphologic features are one of the most important controlling factors for surface water runoff, infiltration, and groundwater recharge. Earth surface features in an area that support stormwater runoff into the groundwater system are considered to have good groundwater potential. This characteristic of geomorphological features led to the classification of a study area into good, moderate, and poor groundwater potential (Figure 6). In this study, 72% of the study area has good groundwater potential, followed by 16% moderate potential, and 12% poor potential (Table 3).

Table 3. Weighting and ranking for each category of thematic layers.

S.No	Parameter	Class	Score	Weights	Area %
1	Geology	Hornblende-biotite gneiss	1	0.404	41.69
2		Quartzo feldspathic gneiss	1	0.404	0.02
3		Argillaceous sandstone with limestone	3	0.404	0.30
4		Quartzite	1	0.404	0.62
5		Laterite (Ferricrete)	2	0.404	29.82
6		Charnockite	1	0.404	0.21
7		Granite	1	0.404	5.98
8		Clay	2	0.404	0.13
9		Sandstone and shales	2	0.404	0.46
10		Sand and silt	3	0.404	0.50
11		Sand, Silt and Clay Partings	2	0.404	6.85
12		Calc-granulite	2	0.404	0.02
13		Clayey sand	3	0.404	11.47
14		Sand, Silt with Clay	2	0.404	2.29
15	Geomorphology	Shallow and moderately Weathered Pediplain	3	0.243	66.50
16		Upland	1	0.243	10.11
17		Pediment	1	0.243	0.81
18		Shallow Flood Plain	3	0.243	4.90
19		Inselberg	1	0.243	0.02
20		Linear Ridge/Dyke	1	0.243	0.11
21		Bazada	3	0.243	0.15
22		Pediment-InselbergComplex	1	0.243	0.47
23		Pediplain Canal Command	3	0.243	0.21
24		Channel bar	1	0.243	0.17
25		Structural Hills	1	0.243	0.06
26		Shallow alluvial plain	2	0.243	13.43
27		Coastal Plain	2	0.243	2.20
28		Lateritic	3	0.243	0.04
29		Salt flat	2	0.243	0.07
30		Brackish water creeks	1	0.243	0.02
31		Beach ridge complex	1	0.243	0.24
32		Dune complex	3	0.243	0.05
33	Soil	Clay	1	0.15	45.66
34		Sandysilt	1	0.15	24.13
35		Sandyclay	2	0.15	22.47
36		ClayeySilt	1	0.15	5.03
37		Sandstone	3	0.15	2.66
38		Silty sand	2	0.15	0.022

Table 3. *Cont.*

S.No	Parameter	Class	Score	Weights	Area %
39	Drainage density	Low	3	0.092	82.85
40		Moderate	2	0.092	15.47
41		High	1	0.092	1.68
42	Lineament density	Low	1	0.055	62.34
43		Moderate	2	0.055	26.06
44		High	3	0.055	11.60
45	Slope	0–4.5 (gentle slope)	3	0.032	99.54
46		4.5–7.9 (moderate slope)	2	0.032	0.44
47		>7.9 (steep slope)	1	0.032	0.02
48	Water Level	<10 (Good)	3	0.024	6.56
49		10–50 (Moderate)	2	0.024	17.51
50		>50 (Poor)	1	0.024	75.9

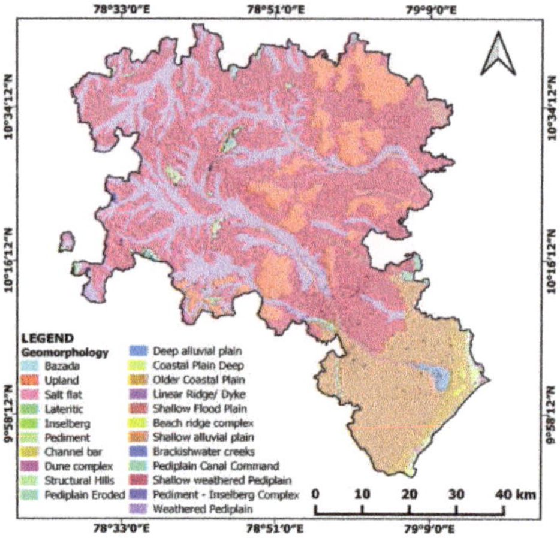

Figure 5. Geomorphological map.

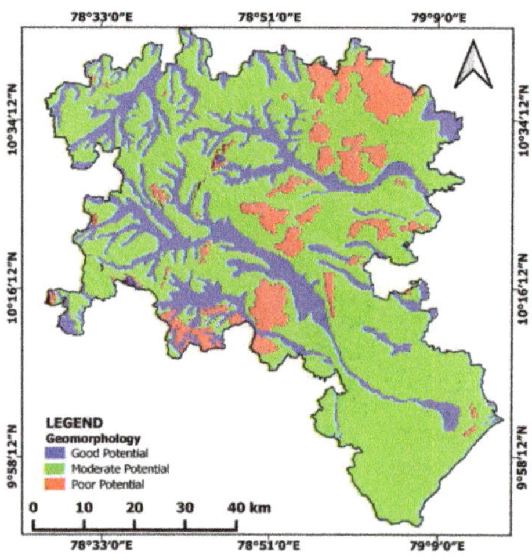

Figure 6. GWPZ map according to the Geomorphology.

4.3. Soil

The study area soils (Figure 7) fall into the categories of black, red, ferruginous, lateritic, alluvial, and beach soils. West of the study area was developed by black soils. Southern parts were constructed by red, ferruginous lateritic soils on the uplands. Alluvial soils are found in the coastal region of the study area. Black and brownish, sandy, and silty soils are found beside the riverine regions such as Vellar, Agniyar, and Ambuliyar. Whereas down the coast are beach sands. The different types of soil appear in the following order: Sand (30.53%) > sandy-loamy (20.27%) > sandy silt (19.28%) > clayey silt (17.66%) > silty sand (6.83%) > clayey (3.13%) > clayey (0.04%). The loamy sand dominates the middle part, while the northern, north-eastern region is dominated by sand. Clay minerals in the south region of the Pudukkottai area are shrinking and swelling. With the exception of the southeastern region, clay soils are found throughout the study area. In general, the infiltration rate of the top sand layer is higher than that of the other soil types. Following is a ranking of the soils based on their infiltration potential: alfisols > entisols > inceptisols > vertisols > silty sands > sandstone [40]. In the identification groundwater potential, soil type is considered an important criterion used in many previous literatures. The infiltration rate in a study area is different if the soil type is different. Different groups of soils have different infiltration rates. The granulometric arrangement of soil types causes them to have different infiltration rates. As mentioned earlier, the study area has different soil types on its surface. Based on the infiltration capacity of these soils, the study area was classified as 75% poor potential, 22% moderate potential, and 3% good potential (Figure 8).

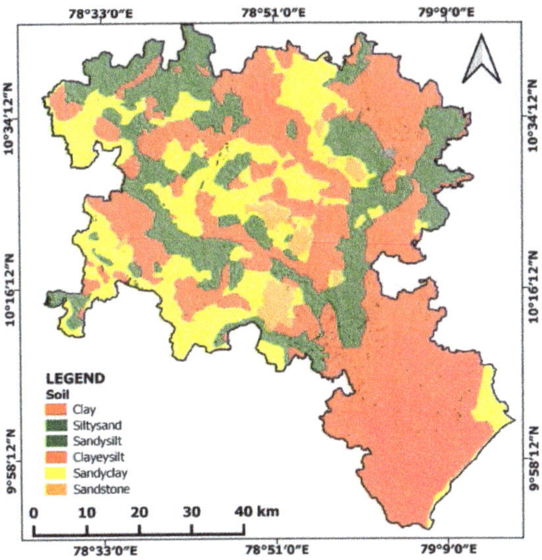

Figure 7. Soil type map.

4.4. Drainage and Drainage Density

Drainage density is divided into three zones on the basis of the proximity to rivers as high, moderate, and low (Figure 9). Pudukottai is a sub-basin of the Cauvery basin. The main river is the Vellar, which flows in a south easterly direction towards the Bay of Bengal [41]. Other rivers that drain in this study area are the Gundar, Pambar, Agniyar, Ambuliyar, and Koraiyar. Most of them are ephemeral rivers and cause structural flooding during the rainy season. Precipitating rainwater forms floods by accumulating in drains. The runoff density of a region controls flood movement at the earth's surface. This is because infiltration and groundwater recharge are relatively slow processes. Dense drainage zones drain floodwaters more quickly, so not enough water remains at the surface for infiltration and groundwater recharge. This phenomenon results in an area with dense drainage systems having low potential because most of the surface water is conveyed through channels or streams. Based on drainage density, the study area was divided into 83% zones with good potential, 15% with moderate potential, and 2% with poor potential (Figure 10).

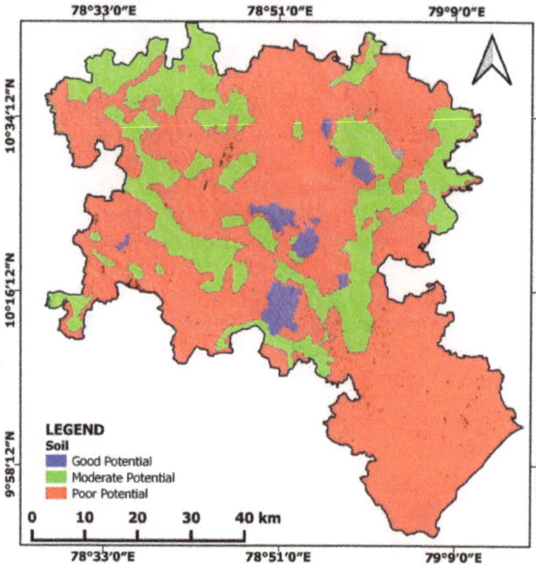

Figure 8. GWPZ map according to the Soil.

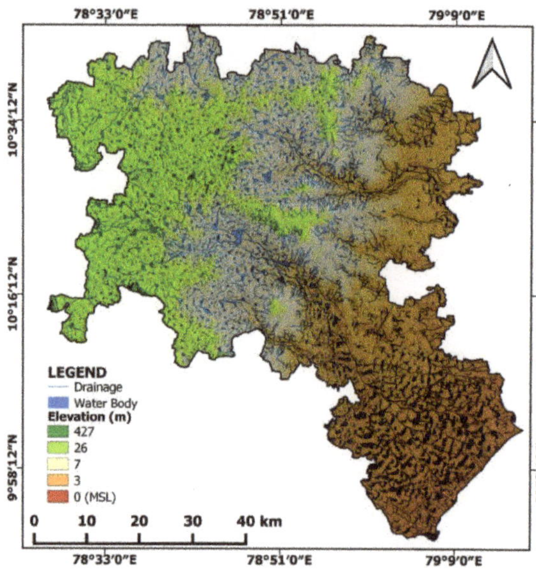

Figure 9. Drainage map.

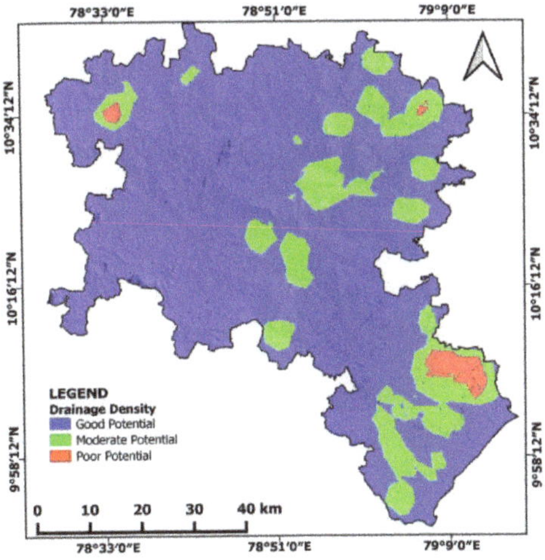

Figure 10. GWPZ map according to the Drainage Density.

4.5. Lineament (L) and Lineament Density (LD)

The study area was divided into low, moderate, and high according to line density and has a uniform distribution of lineaments that serve as conduits for groundwater flow. The highest density category ranks first, followed by moderate density, and low density [42]. The studied area has minimal lineament density throughout. Lineaments are observed in moderate and high density in the center and northeastern part of the study area, respectively (Figure 11). Lineaments act as subsurface conduits that convey infiltrated water to the groundwater regime. Therefore, lineaments are considered good markers for locating potential groundwater zones. Based on lineament density, the study area was divided into 62% low potential, 26% moderate potential, and 12% good potential zones (Figure 12).

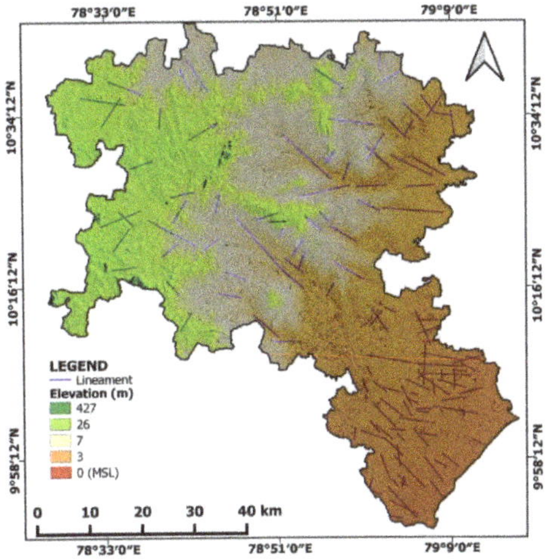

Figure 11. Lineaments density map.

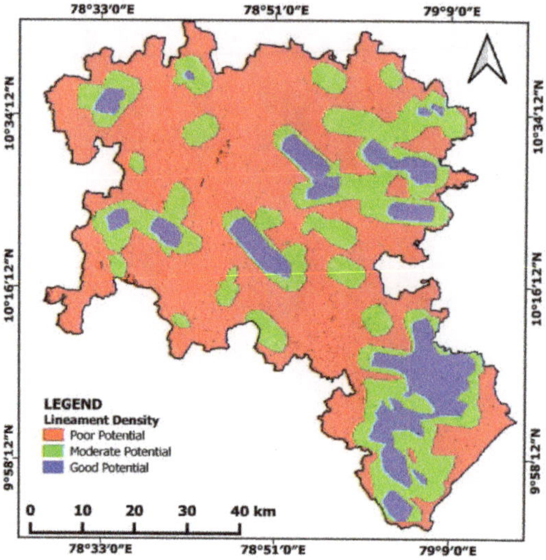

Figure 12. GWPZ map according to the Lineaments.

4.6. Degree of Slope

In contrast to the rest of the study area, which has a gentle (104.5%) to moderate (104.5–457.9%) slope that allows infiltration of rainwater into subsurface aquifers, the hilly terrain in the northwestern section has a steep slope of more than 457.9% (Figure 13) [43]. Consequently, the slope angle of the hillside is important in determining the infiltration and runoff capacity of a site. The amount of runoff increases with the steepness of the slope, but a gentle slope allows more water to infiltrate. Based on the slope factor, 99% of the study area was classified as having good potential, followed by moderate (0.4%), and poor (0.02%) potential (Figure 14).

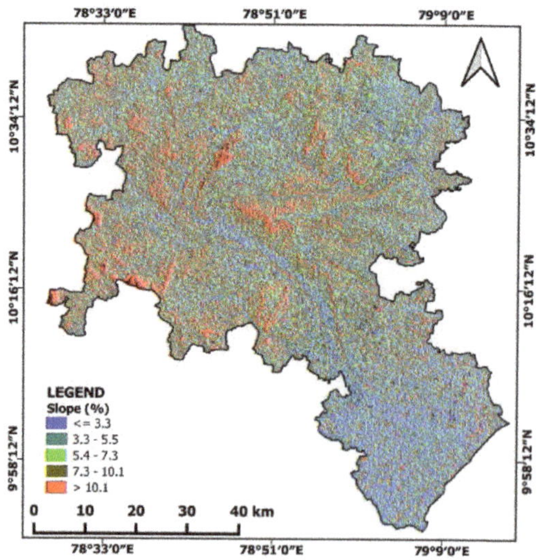

Figure 13. Slope % map of the study region.

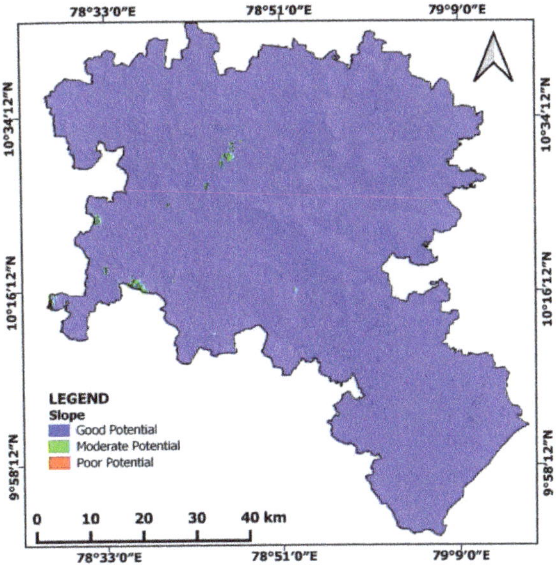

Figure 14. GWPZ map according to the Slope.

4.7. Ground Water Level

Comparison with the groundwater level map of the study area served as validation of the groundwater potential map. (Figure 15), which gives a direct indication of the potential availability of groundwater. The groundwater potential maps were also generated for different well types, and it was found that the GWPZs in 80 wells sampled at different locations correlate well with the groundwater level map. The groundwater potential zone is well associated with the cross-correlation of water level and precipitation [44]. The water table of a near-surface aquifer was measured in sedimentary rocks at depths ranging from 1.80 to 11.5 m and in weathered rocks at depths ranging from 12 to 32 mbar. Hard rock in sedimentary formations ranges from 35 to 125 mbgl, while massive rock ranges from 75 to 152 mbgl. Groundwater levels often provide information about the potential regions of an area. By measuring and monitoring the water table, we can determine its flow path. However, the water table in an area depends on the type or location of the aquifer system. In unconfined areas, the water table is near the surface and may have a low potential, while in confined aquifers it is in deeper horizons but may have a good potential. This leads to uncertainties in determining groundwater potential zones based on water level. In the current study, GWPZs were classified as 76% poor potential, 17% moderate potential, and 7% good potential (Figure 16).

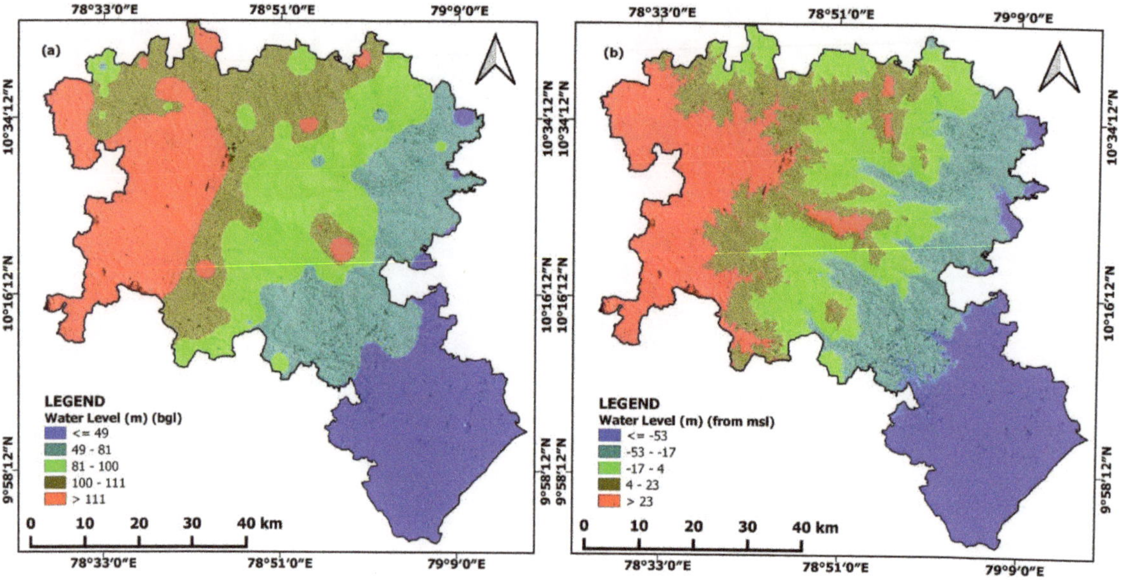

Figure 15. Ground water level of the study area (**a**) measured as below ground level and (**b**) measured as from mean sea level (positive values denote above msl and the negative values denote below msl).

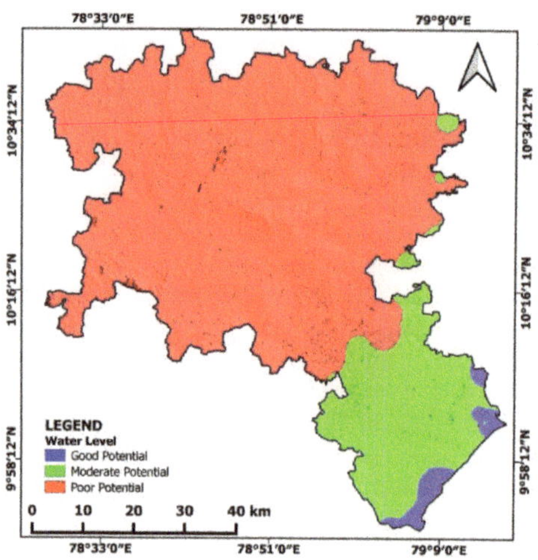

Figure 16. GWPZ map according to the ground water level.

4.8. GWPZ Map

According to the ranking and weighting assigned to the selected thematic layers based on their water storage capacity, they were reclassified (Table 3). The cumulative weighting percentages applied to the geology, geomorphology, soil, drainage density, line density, slope, and water level maps were used in a WOA to create the GWPZ map (Figure 17). The ability to hold the recharging water increases with weighting; conversely, the inability decreases with weighting. Based on the weighted overburden analysis, the groundwater potential map of the study area has three distinct zones of good, moderate, and poor potential. While the peripheral portion of the study area has moderate groundwater potential, the central to northern portions have good potential. Isolated areas in the northwest, southwest, and center have poor GWP. In most of the study region, the groundwater zone potential is excellent, moderate, or fair (Table 4). Validation of the GWPZ map was based on the groundwater level map of the study area, although groundwater level is the most obvious indicator of potential availability [45–47]. Groundwater level data were collected from 80 different well types in 2019. The map of spatial distribution of groundwater level was created using the Inverse Distance Weights (IDW) interpolator in GIS. The map spatially depicts the groundwater level of the study area. The southern and central parts of the study area merge with its northeastern part, which has a narrow groundwater table. The statistical summary of the sensitivity analysis results is provided in Table 5. The relative importance of the selected thematic layers is obtained by comparing the weights determined in the AHP with the effective weights. From the mean of the effective weights, it can be seen that the weights derived from the AHP are consistent in this study. Among the thematic layers, geology is the most important with a mean effective weight of 3.53, followed by geomorphology (2.10), soil (1.18), drainage density (0.87), and line density (0.52). Slope and water table elevations have relatively little influence on the identification of groundwater potential zones, with mean effective weights of 0.30 and 0.26, respectively.

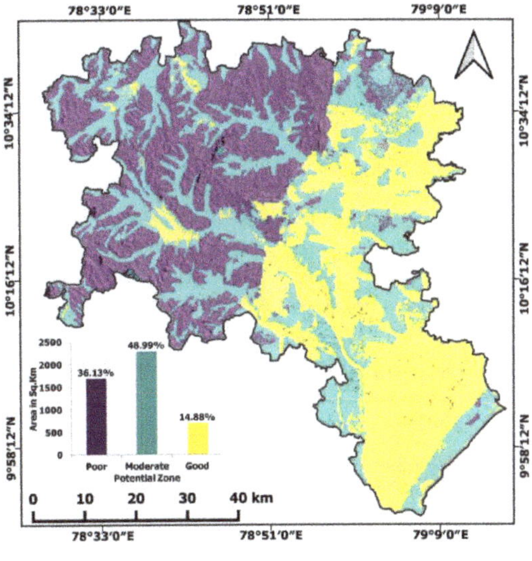

Figure 17. GWPZ classification map.

Table 4. Calculated areas of each zone in groundwater potential GWPZ map.

Zone	Description	Area in km^2	Area in %
1	Poor potential zone	1684.8	36.13
2	Moderate potential zone	2284.5	48.99
3	Good potential zone	693.7	14.88
	Total	4663.0	100

Table 5. Statistical summary of the single-layer sensitivity analysis.

Thematic Layer	AHP Weight	Effective Weight		
		Minimum	Maximum	Mean
Geology	0.404	1.90	5.71	3.53
Geomorphology	0.243	1.14	3.43	2.10
Soil	0.150	0.71	2.12	1.18
Drainage density	0.092	0.43	1.30	0.87

Table 5. *Cont.*

Thematic Layer	AHP Weight	Effective Weight Minimum	Maximum	Mean
Lineament density	0.055	0.26	0.78	0.52
Slope	0.032	0.15	0.45	0.30
Water Level	0.024	0.11	0.34	0.23

4.9. Validation of GWPZ Map

The validation of groundwater potential zone map is significant to evaluate its accuracy and the reliability of adapting the geospatial techniques. There are numerous methods available and widely used by many researchers to validate the GWPZ map [48]. Some of the previous studies compare the results with the bore well yield data [49,50]. However, the availability of these data is very limited in countries such as India. Hence, it is very complex to validate the results of GWPZ. Matching the well inventory data with GWPZ map is one of the validation techniques. In the present study, data from 80 pumping wells were collected through the reconnaissance survey. Among 80 wells, 38% (30 out of 80) fall under the poor potential zone with the average measured depth to water table of 106 m from surface. About 46% (37 out of 80) were found in the moderate potential zone with the average measured depth to water level of 80m. Whereas 16% (13 out of 80) of the wells present in the good potential zone with the average water table measured from the earth's surface at 36 m. Though the proxy of number wells and water table depth gives good matching results with GWPZ, it cannot be taken alone to evaluate the results. Hence, a well density method was used to determine which zone has the maximum number of wells in the total area. Good potential zones possess a greater number of wells in the total area and moderate potential zones contain a relatively fewer number of wells followed by poor potential. As a result, the ratio between the total area and the well count will give the well density. The equation to estimate the well density is given below.

$$Well\ Density = \frac{Total\ Area}{Well\ count}$$

The results of the well density show that in the good potential zone, for every 53 km^2 there should be a well. This is followed by the poor potential zone having a well with every 56 km^{2-} and moderate potential zone having a well in every 62 km^2 (Table 6). It proves that where the good potential of groundwater is present there will be a greater number of wells constructed. However, these findings still depends upon the resolution of the well inventory data. If the well count increases, the interpretation may change accordingly. As a result, it confirms that the GWPZ map produced by the geospatial approach is a reliable and cost effective technique and can be adopted to any part of the world.

Table 6. Summary of the well density estimation.

Zone	Description	Total Area (km^2)	Well Count	Well Density
1	Poor potential zone	1684.8	30	56
2	Moderate potential zone	2284.5	37	62
3	Good potential zone	693.7	13	53
	Total	4663.0	80	58

5. Conclusions

In designated low-potential areas, efforts must be made to improve groundwater levels. In the low-potential groundwater regions, appropriate artificial recharge structures must be created to increase groundwater recharge. Rainwater harvesting in all households must be facilitated in settlements in the low GWP areas. Groundwater pumping needs to be continuously monitored to prevent further drawdown of groundwater levels. Further studies need to be conducted to accurately identify suitable sites for the implementation of appropriate artificial recharge methods. In the critical zones, management strategies should be introduced at the legislative level to prevent further degradation. Geospatial knowledge, remote sensing, and the AHP method are considered the most effective tools for deciphering GWP zones in this region. GIS and the AHP method have been successfully used to create three groups of GWP zones. Thematic layers such as geology map, geomorphology map, soil map, drainage map, drainage and density map, lineament and lineament density map, and slope map are used in groundwater prospecting to distinguish four to three GWP zones as poor, moderate, and good, excellent, medium, moderate, and low potential. The maps were created using pre-processed remote sensing satellite data and data collected from governmental and non-governmental organizations. The results show moderate potential in the peripheral areas, while the central and northern parts have good potential for groundwater. The northwest, southwest, and some coastal areas are classified as low potential groundwater areas. More than 64% of the study area was classified as moderate to good potential for groundwater extraction. The sensitivity analysis for each stratum showed an effective weighting of the selected thematic strata, indicating that geology is the more influential criterion in mapping groundwater potential. Groundwater level maps were prepared and confirmed based on the groundwater potential map. This study shows how effectively GIS and AHP can be used to identify GWP zones. It also shows how GIS and AHP can be used to identify vulnerable, poor potential zones to consider when implementing artificial recharge structures and to reduce uncertainties in defining the boundaries of available groundwater resources. In this drought-prone region, policy makers will use the results of the study to manage water resources responsibly. This technique can also be used by other researchers interested in using spatial methods to map groundwater vulnerability in drought-prone regions.

Author Contributions: Conceptualization and writing—original draft preparation, M.A. and S.K.; Supervision, P.K.; software and visualization, S.K.; validation, K.T.; writing—review and editing, V.S. and S.Y.C.; Data collection support and formal analysis, S.M. (Subagunasekar Muthuramalingam) and M.A.; investigation and visual interpretations, M.R., S.S. and S.M. (Siva Manimuthu). All authors have read and agreed to the published version of the manuscript.

Funding: This article has been written with the financial support of RUSA, Phase 2.0 grant sanctioned vide letter No. 24-51/2014-U, Policy (TNMulti-Gen), Department of Education, Government of India, dated 10 September 2018.

Data Availability Statement: The datasets used and/or analyzed during the current study are available from the corresponding author on reasonable request.

Conflicts of Interest: The authors declare that they have no conflicts of interest.

References

1. Arnell, N.W.; Reynard, N.S. *Impact of Climatic Variability and Change on River Flow Regimes in the United Kingdom: Report to Department of the Environment, Water Directorate*; Institute of Hydrology, NERC: Wallingford, UK, 1990; 163p.
2. William, A.J.; Vaux, H., Jr. The role of science in solving the world's emerging water problems. *Proc. Natl. Acad. Sci. USA* **2005**, *102*, 157155720. [CrossRef]
3. Muruganantham, A.; Sivakumar, K.; Kongeswaran, T.; Prabakaran, K.; Bangaru Priyanga, S.; Karikalan, R.; Agastheeswaran, V.; Perumal, V. Hydrogeochemical Analysis for Groundwater Suitability Appraisal in Sivagangai, an Economically Backward District of Tamil Nadu. *J. Geol. Soc. India* **2021**, *97*, 789–798. [CrossRef]
4. Kongeswaran, T.; Sivakumar, K. Application of Remote Sensing and GIS in Floodwater Harvesting for Groundwater Development in the Upper Delta of Cauvery River Basin, Southern India. In *Water Resources Management and Sustainability, Advances in Geographical and Environmental Sciences*; Pankaj, K., Gaurav, K.N., Manish Kumar, S., Anju, S., Eds.; Springer Nature: Berlin/Heidelberg, Germany, 2022; pp. 257–280. [CrossRef]
5. Sivakumar, K.; Prabakaran, K.; Saravanan, P.K.; Muthusamy, S.; Kongeswaran, T.; Muruganantham, A.; Gnanachandrasamy, G. Agriculture Drought Management in Ramanathapuram District of Tamil Nadu, India. *J. Clim. Change* **2022**, *8*, 59–65. [CrossRef]
6. Muruganantham, A.; Sivakumar, K.; Prabakaran, K.; Kongeswaran, T.; Venkatramanan, S.; Bangaru, P.S.; Agastheeswaran, V.; Perumal, V. Role of GIS in deciphering hydrogeochemical processes and quality in Pudukottai district, Tamil Nadu, India. *Environ. Earth Sci.* **2022**, *81*, 445. [CrossRef]

7. Chandra, S.; Auken, E.; Maurya, P.K.; Ahmed, S.; Verma, S.K. Large Scale Mapping of Fractures and Groundwater Pathways in Crystalline Hardrock By AEM. *Sci Rep.* **2019**, *9*, 398. [CrossRef]
8. Saaty, T.L. A scaling method for priorities in hierarchical structures. *J. Math. Psychol.* **1977**, *15*, 234–281. [CrossRef]
9. Saaty, T.L. *The Analytic Hierarchy Process*; McGraw-Hill: New York, NY, USA, 1980.
10. Saaty, T.L. Axiomatic foundation of the Analytic Hierarchy Process. *Manag. Sci.* **1986**, *32*, 841–855. [CrossRef]
11. Rozann, W. The Analytic Hierarchy Process—What It Is and How It Is Used. *Math. Model.* **1987**, *9*, 161–176. [CrossRef]
12. Saaty, T.L. An exposition of the AHP in reply to the paper Remarks on the analytic hierarchy process. *Manag. Sci.* **1990**, *36*, 259–268. [CrossRef]
13. Kumar, A.; Sah, B.; Singh, A.; Deng, Y.; He, X.; Kumar, P.; Bansal, R. A review of multi criteria decision making (MCDM) towards sustainable renewable energy development. *Renew. Sustain. Energy Rev.* **2017**, *69*, 596–609. [CrossRef]
14. Sivakumar, K.; Shanmugasundaram, A.; Jayaprakash, M.; Prabakaran, K.; Muthusamy, S.; Ramachandran, A.; Venkatramanan, S.; Selvam, S. Causes of heavy metal contamination in groundwater of Tuticorin industrial block, Tamil Nadu, India. *Environ. Sci. Pollut. Res.* **2021**, *28*, 18651–18666. [CrossRef]
15. Prabakaran, K.; Sivakumar, K.; Aruna, C. Use of GIS-AHP tools for potable groundwater potential zone investigations-a case study in Vairavanpatti rural area, Tamil Nadu, India. *Arab. J. Geosci.* **2020**, *13*, 866. [CrossRef]
16. Ramachandran, A.; Sivakumar, K.; Shanmugasundharam, A.; Sangunathan, U.; Krishnamurthy, R.R. Evaluation of potable groundwater zones identification based on WQI and GIS techniques in Adyar River basin, Chennai, Tamilnadu, India. *Acta Ecol. Sin.* **2020**, *41*, 285–295. [CrossRef]
17. Venkatramanan, S.; Chung, S.Y.; Ramkumar, T.; Gnanachandrasamy, G.; Vasudevan, S. A multivariate statistical approaches on physicochemical characteristics of ground water in and around Nagapattinam district, Cauvery deltaic region of Tamil Nadu, India. *Earth Sci. Res. J.* **2013**, *17*, 97–103.
18. Hilletofth, P.; Sequeira, M.; Adlemo, A. Three novel fuzzy logic concepts applied to reshoring decision-making. *Expert Syst. Appl.* **2019**, *126*, 133–143. [CrossRef]
19. Sivakumar, K.; Prabakaran, K.; Venkatramanan, S.; Chung, S.Y.; Kongeswaran, T.; Muruganantham, A.; Sathish, S.; Ho-Na, S. Hydrogeochemical Survey along the Northern Coastal Region of Ramanathapuram District, Tamilnadu, India. *Appl. Sci.* **2022**, *12*, 5595. [CrossRef]
20. Fathy, A.; Karem, M.; Mohamed, A. Groundwater potential mapping using GIS, linear weighted combination techniques and geochemical processes identification, west of the Qena area, Upper Egypt. *J. Taibah Univ. Sci.* **2020**, *14*, 1350–1362. [CrossRef]
21. Saaty, T.L.; Takizawa, M. Dependence and independence: From linear hierarchies to nonlinear networks. *Eur. J. Oper. Res.* **1986**, *26*, 229–237. [CrossRef]
22. Agastheeswaran, V.; Udayaganesan, P.; Sivakumar, K.; Venkatramanan, S.; Prasanna, M.V.; Selvam, S. Identification of groundwater potential zones using geospatial approach in Sivagangai district, South India. *Arab. J. Geosci.* **2021**, *14*, 8. [CrossRef]
23. Kaliraj, S.; Chandrasekar, N.; Magesh, N.S. Identification of potential groundwater recharge zones in Vaigai upper basin, Tamil Nadu, using GIS-based analytical hierarchical process (AHP) technique. *Arab. J. Geosci.* **2014**, *7*, 1385–1401. [CrossRef]
24. Senapati, U.; Das, T. Assessment of basin-scale groundwater potentiality mapping in drought-prone upper Dwarakeshwar River basin, West Bengal, India, using GIS-based AHP techniques. *Arab. J. Geosci.* **2021**, *14*, 960. [CrossRef]
25. Ahmed, A.; Alrajhi, A.; Alquwaizany, A.S. Identification of Groundwater Potential Recharge Zones in Flinders Ranges, South Australia Using Remote Sensing, GIS, and MIF Techniques. *Water* **2021**, *13*, 2571. [CrossRef]
26. Minh, H.V.T.; Ram, A.; Pankaj, K.; Dat, Q.T.; Tran, V.T.; Hari, C.B.; Masaaki, K. Groundwater Quality Assessment Using Fuzzy-AHP in a Giang Province of Vietnam. *Geosciences* **2019**, *9*, 330. [CrossRef]
27. Chandrasekar, N.; Prince, J. Delineation of groundwater potential zones in Theni district, Tamil Nadu, using remote sensing, GIS and MIF techniques. *Geosci. Front.* **2012**, *3*, 189–196. [CrossRef]
28. Gdoura, K.; Anane, M.; Jellali, S. Geospatial and AHP-multicriteria analyses to locate and rank suitable sites for groundwater recharge with reclaimed water. *Resour. Conserv. Recycl.* **2015**, *104*, 19–30. [CrossRef]
29. Shashank, S.; Arvind, C.P. Delineation of groundwater potential zone in hard rock terrain of India using remote sensing, geographical information system (GIS) and analytic hierarchy process (AHP) techniques. *Geocarto Int.* **2015**, *30*, 402–421. [CrossRef]
30. Tanveer, D.; Nachiketa, R.; Aadil, B. Delineation of potential groundwater recharge zones using analytical hierarchy process (AHP). *Geol. Ecol. Landsc.* **2021**, *5*, 292–307. [CrossRef]
31. Mallick, J.; Khan, R.A.; Ahmed, M.; Alqadhi, S.D.; Alsubih, M.; Falqi, I.; Hasan, M.A. Modeling Groundwater Potential Zone in a Semi-Arid Region of Aseer Using Fuzzy-AHP and Geoinformation Techniques. *Water* **2019**, *11*, 2656. [CrossRef]
32. Cabrera-Barona, P.; Ghorbanzadeh, O. Comparing Classic and Interval Analytical Hierarchy Process Methodologies for Measuring Area-Level Deprivation to Analyze Health Inequalities. *Int. J. Environ. Res. Public Health* **2018**, *15*, 140. [CrossRef]
33. Navane, V.S.; Sahoo, S.N. Identification of groundwater recharge sites in Latur district of Maharashtra in India based on remote sensing, GIS and multi-criteria decision tools. *Water Environ. J.* **2020**, *35*, 544–559. [CrossRef]
34. Haile, M.; Yadeta, J.; Gambe, L. Delineation of groundwater potential zones using geo-spatial and AHP techniques: Ambo area, Blue Nile Basin, Ethiopia. *Sustain. Water Resour. Manag.* **2022**, *8*, 36. [CrossRef]
35. Song, B.; Kang, S. A Method of Assigning Weights Using a Ranking and Nonhierarchy Comparison. *Adv. Decis. Sci.* **2016**, *2016*, 8963214. [CrossRef]

36. Recep, Ç. Evaluation of Groundwater Potential by GIS-Based Multicriteria Decision Making as a Spatial Prediction Tool: Case Study in the Tigris River Batman-Hasankeyf Sub-Basin, Turkey. *Water* **2019**, *11*, 2630. [CrossRef]
37. Chakraborty, B.; Roy, S.; Bera, A.; Adhikary, P.P.; Bera, B.; Sengupta, D.; Bhunia, G.S.; Shit, P.K. Groundwater vulnerability assessment using GIS-based DRASTIC model in the upper catchment of Dwarakeshwar river basin, West Bengal, India. *Environ. Earth Sci.* **2022**, *81*, 112085. [CrossRef]
38. Thangavelu, M.; Kumanan, C. Remote Sensing and GIS Based Study of Land Suitability Evaluation in Coastal Domains of Pudukkottai District, Tamil Nadu, India. *Int. J. Geol. Earth Sci.* **2016**, *2*, 25–40.
39. Premkumar, M.; Kongeswaran, T.; Sivakumar, K.; Muruganantham, A.; Muthuramalingam, R.; Chandramohan, S.; Vasanthavigar, M. Spit Evolution and Shoreline Changes Along Manamelkudi Coast Using Geo-Spatial Techniques and Statistical Approach. *J. Clim. Change* **2022**, *8*, 59–67. [CrossRef]
40. Yuvaraj, R.M.; Rajeswari, M. GIS Based Soil Fertility Status of Pudukkottai District. *Int. J. Interdiscip. Res. Innov.* **2020**, *6*, 25–35.
41. Ramachandran Muthulakshmi, Y. Geo-spatial analysis of irrigation water quality of Pudukkottai district. *Appl. Water Sci.* **2020**, *10*, 82. [CrossRef]
42. Gunasekaran, S.; Ramasamy, S.M.; Palanivel, K.; Saravanavel, J.; Selvakumar, R. Remote Sensing and GIS Based 3d Visualization of Groundwater Level Modification and Their Signatures: Pudukkottai District, Tamil Nadu. *Asian J. Appl. Sci.* **2014**, *7*, 403–413. [CrossRef]
43. Yuvaraj, R. M Assessment of Land Use/Land Cover and Its Dynamic Using Geospatial Techniques in Pudukkottai District of Tamil Nadu. *Environ. Ecol. Res.* **2020**, *8*, 85–99. [CrossRef]
44. Monisha, N.; Balasubramaniam, P.; Janaki, D.; Ramesh, T.; Mahendran, P.P. Assessment of groundwater quality and mapping in coastal blocks of Pudukkottai district, Tamil Nadu. *Pharma Innov. J.* **2021**, *SP-10*, 777–781.
45. Pinto, D.; Shrestha, S.; Babel, M.S.; Ninsawat, S. Delineation of groundwater potential zones in the Comoro watershed, Timor Leste using GIS, remote sensing and analytic hierarchy process (AHP) technique. *Appl. Water Sci.* **2017**, *7*, 503–519. [CrossRef]
46. Rajasekhara, M.; Sudarsana, R.G.; Sreenivasulub, Y.; Siddi, R. Delineation of groundwater potential zones in semi-arid region of Jilledubanderu river basin, Anantapur District, Andhra Pradesh, India using fuzzy logic, AHP and integrated fuzzy-AHP approaches. *HydroResearch* **2019**, *2*, 97–108. [CrossRef]
47. Sener, E.; Davraz, A. Assessment of groundwater vulnerability based on a modified DRASTIC model, GIS and an analytic hierarchy process (AHP) method: The case of Egirdir Lake basin (Isparta, Turkey). *Hydrogeol. J.* **2013**, *21*, 701–714. [CrossRef]
48. Pathmanandakumar, V.; Thasarathan, N.; Ranagalage, M. An approach to delineate potential groundwater zones in Kilinochchi district, Sri Lanka, using GIS techniques. *ISPRS Int. J. Geo-Inf.* **2021**, *10*, 730. [CrossRef]
49. Mengistu, T.D.; Chang, S.W.; Kim, I.H.; Kim, M.G.; Chung, I.M. Determination of Potential Aquifer Recharge Zones Using Geospatial Techniques for Proxy Data of Gilgel Gibe Catchment, Ethiopia. *Water* **2022**, *14*, 1362. [CrossRef]
50. Sajil Kumar, P.J.; Elango, L.; Schneider, M. GIS and AHP Based Groundwater Potential Zones Delineation in Chennai River Basin (CRB), India. *Sustainability* **2022**, *14*, 1830. [CrossRef]

water

MDPI

Article

Impacts and Implications of Land Use Land Cover Dynamics on Groundwater Recharge and Surface Runoff in East African Watershed

Tarekegn Dejen Mengistu [1,2], Il-Moon Chung [1,2,*], Min-Gyu Kim [2], Sun Woo Chang [1,2] and Jeong Eun Lee [2,*]

1 Construction Environment Engineering Department, University of Science and Technology, Daejeon 34113, Korea; tarekegnmengistu@kict.re.kr (T.D.M.); chang@kict.re.kr (S.W.C.)
2 Water Resources and River Research Department, Korea Institute of Civil Engineering and Building Technology, Goyang 10223, Korea; kimmingyu@kict.re.kr
* Correspondence: imchung@kict.re.kr (I.-M.C.); jeus22@kict.re.kr (J.E.L.)

Abstract: Assessing the spatiotemporal dynamics of land use land cover (LULC) change on water resources is vital for watershed sustainability and developing proper management strategies. Evaluating LULC scenarios synergistically with hydrologic modeling affords substantial evidence of factors that govern hydrologic processes. Hence, this study assessed the spatiotemporal effects and implications of LULC dynamics on groundwater recharge and surface runoff in Gilgel Gibe, an East African watershed, using the Soil and Water Assessment Tool (SWAT) model. Three different LULC maps (2000, 2010, and 2020) were derived from Landsat images, and the comparisons pointed out that the land-use pattern had changed significantly. The agricultural land and grassland cover increased by 3.76% and 1.36%, respectively, from 2000 to 2020. The implications acquired for 2000 show that forested land covers decreased by 5.49% in 2020. The SWAT simulation process was executed using a digital elevation model, soil, LULC, and weather data. The model was calibrated and validated using streamflow data to understand the surface runoff and groundwater recharge responses of each Hydrologic Response Units on reference simulation periods using the Calibration and Uncertainty Program (SWAT-CUP), Sequential Uncertainty Fitting (SUFI-2) algorithm. The observed and simulated streamflows were checked for performance indices of coefficient of determination (R^2), Nash–Sutcliffe model efficiency (NSE), and percent bias (PBIAS) on monthly time steps. The results show that there is good agreement for all LULC simulations, both calibration and validation periods (R^2 & NSE ≥ 0.84, $-15 < \text{PBIAS} < +15$). This reveals that for the LULC assessment of any hydrological modeling, the simulation of each reference period should be calibrated to have reasonable outputs. The study indicated that surface runoff has increased while groundwater decreased over the last two decades. The temporal variation revealed that the highest recharge and runoff occurred during the wet seasons. Thus, the study can support maximizing water management strategies and reducing adverse driving environmental forces.

Keywords: groundwater recharge; Gilgel Gibe watershed; LULC; Soil and Water Assessment Tool (SWAT); SWAT-CUP; surface runoff

Citation: Mengistu, T.D.; Chung, I.-M.; Kim, M.-G.; Chang, S.W.; Lee, J.E. Impacts and Implications of Land Use Land Cover Dynamics on Groundwater Recharge and Surface Runoff in East African Watershed. *Water* **2022**, *14*, 2068. https://doi.org/10.3390/w14132068

Academic Editors: Sang Yong Chung, Gyoo-Bum Kim and Venkatramanan Senapathi

Received: 23 May 2022
Accepted: 27 June 2022
Published: 28 June 2022

1. Introduction

Land and water resources are dynamically hindered by uncertainties of climate and land use land cover (LULC) changes, aggravating the water crisis worldwide [1,2]. Inappropriate exploitation and poor management systems increasingly threaten land and water resources, changing the natural landscapes for human use [3–5]. Examining the possible effects of LULC change on the hydrologic cycle under natural and human activities is required to manage available resources properly [6]. However, LULC change responses to environmental and socio-economic drivers continue to be a significant scientific hindrance in assessing its effect on water availability [7]. In different climate regions, groundwater is the main fresh water supply; its utilization and management are thoroughly associated with the Sustainable Development Goals [8–10]. Conversely, in most African countries, the salient use of groundwater for domestic and agricultural purposes significantly affects groundwater recharge and food security [5,8,11].

Human activities incur massive changes in the terrestrial environment, but the possible effects and implications of such changes on groundwater recharge and surface runoff (GRSURQ) are poorly understood. In Africa, the influence of LULC is much larger than climate variability [12]. The spatiotemporal variability of groundwater recharge and streamflow is essential in understanding the abstraction scenarios of water management [13,14]. Groundwater is a preferable water supply over surface water, while the extreme inter-annual rainfall variations stress surface water availability [15]. However, the impacts of LULC change on groundwater recharge are not sufficiently recognized, though it is a substantial freshwater source for domestic, agricultural, and commercial uses, which results in groundwater depletion [16]. The main implications are overexploitation of natural resources [17], ecosystem service [18], erosion, land degradation, and deforestation [19,20], which affect ecological and economic sustainability [21].

In developing countries such as Ethiopia, estimating GRSURQ is challenging because of the scarcity of relevant data [13]. Ethiopia has significant land and water resources, playing a minimum role in developing the national economy. It is being affected by various environmental challenges in order to enhance agricultural productivity for food security. Over the past decades, an extensive LULC change has been observed mainly from man-made and natural forces [22,23]. Both natural and anthropogenic activities adversely affect watershed hydrology, causing water stress, which shifts ecosystem biodiversity. Many studies have assessed drivers of LULC change affecting environmental resources [24]. Of which, cutting of trees for fuelwood and charcoal [25,26], resettlement, land tenure policy, population growth, poverty, intensive agricultural practice [27,28], and drought occurrence [29,30] have been recognized as primary drivers in different parts of the country. Poor LULC practices in highland areas have resulted in substantial soil loss, reduced agricultural production, and groundwater depletion [31]. This adversely affects essential aspects of the environment, geomorphologic patterns, flora and fauna [32], habitat fragmentation, depletion of biodiversity [33], and changes in climatic conditions [34]. Where many drivers can potentially conceal, reducing the impacts of LULC change on groundwater recharge is paramount [35].

LULC change affects spatiotemporal scales of land surfaces, changing surface runoff, groundwater recharge, evapotranspiration, and river flow [36–38]. However, critical water stress has been faced due to information gaps in the decision-making process in most developing countries, including Ethiopia [13]. Understanding the hydrologic cycle to establish a suitable model for a watershed is crucial in planning and utilizing water resources. Nevertheless, analyzing and quantifying hydrologic components require a realistic hydrologic model representing watershed processes [39]. Watershed models have been used as a dynamic mechanism to address a comprehensive spectrum of environmental problems, advancing estimates' predictive accuracy [28,40,41]. The Soil and Water Assessment Tool (SWAT) [42] is a vigorous hydrologic model to estimate hydrologic fluxes in combating water scarcity issues [43–46]. The model integrates multiple ecological processes supporting management and decision-making scenarios worldwide [39,41,47–53].

The heterogeneity of variations in hydrological fluxes with LULC change over the years challenges how to realize the possible effects [54]. Qualifying and quantifying the spatiotemporal variations of groundwater recharge based on updated spatial information are indispensable for watershed sustainability [13,55]. Nonetheless, basic information on the spatiotemporal patterns of LULC change and its effect on GRSURQ in Ethiopia is rarely understood, while water and food security are central issues. GRSURQ is challenging because of drastic changes in vast anthropogenic activities. Controlling undesirable surface runoff in watersheds preserves soil nutrients and water availability, maximizing crop yields to reduce food security issues [56]. Furthermore, surface runoff is the driving energy of soil erosion to reservoirs [57], which might result in sediment deposition, directly affecting groundwater recharge if not adequately regulated.

In the Gilgel Gibe watershed, where the surface water is intermittent due to the erratic nature of rainfall, groundwater is becoming a critical source of fresh water, as locating sustainable and productive aquifers is challenging [58]. Hence, modeling efforts in LULC change are fundamental to developing water infrastructure supporting sustainable watershed management strategies in data-sparse environments. Sustainable water resources management requires a significant understanding of the effects and implications of LULC and watershed monitoring strategies. However, no exclusive evidence exists on how changes to LULC affect GRSURQ in the Gilgel Gibe watershed. Thus, the present study aims to assess the spatiotemporal effects and implications of LULC dynamics on groundwater recharge and surface runoff in the Gilgel Gibe watershed, Ethiopia, in East Africa, using the SWAT model over the past two decades. The SWAT model was successfully applied to assess the exclusive effects of LULC changes on GRSURQ to verify the viability of predicting streamflow with independent calibration and validation for each LULC period simulation. This study is robustly helpful in promoting sustainable watershed management and resource development programs under data-scarce conditions. This will enable planners and policy-makers to apply effective management strategies to diminish the undesirable effects of natural and uncontrolled human activities.

2. Materials and Methods

2.1. Description of Study Area

The Gilgel Gibe watershed is situated in Eastern Africa, the Omo-Gibe River Basin, and a semi-arid southwestern region of Ethiopia. The watershed reaches through the Addis Ababa–Jimma–Gambella asphalt road to the west of the Main Ethiopian Rift. The dominant soil types are alisols (45.11%), vertisols (40.79%), nitisols (13.73%), and leptosols (0.37%), with textures ranging from clay to sandy loam [58]. The areal coverage of slope in percentage is 0–5 (15.41%), 5–10 (15.63%), 10–15 (16.55%), 15–35 (40.75%), and 11.65% for slopes above 35 degrees [58]. The geological structures are columnar joints in basalts and flow bands in trachytes and rhyolites. The area reveals Tertiary volcanic rocks, Precambrian basements, Mesozoic rocks, Pliocene age volcanics, quaternary pyroclastics, and alluvial sediments [59].

2.2. SWAT Model Description

The Soil and Water Assessment Tool (SWAT) [42] is a continuous, physically based, and lumped model capable of simulating water management environments [45,60]. The model is developed to simulate runoff and nutrient losses with readily available input data to assess management practices [42,61]. SWAT has usually been used to evaluate the impacts and implications of LULC change on watershed hydrology [17,23,36,48,62]. It simulates a shallow unconfined aquifer that donates water to the mainstream or reaches the sub-basin and a confined deep aquifer [61]. This enhances the precision of the water balance, providing robust physical meaning. In addition, SWAT simulates the hydrological cycle [43,63] using the water balance Equation (1).

$$SW_t = SW_o + \sum_{i=1}^{t} \left(R_{day} - Q_{surf} - E_a - W_{seep} - Q_{gw} \right) \quad (1)$$

where: SW_t final water content (mmH_2O), SW_o initial soil water content (mmH_2O), t time (days), Rday precipitation amount (mmH_2O), Q_{surf} surface runoff (mmH_2O), E_a actual evapotranspiration (mmH_2O), *Wseep* water entering vadose zone from soil profile (mmH_2O), Q_{gw} return flow amount (mmH_2O).

2.3. Model Input Data Preparation

The SWAT model needs topography, LULC, soil, and daily weather data to simulate hydrologic processes. Digital elevation model (DEM) represents watershed topography attained from the Shuttle Radar Topography Mission of the US Geological Survey (Figure 1a). The soil data were acquired from Food and Agriculture Organization (FAO), Harmonized World Soil Database (HWSD) [64]. HWSD Viewer Version 1.21 [65] was used to generate a code for soil properties. Different types of soil texture and physical-chemical properties of soils are required for SWAT simulations [61,66]. The dominant soil groups were identified, the soil map was attached to the soil database (Figure 1b), and a lookup table was prepared that links the soil class and input soil map to the SWAT model database.

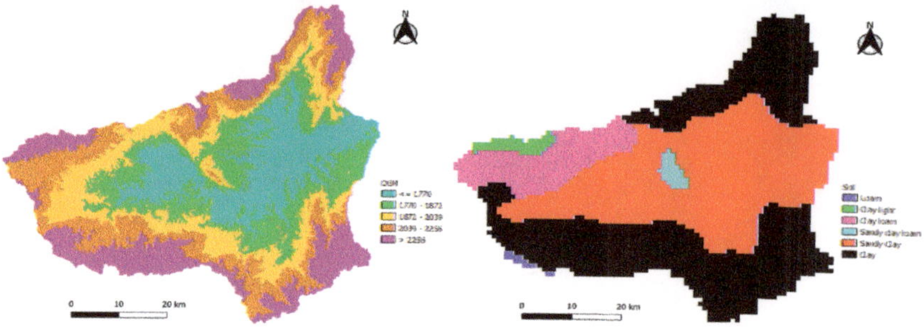

(a) (b)

Figure 1. Study area maps of (**a**) digital elevation model (DEM), and (**b**) soil.

LULC data were retrieved from accurate and high-resolution Landsat images [67], Global Land Cover Datasets of GlobeLand30 [68]. This study considered LULC of the years 2000 (Figure 2a), 2010 (Figure 2b), and 2020 (Figure 2c). LULC accuracy assessment was executed using Semi-Automatic Classification Plugin (SCP) for the Quantum Geographic Information System (QGIS) environment. It was accomplished by comparing a sample of points (ground truth) to assess the dependability and precision of the classified map. A confusion matrix was used to categorize the accuracy derived from the user's accuracy, producer's accuracy, and the kappa information. A kappa distribution rate of more than zero is considered to be good agreement [69]. Kappa coefficient measures how the classification results compare to values assigned randomly. If the kappa coefficient equals one, the classified and ground truth images are identical. Finally, the accuracy of LULC maps was checked to gain confidence for test applications.

The GlobeLand30 Classification System has specifically given details for each LULC type. Cultivated land (AGRC) is used to produce crops; it comprises irrigated upland, vegetable land, cultivated pasture, and coffee garden. Forest (FRST) is land covered with trees, mixed forests, and sparse woodland, occupying a maximum density of over 30%. Grassland (RNGE) is land protected by natural grass with a cover density of over 10%. Shrubland (SHRB) is the land covered with shrubs, and the cover density is over 30%. Wetland (WETN) denotes the junction of land and water areas covered by hygrophyte and wet soils. In contrast, water bodies (WATR) refer to the area's liquid water-covered river, reservoirs, and pit-pond. Artificial surfaces (URBN) denote surfaces made by man-built activities in urban and rural areas and industrial and transportation facilities. Bare land (RNGB) includes naturally covered lands such as desert, sand, gravel ground, bare rocks, saline, and alkaline lands with a cover density lower than 10%. SWAT uses these codes to link the LULC of the study watershed to the SWAT land use database. Subsequently, a user lookup table was prepared that identifies SWAT code (AGRC, FRST, RNGE, SHRB, WATR, URBN, and RNGB) for each LULC category to simulate the model. Finally, each LULC (2000, 2010, and 2020) period map was used for a separate SWAT model (1988–2018) simulation of hydrologic processes.

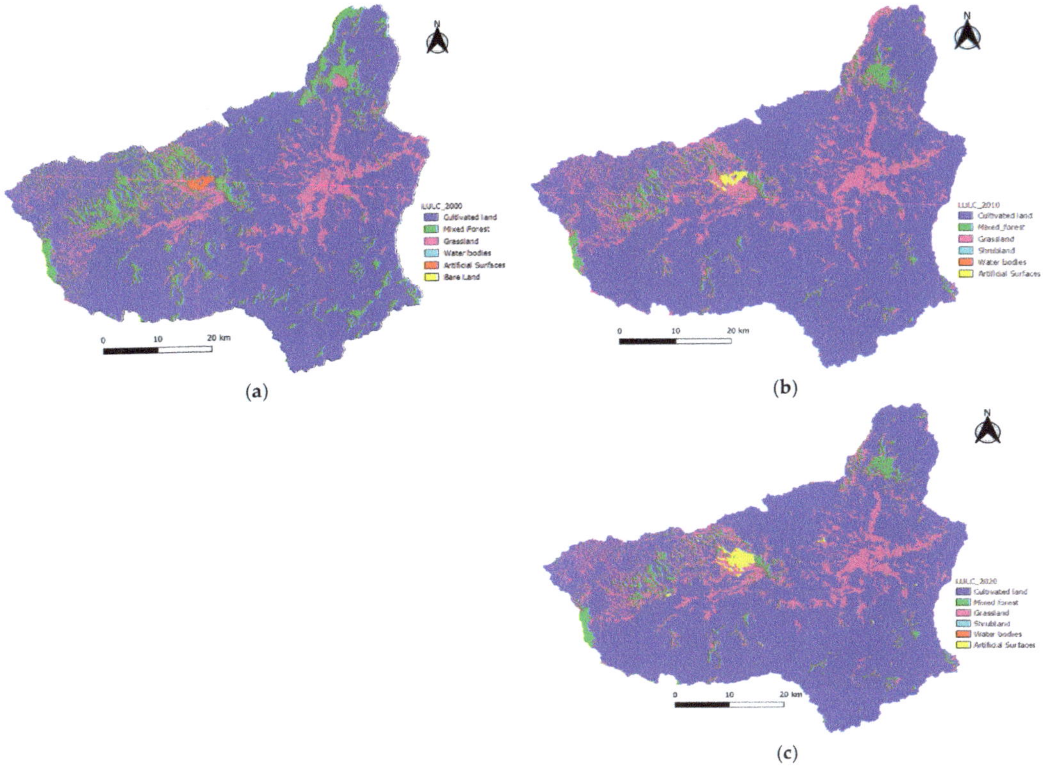

Figure 2. The study area LULC maps of (**a**) 2000, (**b**) 2010, (**c**) 2020.

2.4. SWAT Model Setup and Simulation

SWAT model setup was developed to simulate the watershed hydrological process and evaluate GRSURQ. First, the stream definition was carefully determined by selecting the threshold required to form the origin of streams. In this present study, the slope map was produced from DEM data. The study established the multiple slope option, considering different slope classes to create hydrologic response units (HRUs). Then, the land area was distributed into HRUs with exclusive evidence related to various features of land use, management, and soil attributes [61,70]. HRUs are defined in the QSWAT [71] interface of the QGIS environment. QSWAT is written in a robust programming language, Python, with anticipated supporting code readability [71]. HRUs increase accuracy in predicting loadings from the sub-basin. The LULC and soil data are used to delimit HRUs and tie the DEM with the crop and soil databases. A 5–10% threshold value is used to avoid small HRUs, reduce the total number of HRUs, and increase model efficiency [61]. All soil water balance constituents are computed on an HRU basis as comparable HRUs would have analogous hydrologic features [42].

Weather data definition is an essential requirement for SWAT simulation. The SWAT 2012 requires daily precipitation (mm), temperature (°C), relative humidity (percentage), solar radiation (MJ/m^2), and wind speed (m/s) of meteorological stations inside and in the buffer zone of the watershed. The present study used Jimma, Sekoru, Shebe, Assendabo, Busa, Dedo, and Omo Nada weather stations collected from the National Meteorological Agency to simulate the hydrology of the watershed. The streamflow data were collected from the Ministry of Water, Irrigation, and Electricity. Weather Generator (WGN) Parameters Estimation Tool interpolates missing data from the synoptic stations for specific simulation periods. Hence, WGN provided all necessary statistical information of synthetic daily meteorological records to fill in missing data appropriately. WGN generates wind speed, solar radiation, and relative humidity from precipitation and temperature [61]. After computing the WGN parameter, the corresponding lookup table was prepared according to the SWAT model format, including the two-year warmup periods. The warmup period was taken to ensure no effects from the initial conditions. Then, the LULC, soil, and slope layers were overlaid, basin-wise HRUs were created, weather data were defined, and SWAT2012 was simulated with the total simulation period from 1988 to 2018. The general outline of the workflow structure used in this assessment is shown in Figure 3.

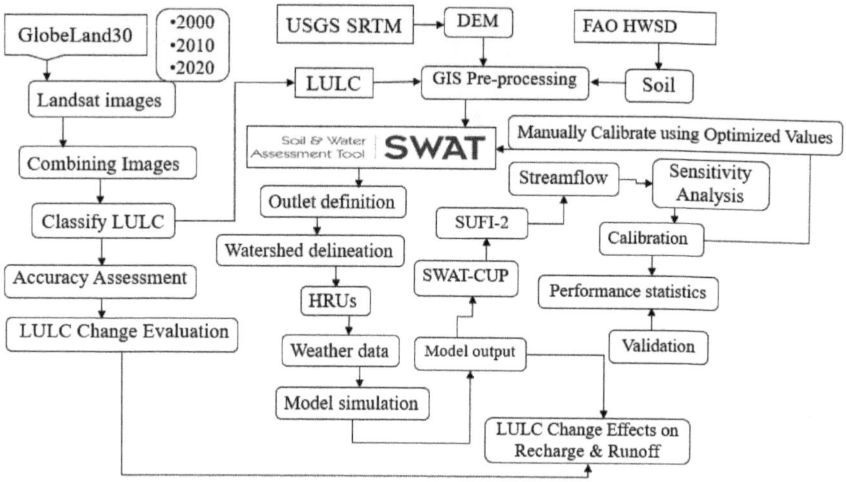

Figure 3. The general workflow design of the study methodology.

2.5. Model Calibration and Validation

SWAT is a sophisticated, robust, and interdisciplinary watershed modeling tool. However, the practical application of the hydrologic model to predict streamflow is determined by calibration and validation techniques [72–75]. It is helpful to find optimal parameters set with the optimum objective functions [74,76]. In the calibration process, parameters transferred from gauged to ungauged sites in data-scarce conditions could be affected by human activities [77]. Correct parameters make model calibration faster, more accurate, with low prediction uncertainty [66], and describe comprehensive hydrological processes [78]. However, model applications in different climatic regions are still challenged by a lack of historical data through the most commonly used watershed response variable in performance evaluation [44].

The SWAT Calibration and Uncertainty Program (SWAT-CUP) was developed to quantify SWAT model calibration, validation, and sensitivity analysis [63]. The capability of the SWAT-CUP encompasses an automated approach to conduct performance analysis more rigorously [66,72]. SWAT-CUP integrates various techniques in one interface that is easy to use and efficient [79]. The SWAT-CUP is an open-access program that connects the Sequential Uncertainty Fitting Version 2 (SUFI-2) algorithm to the SWAT model output [44]. The SUFI-2 accounts for all causes of uncertainty driving factors in water resources [72,79]. Sensitivity analysis determines model output changes regarding model input changes [65]. This study supported the auto analysis using SWAT-CUP and SUFI-2. An auto analysis attains suitable parameter estimates consistent with historical data to decide and refine estimates [80].

A sensitivity analysis affords reflections that limit output variances due to input variability [72]. The t-stat measures the sensitivity, and the *p*-values decide the implications [75]. A *p*-value close to zero has a meaningful value: the larger the t-stat value, the lesser the *p*-value, and the more sensitive the parameter [72]. The practical optimized value within a given bound minimizes the relative error [66]. Model validation indicates that a given model can produce accurate predictions without changing parameter values during calibration [81]. The observed and simulated annual streamflow was used to calibrate (1997–2006) and validate the (2007–2014) effects of LULC changes on GRSURQ. After calibration and validation, the impacts and implications of LULC change on GRSURQ have been analyzed.

2.6. Estimation of Model Predictive Accuracy

Statistical indicators were evaluated using SWAT-CUP to measure the best parameter [77]. The coefficient of determination (R^2), Nash–Sutcliffe efficiency coefficient (NSE), and percent bias (PBIAS) were used to define the dependability of forecasts compared to experimental values of SWAT model performance. R^2 describes the percentage of variance to indicate the correlations between simulated and observed values. R^2 ranges from 0 to 1, where close to 1 and 0 show excellent and poor data presentation [80,82–84]. NSE assesses the predictive power and overall agreement of simulated and observed hydrographs. For acceptable model performance, NSE should be close to 1. PBIAS measures the normal tendency of simulated data with observed data [80]. The model performance ratings were evaluated based on a range of values for R^2, NSE, and PBIAS, as shown by Equations (2), (3), and (4), respectively.

$$R^2 = \frac{\left[\sum_{i=1}^{n}(Q_{si} - \overline{Q}_{si})\,(Q_{oi} - \overline{Q}_{o})\right]^2}{\sum_{i=1}^{n}(Q_{si} - \overline{Q}_{si})^2 \sum_{i=1}^{n}(Q_{oi} - \overline{Q}_{o})^2} \tag{2}$$

$$\text{NSE} = 1 - \frac{\sum_{i=1}^{n}(Q_{oi} - Q_{si})^2}{\sum_{i=1}^{n}(Q_{oi} - \overline{Q}_{oi})^2} \tag{3}$$

$$\text{PBIAS} = \left[\frac{\sum_{i=1}^{n} Q_{oi} - \sum_{i=1}^{n} Q_{si}}{\sum_{i=1}^{n} Q_{oi}}\right] \times 100 \tag{4}$$

where: Q_{si} (simulated value),

Q_{oi} (measured value), $Q_{\ si}$ and $Q_{\ oi}$ (mean of simulated and observed discharge, respectively).

3. Results and Discussion

3.1. SWAT Model Sensitivity Analysis

Hydrologic modeling and simulation were performed using the QSWAT in the QGIS environment. After preparing all model inputs, the model was simulated. Then, the sensitivity analysis was performed at the Asendabo station of the main Gilgel Gibe River. The sensitivity analysis identifies the most responsive hydrological parameters that significantly influence specific model output to enhance the reliability of results. The monthly streamflow data simulation was performed for 12 years, from 1997 to 2006. The sensitivity was determined using the t-stat and *p*-values provided by the SUFI-2 program. In these statistics, the higher the t-stat values, the greater the relative sensitivity. The *p*-values were used to fix the sensitivity implication so that the closer the *p*-values to zero, the more critical the parameters become. Before calibration and validation, many more parameters were used to identify the most sensitive parameters that govern streamflow generation. Then, the 14 most sensitive parameters were selected to calibrate and validate model predictive capability based on sensitivity evaluation criteria. The parameter values were adjusted by changing parameters at a time within acceptable ranges until the best simulation was attained [70]. The selected most sensitive parameters and their relatively optimized values are indicated in Table 1 with qualifiers. The parameter qualifier (R_) refers to the default value multiplied by (1 plus an optimized value), and (V_) refers to the replacement of the default value with an optimized one. The extensions .hru, .mgt, .bsn, .gw, .sol, and .rte indicate the SWAT parameter family of HRU, management, basin, groundwater, soil, and route, respectively.

Table 1. Sensitive model calibration parameters and optimized values for LULC-2000 simulation.

Parameter Name	Description	Range	Optimized Value
R_CN2.mgt	Curve number for moisture condition II	−0.02–0.2	0.077
V_ALPHA_BF.gw	Coefficient of depletion of groundwater	0.01–1	0.120
V_ESCO.hru	Soil evaporation compensation factor	0–1	0.069
V_GW_DELAY.gw	Groundwater delay	0–350	195.650
V_GW_REVAP.gw	Groundwater "revap" coefficient	0.02–0.2	0.091
V_CANMX.hru	Maximum canopy storage	10–100	70.570
V_EPCO.hru	Plant uptake compensation factor	0–1	0.649
V_CH_K2.rte	Effective hydraulic conductivity	0.01–150	118.052
R_SOL_AWC().sol	Available water capacity of the soil	−0.5–0.5	0.403
R_SOL_K().sol	Saturated hydraulic conductivity	−0.5–0.5	−0.443
V_SURLAG.bsn	Surface runoff lag coefficient	0–24	12.744
V_SHALLST.gw	Initial depth of the shallow aquifer	0–500	404.450
R_SOL_Z().sol	Soil depth	0.5–1	0.681
R_GWQMN.gw	Threshold depth of shallow water aquifer	0–2	0.242

The model was calibrated and validated to evaluate model simulation performance in the watershed under changing environments due to extensive human-induced factors. LULC change uncertainty substantially influences hydrologic cycles, complicating groundwater recharge and surface runoff modeling outputs [8]. Model parameters are usually calibrated under one LULC scenario, and are then supposed to be time-invariant while simulating different hydrological LULC scenarios. However, other variables of LULC affect simulations to quantify streamflow variability for a new LULC scenario. Through anthropogenic activities, ensuing changes in the watershed feature that model parameters are not appropriate for other LULC situations [5,8,17]. As calibration parameters differ while LULC changes, using the same parameters may not always be feasible. The simulations change when LULC adjustments affect HRU configurations; this no longer affects the new LULC simulation. The attributes of HRUs are the significant influences affecting streamflow and other hydrologic components.

Consequently, in the present study, simulations of each reference LULC period were calibrated using an auto-calibration technique to optimize the values of sensitive parameters. In addition, new parameters were included to check whether the previously applied parameter could represent the hydrologic simulation process or not. Nevertheless, optimal values and parameter substitutions were observed regardless of the similarity of parameters for the 2010 and 2020 LULC simulation periods. Hence, the calibration for simulation of 2010 and 2020 LULC indicates that CN2, GW_DELAY, GW_REVAP, ALPHA_BF, ESCO, EPCO, CH_K2, RCHRG_DP, SOL_AWC, SOL_K, and GWQMN were newly optimized best parameters with replacement and multiplication of actual simulation results. This might support reliable land and water resource development plans.

3.2. *Assessing Hydrological Model Performance on Streamflow*

SWAT-CUP application with SUFI-2 set of rules has been employed for calibration, validation, and uncertainty assessment of SWAT output. This study calibrated the model to make the simulation result more realistic for the independent calibration time steps. The statistically significant model performance among time intervals was judged based on recommendations given in Moriasi et al. [75,76], the coefficient of determination (R^2), Nash and Sutcliffe's model efficiency (NSE), and percentage bias (PBIAS). The hydrology is well-simulated and is considered representative of the watershed if the statistical indicators R^2, NSE, PBIAS, and graphical fitness are satisfied. The calibration period of the SWAT model was (1997–2006), excluding two years of model warmup periods and its validation period (2007–2014). R^2, NSE, and PBIAS compared the model's applicability. R^2 suggests that the observed and simulated values are in good agreement as it is close to one. PBIAS characterizes the error among the experimental and simulated values as a percentage. The objective functions were within the acceptable range of goodness of fit tests [75]. The model performance showed that statistical values simulating monthly streamflow were R^2, NSE, and PBIAS of 0.88, 0.87, and −7.9%, respectively, during the calibration time steps of the 2000 land use simulation. Model validation results should increase user confidence in model predictive capabilities [76]. The model was validated with observed flow data for eight years (2007–2014) without further adjusting calibration parameters. The SWAT overall performance for the 2000 LULC simulation during validation was 0.85, 0.85, and −4.0% for R^2, NSE, and PBIAS, respectively.

As depicted in Figure 4a–c, the hydrographs indicated that the results were in good agreement with measured and simulated streamflow data. The selected statistical performance indicators show that there is good agreement for both calibration and validation periods (R^2 & NSE > 0.84, −15 < PBIAS < +15), which are in reasonably acceptable ranges as in [75,76]. The positive and negative results of PBIAS showed underestimation and overestimation, respectively. Hence, during the model's second simulation, the model overestimated (1%, 3%) calibration and validation time steps, respectively. Moreover, the PBIAS (−7.9% and −4.0%) for calibration and validation, respectively, indicates that the model was underestimated by (7.9%, 4.0%) for 2000 and (14.3%, 9.3%) for 2020 land use periods (Table 2) during calibration and validation time steps. As revealed in Figure 5a–c for calibration and Figure 5d–f for validation, the scattered plot of the observed and simulated streamflow, the best-fit line's correlation coefficient of (0.88, 0.86, 0.87) during calibration and (0.85, 0.85, 0.86) during validation time steps is observed for 2000, 2010, and 2020 land use periods, respectively. The statistical performance evaluation results have been statistically accurate for all calibration and validation time steps.

Table 2. Statistical performance indicators during calibration and validation LULC periods.

Variable	2000		2010		2020	
	Calibration	Validation	Calibration	Validation	Calibration	Validation
R^2	0.88	0.85	0.86	0.85	0.87	0.86
NSE	0.87	0.85	0.86	0.85	0.84	0.85
PBIAS	−7.9	−4.0	1.0	3.0	−14.3	−9.3

Statistical values (Table 2) indicate that objective functions were acceptable for model evaluation [75,76]. Several authors [14,23,36,40,57,62] calibrated the SWAT model in other Ethiopian watersheds. They mainly reported statistical performance indicators of R^2, NSE, and PBIAS based on [75], comparable with the present findings. The resulting statistics were comparable to other worldwide studies [35,48–53,77]. The SURQ relies upon overall precipitation, evaporation, and soil water storage. The most considerable SURQ is derived from high precipitation and elevation. The most increased flows are typical from July to September, and the lowest flows occur during February and May. A regular correlation between rainfall and runoff was observed (Figure 4a–c). Accordingly, the applicability of the simulated SWAT model was found to be in agreement with regional and global studies and was reasonably acceptable in the Gilgel Gibe watershed.

SWAT was found to be reasonably appropriate for estimating the spatiotemporal variability of GRSURQ in the Gilgel Gibe watershed. Therefore, the simulated hydrological parameters can be helpful in the planning and management of water resource projects. The SWAT model was simulated, properly calibrated, validated, and confirmed to assess the impact of LULC on the watershed's hydrology. The performance of simulated and measured results suggests that SWAT properly represents the streamflow modeling. Furthermore, the results indicated that the SWAT model performs well. Hence, the calibrated model can be used to analyze the effect of climate and LULC change on groundwater abstraction and drought extremes. Thus, this can be improved with integrated surface water and groundwater interaction of abstraction scenarios to assess the impact on rivers drying up or groundwater depletion. Consequently, the model output can support the water resource decision-making process.

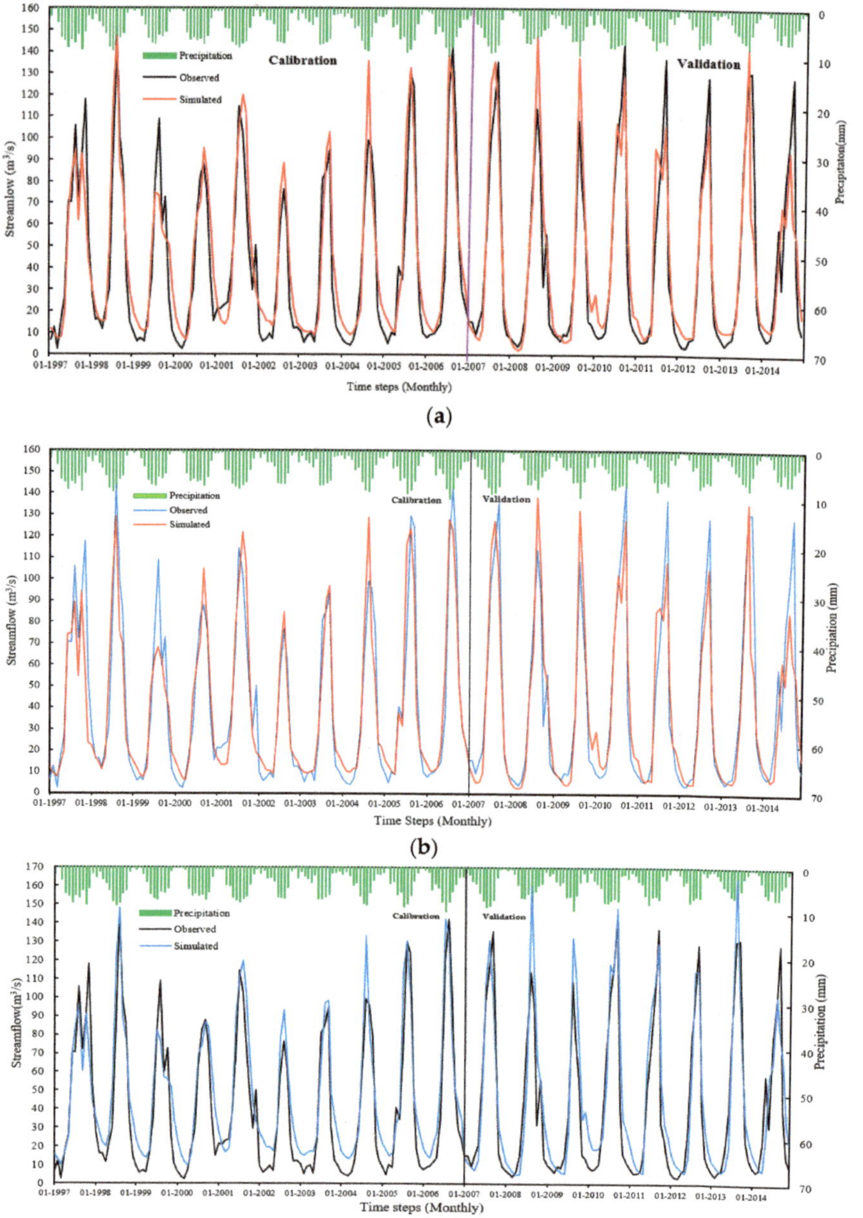

Figure 4. Hydrograph of measured and simulated flow during calibration and validation of (**a**) 2000, (**b**) 2010, and (**c**) 2020 reference periods.

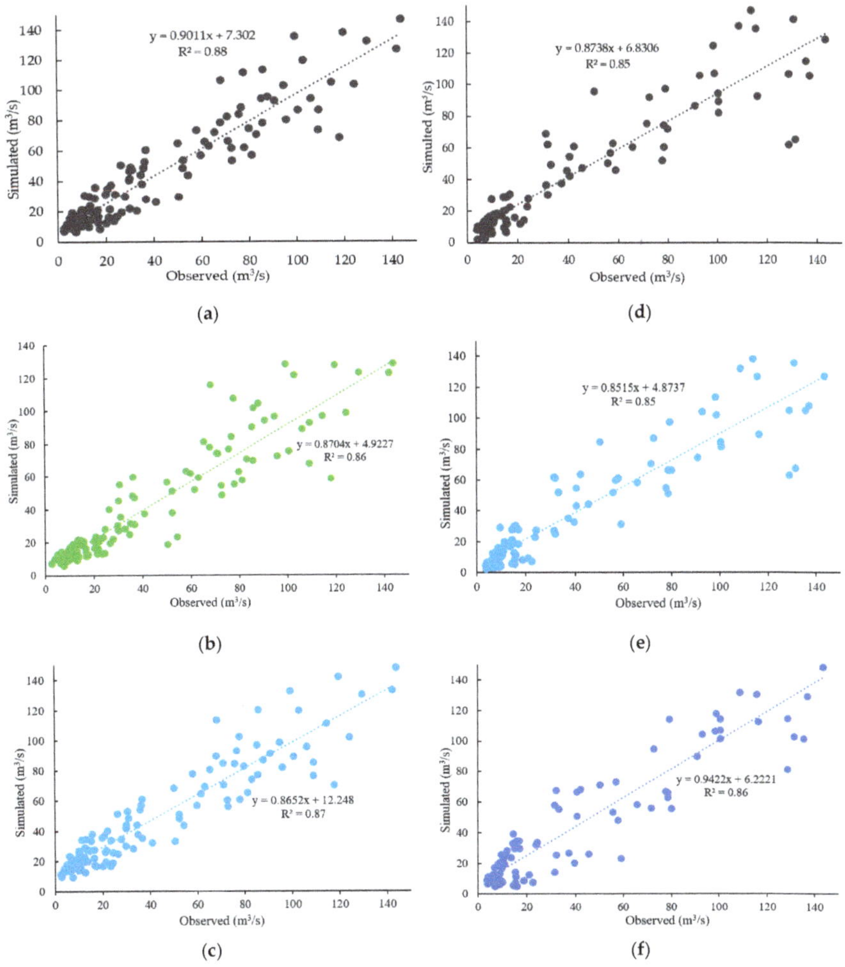

Figure 5. Scatter plot of measured and simulated flow during calibration (**a**–**c**); validation (**d**–**f**) for LULC-2000, -2010, and -2020 simulations, respectively.

3.3. Effects and Implications of LULC Change

In watershed hydrology, understanding LULC change will empower planners and policy-makers to reduce adverse effects. The forest covers have undergone deforestation and decreased by 5.4936%. The barren land has been entirely transformed by 0.001% of other activities. The urban/built-up land classes have increased by 0.4511% over the last two decades. Agricultural land covered about 77.6760%, 80.8945%, and 81.4319% of the entire watershed in 2000, 2010, and 2020. Assessment of LULC over a long time confirmed continuous agricultural activity and deforestation. The agricultural land use was augmented by 3.2184% from 2000 to 2010 and 0.5374% from 2010 to 2020. During the last two decades, agricultural land has increased significantly by 3.7558% from 2000 to 2020. The artificial surfaces have proven to be 0.0416% from 2000 to 2010, 0.4094% from 2010 to 2020, and 0.4510% from 2000 to 2020. An increase in agricultural land results in a reduction of forested areas and grassland. Accordingly, the forested land area changed by −5.6942%, +0.2006%, and −5.4936% from 2000–2010, 2010–2020, and 2000–2020. The grassland change was observed to be +2.5079%, −1.1485%, and +1.3594% from 2000–2010, 2010–2020, and 2000–2020, respectively. The available bare lands in 2000 have completely transformed into other LULC types. The implications acquired for 2000 show that forested area land covers were 9.8708% but decreased by 5.6942% in 2010 and 5.4936% in 2020.

Similarly, grassland cover was 13.360% in 2000, increased by 2.5079% in 2010, but decreased by 1.1485% in 2020. Numerous anthropogenic activities associated with socio-economic and biophysical environments reduced forested areas. The effects showed significant variations in LULC that occurred from 2000 to 2010. From a political atmosphere point of view, it could be regarded that poor development policy, globalization, and market forces of multi-national initiatives drive LULC changes. It has been hypothesized that LULC change results from favorable economic and institutional conditions triggered by the expansion of agriculture and rangelands.

LULC disturbs water budgets by reducing infiltration and increasing surface runoff in the watershed. Human activities in the biophysical environment increase reliance on agriculture. The expansion of cultivation is to produce a crop at the expense of forests and grazing lands. This increases land susceptibility to erosion and sedimentation of water bodies and reservoirs. This is attributed to the loss of fertile topsoil, soil degradation, the decline in organic matter, decreased plant-available water, and nutrient loss, reducing crop yields. Agricultural land increases as the population increases, which is the reason for the cultivation of farming products. LULC changes in the watershed have been altered extensively because of human activities. There is a change in vegetation and forestlands due to agriculture-affected water bodies in the ecosystem and land surfaces. The agricultural land use practice encourages more surface runoff than infiltration. Hence, LULC change impact on spatiotemporal assessment is substantial for socio-economic and environmental development. The findings are in agreement with studies conducted by [6,17,20–23,29,56]. Similar studies confirmed that LULC changed from forest to agricultural land, waterbody, industrial farmland, and built-up areas.

3.4. *Effects and Implications of LULC Change on Surface Runoff*

LULC change usually induces significant changes in flood peak and infiltration properties, thus affecting the entire hydrological condition of the watershed. Assessing the spatiotemporal variability, the effect, and the implications of LULC changes is essential to estimating water balances. Computational reliability of hydrological model simulation increases with well-calibrated model simulation. In the present study, the SWAT model was initially calibrated using the LULC map of 2000, then updated to 2010 and 2020 to examine the effect of LULC changes on GRSURQ. For the first LULC-2000 period, the watershed has a total mean annual value of actual evapotranspiration of 608.61 mm, groundwater contribution to streamflow of 313.28 mm, lateral flow of 75.9 mm, recharge of 348.16 mm, SURQ 347.9 mm, and water yield 754.52 mm. The simulated annual water balances indicated that actual evapotranspiration loses 44% of the yearly precipitation and 56% of the rainfall in the watershed contributes to the streamflow during the simulation period. The average annual contribution of groundwater relative to rainfall is 22.7%. Therefore, 25.22% of precipitation is lost as groundwater recharge.

Surface runoff is the primary integral part of streamflow and is essential in estimating groundwater recharge potential. LULC is a significant characteristic of the surface runoff process that affects soil water content, water yield, infiltration rate, and groundwater flow. In watershed hydrology, surface runoff is the prime streamflow contribution to aquifers. The simulated total average surface runoff in 2000, 2010, and 2020 was 347.9 mm, 599.36 mm, and 282.99 mm, respectively. The results of the LULC changes indicated that surface runoff increased in 2010 and decreased in 2020. In the second decade, a decrease in surface runoff and water yield increased groundwater recharge. In high slope areas, there is no infiltration, and precipitation cannot recharge, resulting in an increased volume of surface runoff. The other reason for reducing surface runoff is the excellent watershed management practice promoted by the government and local administration in the last decade. If not adequately managed, the time to peak flow is reduced, causing flooding and affecting lives and property. The model simulated high water yield from June to September as precipitation increased from June to August in the summer.

Consequently, excess water flows can be stored and used during low flow conditions. The simulated streamflow accounts for the LULC change scenarios classified into moist (August to October) and dry months (February to April). As a result, surface runoff was very high during the wet season and low during the dry. This evaluation showed that the increment of cultivated land causes a direct runoff throughout the wettest months. The increase in surface runoff simulated during the wet season due to LULC changes has broader ecological resource development implications. The increase in runoff may have broader implications for growing soil erosion and sedimentation, if proper control is not implemented. It also removes the top productive soil and causes degradation, affecting agricultural land, natural river banks, and low plain areas. This reduces crop yield and leads to food insecurity and sediment inflow to downstream reservoirs, decreasing the life span of service reservoirs and hydraulic structures. The SURQ map suggests that sub-basins with high rainfall correspond to extreme runoff (Figure 6a–c). For all LULC reference periods, the sub-basin numbers (8, 19, 22, and 39) in 2000, (50–53) in 2010, and 2020 were highly attributed to the surface runoff of 104.29 mm to 905.39 mm annually. High annual surface runoff is observed in the highland elevation areas, attributed to the high rainfall and steep slope topography. In flat-sloping regions of low lands, surface runoff is also higher on cultivated land, resulting in human activities in the watershed and significant driving factors for LULC change. These findings agree with other similar efforts [5,14,34–36,38,85].

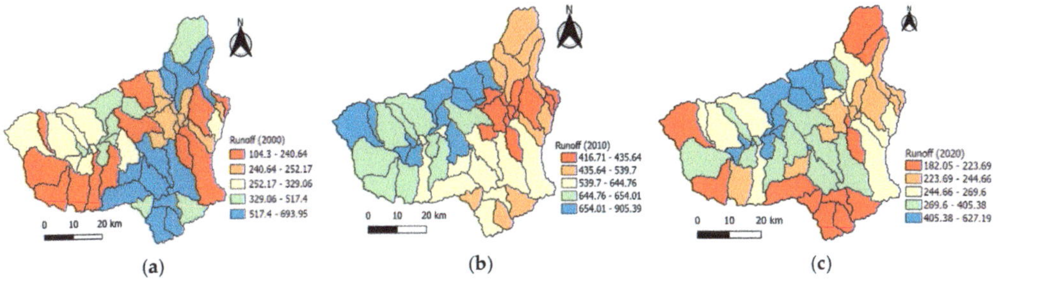

Figure 6. Spatiotemporal patterns of surface runoff during (**a**) 2000, (**b**) 2010, and (**c**) 2020 LULC simulation periods.

3.5. Effects and Implications of LULC Change on Groundwater Recharge

Groundwater recharge is a vital hydrologic cycle for sustaining aquifers recharged through precipitation and recharged artificially through human activity. Hence, developing the best management scenarios would help conserve stream biotas within the ecosystems by increasing recharge and decreasing surface runoff [18]. Best management scenarios increase groundwater recharge and decrease surface runoff, reducing erosion as in-stream sediment loads decrease. For example, converting agricultural land back to natural land cover decreased surface runoff. In addition, it decreased in-stream sediment loads due to reduced erosion. It has been observed that from 2010 to 2020, soils recovering to natural land conditions had a lower bulk density, higher saturated hydraulic conductivity, and decreased surface runoff. In addition, surface runoff was affected by LULC change and increased when the interception decreased as forest cover decreased. Poor land-use practices alter soil structure and porosity, reduce infiltration rate, and increase surface runoff. However, intensive agricultural practices removing vegetation covers exposed dense soils to erosion, decreasing groundwater recharge in the aquifer.

LULC strongly influences groundwater recharge, and it is essential to understand its interactions with increasing natural and human activities. The calibrated SWAT model estimated that the simulated groundwater recharge for the LULC reference periods 2000,

2010, and 2020 averages 348.16 mm, 13.59 mm, and 76.37 mm, respectively. At the sub-basin scale, groundwater recharge varies from 0 to 704.31, 0 to 27.8, and 0 to 190.79 mm for 2000, 2010, and 2020, respectively. The sub-basin numbers (35, 50–53), (50–53), and (35, 27, 44, 51) for 2000, 2010, and 2020 were attributed to high groundwater recharge. Rangeland and sandy, loamy soil allow alluvial deposits to infiltrate the sub-surface. The spatial pattern of GRSURQ showed the direct effects of surface runoff reducing the recharge rate (Figures 6a–c and 7a–c). The lowest recharge values are observed in forest-dominated sub-basins. The decline in streamflow is attributed to lower surface runoff due to increasing forestland, which advances the water holding capacity of the soil, reducing infiltration and recharge conditions. The low recharge estimates in the flat areas could be due to heavy clay soils with low infiltration. The decrease in groundwater recharge could be attributed to high evapotranspiration. It increases recharge and runoff for water resource control strategies tailored to the watershed.

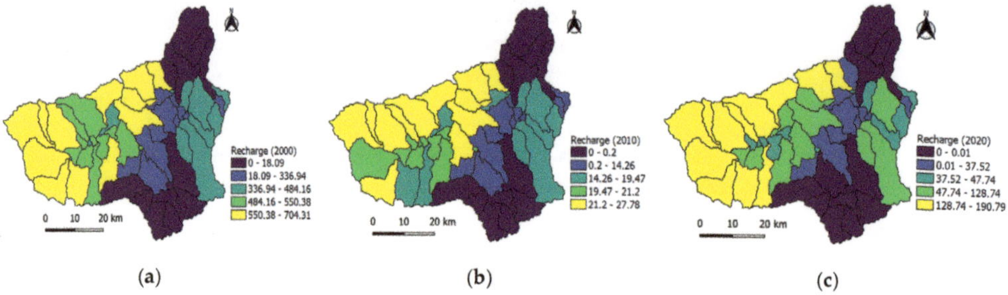

Figure 7. Spatiotemporal pattern of groundwater recharge: (**a**) 2000, (**b**) 2010, and (**c**) 2020 reference LULC periods.

The variabilities of mean yearly groundwater recharge found in the three reference periods of this study were lower than other watershed recharge rates [13]. Moreover, the 2010 LULC simulation groundwater recharge was much lower than the estimated rate [86]. The temporal variation of groundwater recharge showed that the highest value occurred during the wet months (June to August). The LULC simulation scenarios dry season flow (December to April) had shown lowering responses of groundwater recharge during dry periods. Therefore, the study deduced that the SWAT model underestimated and overestimated the high flow rate during low flow. The main reason for reducing the average annual groundwater recharge is the long dry season from December to May, before the wet season. These changes improved the wet seasonal flow and decreased dry seasonal flow. The findings revealed increasing wet flow (June–September) and lower dry flow due to alteration of vegetation cover in agricultural lands over study periods.

Similar efforts in Ethiopia reported that expansion of agricultural land diminishing forest, shrubland, and grasslands increased runoff, decreasing recharge [14,23,26,36,54]. This produces flooding during the rainy season, decreasing low flow during the dry period. Therefore, reduced recharge due to LULC change revealed a probable recurrent hydrological drought. Understanding its effect on GRSURQ is essential to knowing the flow regimes of wet and dry seasons in the watershed. Hence, there should be an effort to enhance watershed management practices for efficient use of resources for national socio-economic development. Hence, sustainable land and water management is crucial to safeguard the environmental and riverine ecosystems.

4. Conclusions

The present study examined spatiotemporal effects and implications of LULC changes on groundwater recharge and surface runoff (GRSURQ) under three reference scenarios in the Gilgel Gibe, an East African watershed. The SWAT model was built and simulated using DEM, LULC, soil, weather, and hydrological data to understand GRSURQ over the watershed. LULC changes derived from satellite images showed an increase in agricultural lands (3.76%) and a decline in forestland (5.49%) and grassland (2.51%) due to population growth and associated human activities. The effects reveal that the forests have undergone deforestation and were reduced due to a change in the considerable agricultural activity in the watershed.

The ability of the SWAT model to satisfactorily simulate stream flows was evaluated using SWAT-CUP, SUFI-2. The coefficient of determination (R^2), Nash–Sutcliffe efficiency (NSE), and percent bias (PBIAS) for calibration and validation showed excellent agreement between observed and simulated hydrographs. The results found that R2, NSE, and PBIAS values were in acceptable ranges for all LULC simulations. The results suggest that LULC changes substantially affect the watershed GRSURQ. The calibrated model confirmed an increase in runoff and a decline in recharge because of changes in LULC. The study scrutinized the impact of LULC changes on GRSURQ, which is mainly due to intensive agricultural expansion and withdrawal of the forestlands. In addition, the increase in agricultural land practice is related to water abstraction for household consumption, resulting in reduction in ground recharge. The increase in surface runoff and the decline of groundwater recharge during the wet season could lead to water scarcity during the dry season, bringing about an aquifer drought. Therefore, developing watershed management scenarios is indispensable in reducing the adverse effect of LULC changes on GRSURQ.

Therefore, an enormous concern with LULC response is imperative for planners and policy-makers of water resource projects to ensure ecosystem sustainability. Furthermore, if adequately assessed, the effects of human-induced LULC changes are indispensable in understanding hydrologic dynamics. Hence, it was found that providing insights into calibrated results helps to contribute to a concrete plan for future management strategies on watershed hydrology. Therefore, the calibrated model setup can be an ensemble for further assessment, integrating with groundwater flow modeling under different scenarios and driving factors that might influence groundwater storage.

Author Contributions: Conceptualization, T.D.M. and I.-M.C.; methodology, T.D.M. and I.-M.C.; software, T.D.M.; calibration and validation, T.D.M.; writing—original draft preparation, T.D.M.; writing—review and editing, T.D.M. and I.-M.C.; resources and data curation, M.-G.K. and J.E.L.; visualization, J.E.L., M.-G.K., and I.-M.C.; supervision, I.-M.C. and S.W.C.; funding acquisition, I.-M.C. and S.W.C. All authors have read and agreed to the published version of the manuscript.

Funding: This research was supported by a grant from a Strategic Research Project (20220178-001) funded by the Korea Institute of Civil Engineering and Building Technology.

Institutional Review Board Statement: Not applicable.

Informed Consent Statement: Not applicable.

Data Availability Statement: Not applicable.

Conflicts of Interest: The authors declare no conflict of interest.

References

1. Lal, R. World Water Resources and Achieving Water Security. *Agron. J.* **2015**, *107*, 1526–1532. [CrossRef]
2. Scanlon, B.R.; Jolly, I.; Sophocleous, M.; Zhang, L. Global impacts of conversions from natural to agricultural ecosystems on water resources: Quantity versus quality. *Water Resour. Res.* **2007**, *43*, W03437. [CrossRef]
3. Foley, J.A.; DeFries, R.; Asner, G.P.; Barford, C.; Bonan, G.; Carpenter, S.R.; Chapin, F.S.; Coe, M.T.; Daily, G.C.; Gibbs, H.K.; et al. Global Consequences of Land Use. *Science* **2005**, *309*, 570–574. [CrossRef] [PubMed]
4. Scanlon, B.R.; Reedy, R.C.; Stonestrom, D.A.; Prudic, D.E.; Dennehy, K.F. Impact of land use and land cover change on groundwater recharge and quality in the southwestern US. *Glob. Chang. Biol.* **2005**, *11*, 1577–1593. [CrossRef]
5. Guzha, A.C.; Rufino, M.C.; Okoth, S.; Jacobs, S.; Nóbrega, R.L.B. Impacts of land use and land cover change on surface runoff, discharge and low flows: Evidence from East Africa. *J. Hydrol. Reg. Stud.* **2018**, *15*, 49–67. [CrossRef]
6. Bewket, W.; Sterk, G. Dynamics in land cover and its effect on stream flow in the Chemoga watershed, Blue Nile basin, Ethiopia. *Hydrol. Process.* **2005**, *19*, 445–458. [CrossRef]
7. Owuor, S.O.; Butterbach-Bahl, K.; Guzha, A.C.; Rufino, M.C.; Pelster, D.E.; Díaz-Pinés, E.; Breuer, L. Groundwater recharge rates and surface runoff response to land use and land cover changes in semi-arid environments. *Ecol. Process.* **2016**, *5*, 16. [CrossRef]

8. Mensah, J.K.; Ofosu, E.A.; Yidana, S.M.; Akpoti, K.; Kabo-bah, A.T. Integrated modeling of hydrological processes and groundwater recharge based on land use land cover, and climate changes: A systematic review. *Environ. Adv.* **2022**, *8*, 100224. [CrossRef]
9. Taylor, R.G.; Scanlon, B.; Döll, P.; Rodell, M.; van Beek, R.; Wada, Y.; Longuevergne, L.; Leblanc, M.; Famiglietti, J.S.; Edmunds, M.; et al. Ground water and climate change. *Nat. Clim. Chang.* **2013**, *3*, 322–329. [CrossRef]
10. Coelho, V.H.R.; Montenegro, S.; Almeida, C.N.; Silva, B.; Oliveira, L.M.; Gusmão, A.C.V.; Freitas, E.S.; Montenegro, A.A.A. Alluvial groundwater recharge estimation in semi-arid environment using remotely sensed data. *J. Hydrol.* **2017**, *548*, 1–15. [CrossRef]
11. Pavelic, P. *Groundwater Availability and Use in Sub-Saharan Africa: A Review of 15 Countries*; International Water Management Institute (IWMI): Colombo, Sri Lanka, 2012; ISBN 9789290907589.
12. Scanlon, B.R.; Keese, K.E.; Flint, A.L.; Flint, L.E.; Gaye, C.B.; Edmunds, W.M.; Simmers, I. Global synthesis of groundwater recharge in semiarid and arid regions. *Hydrol. Process.* **2006**, *20*, 3335–3370. [CrossRef]
13. Mengistu, T.D.; Chung, I.-M.; Chang, S.W.; Yifru, B.A.; Kim, M.-G.; Lee, J.; Ware, H.H.; Kim, I.-H. Challenges and Prospects of Advancing Groundwater Research in Ethiopian Aquifers: A Review. *Sustainability* **2021**, *13*, 11500. [CrossRef]
14. Gessesse, A.A.; Melesse, A.M.; Abera, F.F.; Abiy, A.Z. Modeling Hydrological Responses to Land Use Dynamics, Choke, Ethiopia. *Water Conserv. Sci. Eng.* **2019**, *4*, 201–212. [CrossRef]
15. Carter, R.C.; Parker, A. Climate change, population trends and groundwater in Africa. *Hydrol. Sci. J.* **2009**, *54*, 676–689. [CrossRef]
16. Wada, Y.; van Beek, L.P.H.; van Kempen, C.M.; Reckman, J.W.T.M.; Vasak, S.; Bierkens, M.F.P. Global depletion of groundwater resources. *Geophys. Res. Lett.* **2010**, *37*, L20402. [CrossRef]
17. Woldesenbet, T.A.; Elagib, N.A.; Ribbe, L.; Heinrich, J. Hydrological responses to land use/cover changes in the source region of the Upper Blue Nile Basin, Ethiopia. *Sci. Total Environ.* **2017**, *575*, 724–741. [CrossRef]
18. Guida-Johnson, B.; Zuleta, G.A. Land-use land-cover change and ecosystem loss in the Espinal ecoregion, Argentina. *Agric. Ecosyst. Environ.* **2013**, *181*, 31–40. [CrossRef]
19. Haregeweyn, N.; Tsunekawa, A.; Poesen, J.; Tsubo, M.; Meshesha, D.T.; Fenta, A.A.; Nyssen, J.; Adgo, E. Comprehensive assessment of soil erosion risk for better land use planning in river basins: Case study of the Upper Blue Nile River. *Sci. Total Environ.* **2017**, *574*, 95–108. [CrossRef] [PubMed]
20. Zeleke, G.; Hurni, H. Implications of land use and land cover dynamics for mountain resource degradation in the Northwestern Ethiopian highlands. *Mt. Res. Dev.* **2001**, *21*, 184–191. [CrossRef]
21. Moges, D.M.; Bhat, H.G. An insight into land use and land cover changes and their impacts in Rib watershed, north-western highland Ethiopia. *Land Degrad. Dev.* **2018**, *29*, 3317–3330. [CrossRef]
22. Demissie, F.; Yeshitila, K.; Kindu, M.; Schneider, T. Land use/Land cover changes and their causes in Libokemkem District of South Gonder, Ethiopia. *Remote Sens. Appl. Soc. Environ.* **2017**, *8*, 224–230. [CrossRef]
23. Getu Engida, T.; Nigussie, T.A.; Aneseyee, A.B.; Barnabas, J. Land Use/Land Cover Change Impact on Hydrological Process in the Upper Baro Basin, Ethiopia. *Appl. Environ. Soil Sci.* **2021**, *2021*, 6617541. [CrossRef]
24. Regasa, M.S.; Nones, M.; Adeba, D. A Review on Land Use and Land Cover Change in Ethiopian Basins. *Land* **2021**, *10*, 585. [CrossRef]
25. Hailu, A.; Mammo, S.; Kidane, M. Dynamics of land use, land cover change trend and its drivers in Jimma Geneti District, Western Ethiopia. *Land Use Policy* **2020**, *99*, 105011. [CrossRef]
26. Birhanu, A.; Masih, I.; van der Zaag, P.; Nyssen, J.; Cai, X. Impacts of land use and land cover changes on hydrology of the Gumara catchment, Ethiopia. *Phys. Chem. Earth* **2019**, *112*, 165–174. [CrossRef]
27. Dibaba, W.T.; Demissie, T.A.; Miegel, K. Drivers and Implications of Land Use/Land Cover Dynamics in Finchaa Catchment, Northwestern Ethiopia. *Land* **2020**, *9*, 113. [CrossRef]
28. Zewdie, M.; Worku, H.; Bantider, A. Temporal Dynamics of the Driving Factors of Urban Landscape Change of Addis Ababa During the Past Three Decades. *Environ. Manag.* **2018**, *61*, 132–146. [CrossRef]
29. Tsegaye, D.; Moe, S.R.; Vedeld, P.; Aynekulu, E. Land-use/cover dynamics in Northern Afar rangelands, Ethiopia. *Agric. Ecosyst. Environ.* **2010**, *139*, 174–180. [CrossRef]
30. Biazin, B.; Sterk, G. Drought vulnerability drives land-use and land cover changes in the Rift Valley dry lands of Ethiopia. *Agric. Ecosyst. Environ.* **2013**, *164*, 100–113. [CrossRef]
31. Dile, Y.T.; Tekleab, S.; Ayana, E.K.; Gebrehiwot, S.G.; Worqlul, A.W.; Bayabil, H.K.; Yimam, Y.T.; Tilahun, S.A.; Daggupati, P.; Karlberg, L.; et al. Advances in water resources research in the Upper Blue Nile basin and the way forward: A review. *J. Hydrol.* **2018**, *560*, 407–423. [CrossRef]
32. Githui, F.; Mutua, F.; Bauwens, W. Estimating the impacts of land-cover change on runoff using the soil and water assessment tool (SWAT): Case study of Nzoia catchment, Kenya/Estimation des impacts du changement d'occupation du sol sur l'écoulement à l'aide de SWAT: Étude du cas du bassi. *Hydrol. Sci. J.* **2009**, *54*, 899–908. [CrossRef]
33. Hailemariam, S.; Soromessa, T.; Teketay, D. Land Use and Land Cover Change in the Bale Mountain Eco-Region of Ethiopia during 1985 to 2015. *Land* **2016**, *5*, 41. [CrossRef]
34. Mango, L.M.; Melesse, A.M.; McClain, M.E.; Gann, D.; Setegn, S.G. Land use and climate change impacts on the hydrology of the upper Mara River Basin, Kenya: Results of a modeling study to support better resource management. *Hydrol. Earth Syst. Sci.* **2011**, *15*, 2245–2258. [CrossRef]

35. Gyamfi, C.; Ndambuki, J.M.; Anornu, G.K.; Kifanyi, G.E. Groundwater recharge modelling in a large scale basin: An example using the SWAT hydrologic model. *Model. Earth Syst. Environ.* **2017**, *3*, 1361–1369. [CrossRef]
36. Gashaw, T.; Tulu, T.; Argaw, M.; Worqlul, A.W. Modeling the hydrological impacts of land use/land cover changes in the Andassa watershed, Blue Nile Basin, Ethiopia. *Sci. Total Environ.* **2018**, *619–620*, 1394–1408. [CrossRef]
37. Sterling, S.; Ducharne, A. Comprehensive data set of global land cover change for land surface model applications. *Glob. Biogeochem. Cycles* **2008**, *22*, GB3017. [CrossRef]
38. Jin, X.; Jin, Y.; Yuan, D.; Mao, X. Effects of land-use data resolution on hydrologic modelling, a case study in the upper reach of the Heihe River, Northwest China. *Ecol. Modell.* **2019**, *404*, 61–68. [CrossRef]
39. Santhi, C.; Allen, P.M.; Muttiah, R.S.; Arnold, J.G.; Tuppad, P. Regional estimation of base flow for the conterminous United States by hydrologic landscape regions. *J. Hydrol.* **2008**, *351*, 139–153. [CrossRef]
40. Setegn, S.G.; Srinivasan, R.; Dargahi, B. Hydrological Modelling in the Lake Tana Basin, Ethiopia Using SWAT Model. *Open Hydrol. J.* **2008**, *2*, 49–62. [CrossRef]
41. Suryavanshi, S.; Pandey, A.; Chaube, U.C. Hydrological simulation of the Betwa River basin (India) using the SWAT model. *Hydrol. Sci. J.* **2017**, *62*, 960–978. [CrossRef]
42. Arnold, J.G.; Srinivasan, R.; Muttiah, R.S.; Williams, J.R. Large Area Hydrologic Modeling and Assessment Part I: Model Development. *J. Am. Water Resour. Assoc.* **1998**, *34*, 73–89. [CrossRef]
43. Gassman, P.W.; Reyes, M.R.; Green, C.H.; Arnold, J.G. The Soil and Water Assessment Tool: Historical Development, Applications, and Future Research Directions. *Trans. ASABE* **2007**, *50*, 1211–1250. [CrossRef]
44. Abbaspour, K.C.; Yang, J.; Maximov, I.; Siber, R.; Bogner, K.; Mieleitner, J.; Zobrist, J.; Srinivasan, R. Modelling hydrology and water quality in the pre-alpine/alpine Thur watershed using SWAT. *J. Hydrol.* **2007**, *333*, 413–430. [CrossRef]
45. Srinivasan, R.; Ramanarayanan, T.S.; Arnold, J.G.; Bednarz, S.T. Large Area Hydrologic Modeling and Assessment Part II: Model Application. *J. Am. Water Resour. Assoc.* **1998**, *34*, 91–101. [CrossRef]
46. Arnold, J.; Muttiah, R.; Srinivasan, R.; Allen, P. Regional estimation of base flow and groundwater recharge in the Upper Mississippi river basin. *J. Hydrol.* **2000**, *227*, 21–40. [CrossRef]
47. Akoko, G.; Le, T.H.; Gomi, T.; Kato, T. A Review of SWAT Model Application in Africa. *Water* **2021**, *13*, 1313. [CrossRef]
48. Gyamfi, C.; Ndambuki, J.; Salim, R. Hydrological Responses to Land Use/Cover Changes in the Olifants Basin, South Africa. *Water* **2016**, *8*, 588. [CrossRef]
49. Chen, Y.; Niu, J.; Sun, Y.; Liu, Q.; Li, S.; Li, P.; Sun, L.; Li, Q. Study on streamflow response to land use change over the upper reaches of Zhanghe Reservoir in the Yangtze River basin. *Geosci. Lett.* **2020**, *7*, 6. [CrossRef]
50. Chen, Y.; Nakatsugawa, M. Analysis of Changes in Land Use/Land Cover and Hydrological Processes Caused by Earthquakes in the Atsuma River Basin in Japan. *Sustainability* **2021**, *13*, 13041. [CrossRef]
51. Astuti, I.S.; Sahoo, K.; Milewski, A.; Mishra, D.R. Impact of Land Use Land Cover (LULC) Change on Surface Runoff in an Increasingly Urbanized Tropical Watershed. *Water Resour. Manag.* **2019**, *33*, 4087–4103. [CrossRef]
52. Huang, T.C.C.; Lo, K.F.A. Effects of land use change on sediment and water yields in yang ming shan national park, taiwan. *Environments* **2015**, *2*, 32–42. [CrossRef]
53. Ghoraba, S.M. Hydrological modeling of the Simly Dam watershed (Pakistan) using GIS and SWAT model. *Alexandria Eng. J.* **2015**, *54*, 583–594. [CrossRef]
54. Tekleab, S.; Mohamed, Y.; Uhlenbrook, S.; Wenninger, J. Hydrologic responses to land cover change: The case of Jedeb mesoscale catchment, Abay/Upper Blue Nile basin, Ethiopia. *Hydrol. Process.* **2014**, *28*, 5149–5161. [CrossRef]
55. Qiu, W.; Ma, T.; Wang, Y.; Cheng, J.; Su, C.; Li, J. Review on status of groundwater database and application prospect in deep-time digital earth plan. *Geosci. Front.* **2022**, *13*, 101383. [CrossRef]
56. Gessesse, B.; Bewket, W.; Bräuning, A. Model-Based Characterization and Monitoring of Runoff and Soil Erosion in Response to Land Use/land Cover Changes in the Modjo Watershed, Ethiopia. *Land Degrad. Dev.* **2015**, *26*, 711–724. [CrossRef]
57. Sime, C.H.; Demissie, T.A.; Tufa, F.G. Surface runoff modeling in Ketar watershed, Ethiopia. *J. Sediment. Environ.* **2020**, *5*, 151–162. [CrossRef]
58. Mengistu, T.D.; Chang, S.W.; Kim, I.-H.; Kim, M.; Chung, I. Determination of Potential Aquifer Recharge Zones Using Geospatial Techniques for Proxy Data of Gilgel Gibe Catchment, Ethiopia. *Water* **2022**, *14*, 1362. [CrossRef]
59. Tefera, M.; Cherinet, T.; Haro, W. *Explanation to the Geological Map of Ethiopia. Ministry of Mines and Energy;* Ethiopian Institute of Geological Surveys: Addis Ababa, Ethiopia, 1996.
60. Tuppad, P.; Mankin, K.R.D.; Lee, T.; Srinivasan, R.; Arnold, J.G. Soil and Water Assessment Tool (SWAT) Hydrologic/Water Quality Model: Extended Capability and Wider Adoption. *Am. Soc. Agric. Biol. Eng.* **2011**, *54*, 1677–1684. [CrossRef]
61. Neitsch, S.L.; Arnold, J.G.; Kiniry, J.R.; Williams, J.R. *Soil and Water Assessment Tool Theoretical Documentation Version 2009*; Texas Water Resources Institute: Temple, TX, USA, 2011; Volume 543.
62. Leta, M.K.; Demissie, T.A.; Tränckner, J. Hydrological Responses of Watershed to Historical and Future Land Use Land Cover Change Dynamics of Nashe Watershed, Ethiopia. *Water* **2021**, *13*, 2372. [CrossRef]
63. Arnold, J.G.; Moriasi, D.N.; Gassman, P.W.; Abbaspour, K.C.; White, M.J.; Srinivasan, R.; Santhi, C.; Harmel, R.D.; van Griensven, A.; Van Liew, M.W.; et al. SWAT: Model Use, Calibration, and Validation. *Trans. ASABE* **2012**, *55*, 1491–1508. [CrossRef]

64. Nachtergaele, F.; Velthuizen, H.V.; Verelst, L.; Batjes, N.; Dijkshoorn, K.; Engelen, V.V.; Fischer, G.; Jones, A.; Montanarella, L.; Petri, M.; et al. Harmonized World Soil Database (version 1.2). Food and Agriculture Organization of the UN, International Institute for Applied Systems Analysis, ISRIC-World Soil Information, Institute of Soil Science-Chinese Academy of Sciences, Joint Research Centre of the EC. Available online: http://www.iiasa.ac.at/Research/LUC/External-World-soil-database/HWSD_Documentation (accessed on 20 June 2022).
65. Abbaspour, K.C.; Vaghefi, S.A.; Yang, H.; Srinivasan, R. Global soil, landuse, evapotranspiration, historical and future weather databases for SWAT Applications. *Sci. Data* **2019**, *6*, 263. [CrossRef]
66. Arnold, J.; Kiniry, R.; Williams, E.; Haney, S.; Neitsch, S. *Soil & Water Assessment Tool*; Texas Water Resources Institute: Temple, TX, USA, 2012.
67. Chen, J.; Chen, J.; Liao, A.; Cao, X.; Chen, L.; Chen, X.; He, C.; Han, G.; Peng, S.; Lu, M.; et al. Global land cover mapping at 30m resolution: A POK-based operational approach. *ISPRS J. Photogramm. Remote Sens.* **2015**, *103*, 7–27. [CrossRef]
68. Jun, C.; Ban, Y.; Li, S. Open access to Earth land-cover map. *Nature* **2014**, *514*, 434. [CrossRef] [PubMed]
69. Anthony, J.; Viera, A.J.V. The kappa statistic. *JAMA J. Am. Med. Assoc.* **1992**, *268*, 2513–2514. [CrossRef]
70. Arnold, J.G.; Kiniry, J.R.; Srinivasan, R.; Williams, J.R.; Haney, E.B.; Neitsch, S.L. Input/Output Documentation Soil & Water Assessment Tool. 2012. Available online: https://swat.tamu.edu/media/69296/swat-io-documentation-2012.pdf (accessed on 20 June 2022).
71. Dile, Y.T.; Daggupati, P.; George, C.; Srinivasan, R.; Arnold, J. Introducing a new open source GIS user interface for the SWAT model. *Environ. Model. Softw.* **2016**, *85*, 129–138. [CrossRef]
72. Abbaspour, K.C.; Rouholahnejad, E.; Vaghefi, S.; Srinivasan, R.; Yang, H.; Kløve, B. A continental-scale hydrology and water quality model for Europe: Calibration and uncertainty of a high-resolution large-scale SWAT model. *J. Hydrol.* **2015**, *524*, 733–752. [CrossRef]
73. Setegn, S.G.; Srinivasan, R.; Melesse, A.M.; Dargahi, B. SWAT model application and prediction uncertainty analysis in the Lake Tana Basin, Ethiopia. *Hydrol. Process.* **2009**, *24*, 357–367. [CrossRef]
74. Kouchi, D.H.; Esmaili, K.; Faridhosseini, A.; Sanaeinejad, S.H.; Khalili, D.; Abbaspour, K.C. Sensitivity of Calibrated Parameters and Water Resource Estimates on Different Objective Functions and Optimization Algorithms. *Water* **2017**, *9*, 384. [CrossRef]
75. Moriasi, D.N.; Arnold, J.G.; Van Liew, M.W.; Bingner, R.L.; Harmel, R.D.; Veith, T.T.; Moriasi, D.N.; Arnold, J.G.; Van Liew, M.W.; Bingner, R.L.; et al. Model Evaluation Guidelines for Systematic Quantification of Accuracy in Watershed Simulations. *Colomb. Med.* **2007**, *50*, 885–900. [CrossRef]
76. Moriasi, D.N.; Gitau, M.W.; Pai, N.; Daggupati, P. Hydrologic and Water Quality Models: Performance Measures and Evaluation Criteria. *Trans. ASABE* **2015**, *58*, 1763–1785. [CrossRef]
77. Mengistu, A.G.; van Rensburg, L.D.; Woyessa, Y.E. Techniques for calibration and validation of SWAT model in data scarce arid and semi-arid catchments in South Africa. *J. Hydrol. Reg. Stud.* **2019**, *25*, 100621. [CrossRef]
78. Meaurio, M.; Zabaleta, A.; Uriarte, J.A.; Srinivasan, R.; Antigüedad, I. Evaluation of SWAT models performance to simulate streamflow spatial origin. The case of a small forested watershed. *J. Hydrol.* **2015**, *525*, 326–334. [CrossRef]
79. Abbaspour, K.C. Calibration of hydrologic models: When is a model calibrated? In Proceedings of the MODSIM05: International Congress on Modelling and Simulation: Advances and Applications for Management and Decision Making, Melbourne, Australia, 12–15 December 2005; pp. 2449–2455.
80. Gupta, H.V.; Sorooshian, S.; Yapo, P.O. Status of Automatic Calibration for Hydrologic Models: Comparison with Multilevel Expert Calibration. *J. Hydrol. Eng.* **1999**, *4*, 135–143. [CrossRef]
81. Refsgaard, J.C.; Knudsen, J. Operational Validation and Intercomparison of Different Types of Hydrological Models. *Water Resour. Res.* **1996**, *32*, 2189–2202. [CrossRef]
82. Daggupati, P.; Pai, N.; Ale, S.; Douglas-Mankin, K.R.; Zeckoski, R.W.; Jeong, J.; Parajuli, P.B.; Saraswat, D.; Youssef, M.A. A recommended calibration and validation strategy for hydrologic and water quality models. *Trans. ASABE* **2015**, *58*, 1705–1719. [CrossRef]
83. Krause, P.; Boyle, D.P.; Bäse, F. Comparison of different efficiency criteria for hydrological model assessment. *Adv. Geosci.* **2005**, *5*, 89–97. [CrossRef]
84. Legates, D.R.; McCabe, G.J. Evaluating the use of "goodness-of-fit" Measures in hydrologic and hydroclimatic model validation. *Water Resour. Res.* **1999**, *35*, 233–241. [CrossRef]
85. Guzman, C.D.; Tilahun, S.A.; Dagnew, D.C.; Zimale, F.A.; Zegeye, A.D.; Boll, J.; Parlange, J.Y.; Steenhuis, T.S. Spatio-temporal patterns of groundwater depths and soil nutrients in a small watershed in the Ethiopian highlands: Topographic and land-use controls. *J. Hydrol.* **2017**, *555*, 420–434. [CrossRef]
86. Döll, P.; Fiedler, K. Global-scale modeling of groundwater recharge. *Hydrol. Earth Syst. Sci.* **2008**, *12*, 863–885. [CrossRef]

water

MDPI

Article

Performance Evaluation of Artificial Recharge–Water Intake System Using 3D Numerical Modeling

Jae-Young Lee * and **Tae-Young Woo**

Fusion Research Institute, Sinwoo Engineering Co., Ltd., Seoul 06184, Korea; lovewty@naver.com
* Correspondence: vennard386@gmail.com; Tel.: +82-2-6959-3038

Abstract: In this study, 3D detailed numerical modeling was performed to evaluate the performance of an artificial recharge–water intake system installed to secure agricultural water in drought areas. Using a 3D irregular finite element grid, a conceptual model was constructed that reflected the actual scale of the study area and artificial recharge–water intake system and considered the characteristics of saturated–unsaturated aquifers. The optimal design factors for the artificial recharge system were derived through the constructed conceptual model, and were reflected to evaluate the individual performance of the artificial recharge and water intake system in the study area. Finally, an optimal operating scenario for the artificial recharge and water intake system was developed. The operation scenarios were composed of an appropriate injection rate and water withdrawal for each period from March, when the demand for agricultural water was low, to June, when the dry season and farming season overlapped, considering the target water withdrawal amount (30,000 tons) of the region, derived from water budget analysis. The proposed results are expected to be very useful in the future for the efficient operation and management of artificial recharge–water intake systems installed in drought areas.

Keywords: artificial recharge; water intake; detailed numerical modeling; performance evaluation; operation scenario

Citation: Lee, J.-Y.; Woo, T.-Y. Performance Evaluation of Artificial Recharge–Water Intake System Using 3D Numerical Modeling. *Water* **2022**, *14*, 1974. https://doi.org/10.3390/w14121974

Academic Editors: Sang Yong Chung, Gyoo-Bum Kim and Venkatramanan Senapathi

Received: 26 April 2022
Accepted: 14 June 2022
Published: 20 June 2022

1. Introduction

Due to recurrent drought, it is not easy to secure water resources, so there are many areas that are experiencing difficulties in cultivating crops. South Korea has implemented policies and technologies focusing on surface water resources such as dams and reservoirs, but surface water resources sensitive to climate change have limitations as a solution to drought, and they require long-term preparation. Conversely, groundwater resources have relatively little evaporative loss and are not sensitive to climate change, so they can complement the limitations of surface water resources [1]. Accordingly, the paradigm of domestic water resources policy is changing with the development and expansion of alternative water resources employing groundwater to stably secure high-quality water in the event of an extreme drought. However, groundwater resources are also decreasing due to a decrease in natural recharge, excess extraction, and increased risk of contamination. Therefore, in order for groundwater to become a more effective water resource to respond to climate change, it is necessary to develop artificial recharge technology that can artificially increase high quality groundwater resources [2]. The artificial groundwater recharge technique is a useful method to continuously secure water resources by artificially injecting water into an aquifer by installing an injection well. Compared to surface dams, there are certain hydrological and economic advantages, so the use of the artificial recharge method is increasing for both long-term and short-term underground storage [3].

The artificial groundwater recharge technique was proposed as an effective method of securing water resources in preparation for uncertain future climate change [4], and the range of artificial recharge and appropriate artificial recharge techniques were reviewed considering the characteristics of aquifers [5]. An artificial recharge evaluation model through 3D groundwater modeling was developed for Hancheon basin in Jeju Island using MODFLOW [6], and the optimal location of the artificial recharge well was evaluated by comparing the pumping rate change in the existing pumping well according to the location of the injection well using the groundwater model [7]. In addition, MODFLOW has been used to analyze the groundwater flow in a ditch filled with homogeneous anisotropic soil [8], and a numerical analysis model was developed to examine the application of groundwater artificial recharge in the Yongding River basin in Beijing, China [9]. Meanwhile, an underground drainage system that can mitigate landslide damage was developed and constructed on site; the stability of landslides and slopes was reviewed using a three-dimensional finite element program (Plaxis 3D) [10].

In relation to artificial groundwater recharge, various studies have been conducted internationally, but there are very few case studies based on numerical analysis in which actual water shortage areas are set, the recharge potential is evaluated in consideration of the target water withdrawal in the watershed, and where the operation conditions of the recharge rate and water withdrawal by period are presented in detail. Therefore, in this study, the optimal design factors for artificial recharge and water intake facilities were derived by using detailed numerical modeling for the area where actual artificial recharge–water intake system construction has been completed. The performance of each artificial recharge system and water intake system was then evaluated, and an optimal operation scenario for supplying agricultural water suitable for the region during the drought period was presented.

2. Materials and Methods

2.1. Study Area

The study area is located in Ungok-ri, Hongsung-gun, in the Chungnam province, and consists of agricultural land for paddy and field farming (Figure 1). The geography of the basin is surrounded by mountains with a height of about 350 m, and the large and small tributaries originating from these mountains join together to form two major streams (Shingok stream and Ungok stream). The two streams have a stream width of about 15 m, a channel width of about 2 m, and an ordinary water depth of about 0.1 m; the range of streamflow variability is highly dependent on rainfall variability. During the busy farming season (April to June), there is relatively little rainfall, so there is almost no streamflow. As continuous groundwater extraction and depletion of groundwater are occurring rapidly, artificial groundwater recharge technology is being considered. As a result of analyzing the precipitation status through meteorological data (Seosan meteorological station, 2012~2021) for 10 years in the target area, the average annual precipitation was 1106.8 mm/year. In addition, about 40% of the total annual precipitation was concentrated between July and August, and about 20% occurred between April and June, the farming season, indicating a small amount of precipitation. The main geological status of the study area is mainly composed of granitic gneiss, flaky granite, and migmatite belts of Precambrian. On the east side, biotite granite of the Cretaceous penetrates metamorphic rocks, and on the west side, sedimentary rocks from the Muryangri Formation, presumed to be Permian, are distributed. In addition, the alluvial layer is very developed due to the hilly topography. Although the average annual groundwater level in the study area fluctuates slightly by period, it is distributed at about 2 m below the ground surface. The groundwater level repeatedly rises or falls depending on whether or not there is rainfall, and accordingly, the water level drops to about 5 m below the surface during the dry season and farming season, from April to June.

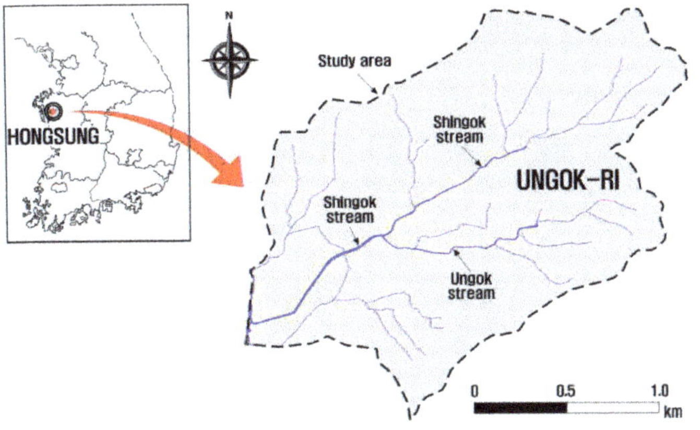

Figure 1. Location map of the study area.

2.2. Methods

For detailed numerical modeling of the artificial recharge–water intake system, a 3D, FEM-based SEEFLOW3D [11] (Cha et al., 2017) model was adopted and an irregular finite element grid was applied to implement artificial injection and water intake facilities more precisely [11]. In consideration of the physical properties, it was attempted to accurately reflect the behavior of groundwater flow in the saturated–unsaturated aquifer. In order to build a conceptual model using Digital Elevation Model (DEM) data of the digital topographic map, a modeling range of 2450 m × 2350 m was built within the boundary of the mountain ridge, which is a watershed, and the stratum information of the drilling column obtained through on-site drilling survey data was utilized. Three numerical layers were composed of an alluvial layer, weathered soil, and bedrock, and input data for the stratum characteristics of the numerical model were composed using the results of field hydraulic tests and grain size analysis in the unsaturated zone (see Table 1). A mixed artificial injection system (1 ditch + 4 vertical wells), an intake system (intake well, horizontal well), and three observation wells near the downstream part of the study area for monitoring groundwater level fluctuations were reflected in the model. A computational mesh with a total of 24,650 nodes and 43,596 elements was built by constructing irregular three-dimensional elements. As shown in Figure 2, the outside of the target area was set as the no-flux boundary, and the rivers within the modeling area were set as the time-series river stage (Dirichlet boundary). In addition, in order to derive the optimal design factors for the artificial injection system to develop an operational scenario for an artificial recharge–water intake system, three major design factors were set and evaluated as follows:

1. Optimal size of media in the ditch of the artificial recharge facilities from various media sizes (10/30/50/70/90 mm)
2. Optimal spacing of vertical wells for artificial recharge facilities from various vertical well intervals (5/10/15 m)
3. Optimal screen opening ratio of horizontal wells for water intake facilities from various opening ratios (10/20/30%)

Table 1. Distribution of Geological Layer and Hydraulic Conductivity.

Media	Layer	Depth (GL.-m)		Hydraulic Conductivity (cm/s)	
		Range	Average	Range	Average
Alluvium	1	1.5~5.8	3.1	4.05×10^{-5}~1.14×10^{-3}	4.11×10^{-4}
Weathered soil	2	1.5~6.5	3.7	5.82×10^{-5}~4.10×10^{-3}	7.00×10^{-4}
Bedrock	3	5.8~12.3	8.7	1.40×10^{-6}~2.13×10^{-4}	7.20×10^{-5}

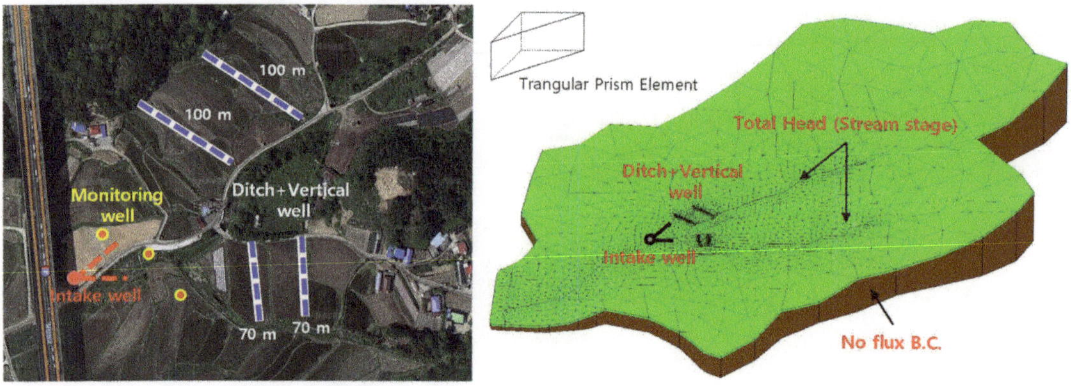

Figure 2. Configuration of conceptual model reflecting the recharge facilities and intake well in the target area.

Based on the optimal design factors, the performance of each artificial injection and intake system in this study area was evaluated, and finally, an optimal operation scenario of the artificial recharge–water intake system was developed.

3. Results

3.1. Evaluation of Optimal Size of Media in the Ditch

Because the ditch of the artificial recharge facility is located 1.0 m below the ground surface, it is necessary to prevent the upper soil from flowing into the artificial recharge facility as much as possible. A filter media is required because the water flow in the ditch should be stable and uniform. Therefore, in order to derive the optimal size of the media in the ditch of the artificial recharge facility, numerical simulations were performed for each media size of 10, 30, 50, 70, and 90 mm, and the optimal size of the media was derived by setting the time when the groundwater level was full in the ditch. The hydraulic conductivity and porosity were used as modeling input data for each media size. The hydraulic conductivity was applied as 35 m/day for 10 mm, 1000 m/day for 30 mm, 3500 m/day for 50 mm, 6000 m/day for 70 mm, and 8600 m/day for 90 mm, and the porosity was used, ranging from 0.34 for 10mm to 0.28 for 90mm, respectively [12]. For the artificial injection conditions according to the media size, the injection location was designated at regular intervals on the upper element of the ditch, and the inflow flux boundary condition was applied at an injection rate of 45 ton/day. The numerical simulations were performed for a total of 1 month, with a computational time step of 1 h, the time for filling the ditch for each media size was estimated, and the amount of penetration of water injected through the ditch into the bottom of the ditch was compared.

As a result of the simulations, the time to fill the groundwater level for each media size was 16 h at 10 mm, 10 h at 30 mm, 5 h at 50 mm, 4 h at 70 mm, and 3 h at 90 mm (Figure 3a). The infiltration rate was 38 ton/day at 10 mm, 41 ton/day at 30 mm, 45 ton/day at 50 mm, 47 ton/day at 70 mm, and 45 ton/day at 90 mm (Figure 3b). In addition, Figure 4 shows the spatial distribution of the groundwater level by time in the ditch according to the artificial injection. Thus, the fill time is fastest at 90 mm and the maximum rate of infiltration is shown at 70 mm, but 50 mm is the optimal size of media considering the economic feasibility because the fill time and infiltration rate are similar at a media size of more than 50 mm.

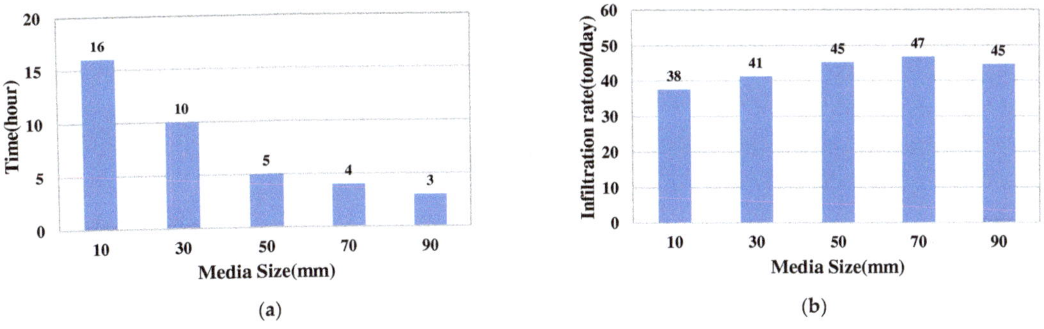

Figure 3. Time for filling the ditch and infiltration rate according to media size: (**a**) Time for filling the ditch; (**b**) Infiltration rate.

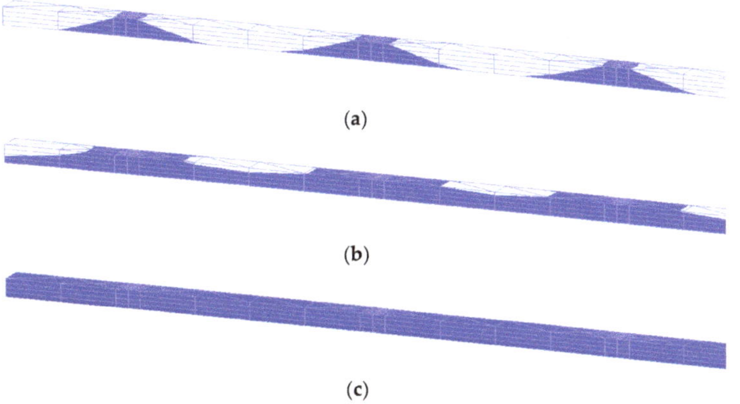

Figure 4. Spatial and temporal variation of the groundwater level in the ditch according to artificial recharge (media size 50 mm): (**a**) 1 h; (**b**) 3 h; (**c**) 5 h.

3.2. Evaluation of Proper Spacing of Vertical Wells in the Ditch

In an artificial injection system, vertical wells are very effective because they can be directly introduced into the lower strata. When multiple vertical wells are installed, a large amount of artificial recharge is possible. However, because the cost of construction and maintenance increases with the number of wells, it is necessary to determine their appropriate spacing and number. Therefore, in order to derive the proper spacing of vertical wells for artificial recharge facilities, the vertical well spacing was divided into 5, 10, and 15 m in mixed structures (ditch + vertical well), and the efficiency for each interval (ratio of aquifer infiltration to injection rate) was calculated after setting the target groundwater level rise of less than 0.1 m below the ground surface, in consideration of the depth of crop growth (Figure 5).

First, a numerical simulation of the steady state was performed to predict the spatial distribution of the groundwater level in the modeling area. As a result of the steady state simulation, the groundwater level distribution in the target watershed was EL. 21.23~318.42 m, and it was simulated with a distribution similar to the topographic elevation in the watershed (Figure 6).

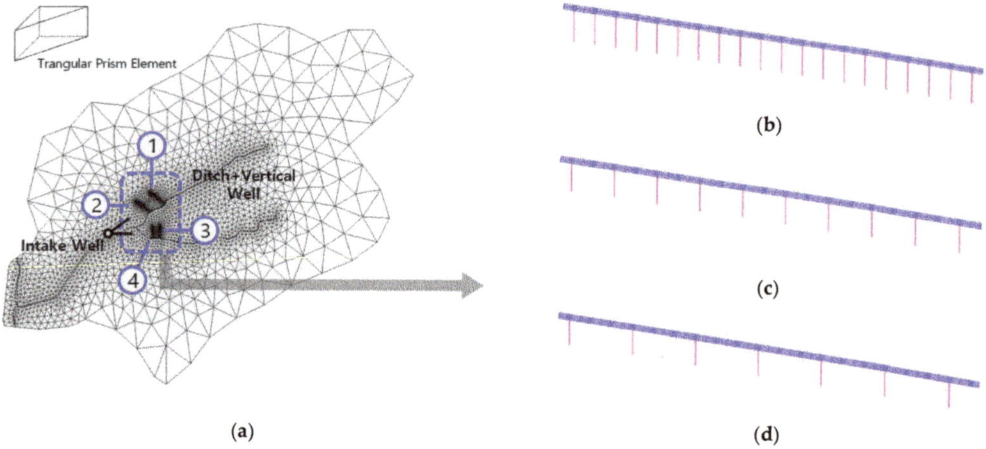

Figure 5. Configuration of 3D conceptual model of vertical well spacing: (**a**) Mesh generation reflecting the recharge facilities and Intake well. (**b**) Vertical well interval: 5 m. (**c**) Vertical well interval: 10 m. (**d**) Vertical well interval: 15 m.

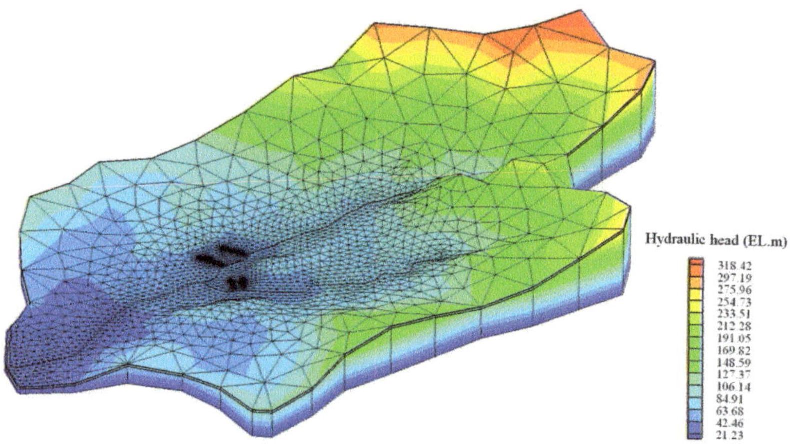

Figure 6. Steady-state simulation result of study domain.

For the artificial injection conditions, the injection rate for each vertical well and the ditch injection rate according to the vertical well spacing calculated from the injection test data by applying the inflow boundary condition to the vertical well and the element at the top of the ditch were used (Table 2) [13]. As a result of the simulation, at an interval of 5 m, the aquifer penetration for the injection rate of 224.8 ton/day was simulated as 184.0 ton/day, indicating an efficiency of 81.9%. At an interval of 10 m, the aquifer penetration for the injection rate of 209.2 ton/day was simulated as 185.2 ton/day, indicating a high efficiency of about 88.5%. On the other hand, intervals of 15 m, the aquifer penetration for the injection rate of 199.6 ton/day was simulated as 158.3 ton/day, which showed the lowest efficiency of 79.3% (Table 3, Figure 7). Figure 8 shows the time-series behavior of the groundwater level within the ditch, including vertical well spacing of 10 m after artificial injection, and all four ditches show that the groundwater level rises gradually after the start of injection and the fill time is similar. In this way, the injection efficiency was simulated, and in consideration of the economic feasibility and maintenance aspects of the vertical well construction, the proper spacing of the vertical wells was determined to be 10 m.

Table 2. Dimensions of artificial recharge facilities and injection rate conditions.

Vertical Well Interval (m)	Vertical Well Number		Vertical Well Injection Rate (ton/day)					Ditch Injection Rate (ton/day)					Total Injection Rate (ton/day)
	100m Ditch (①,②)	70m Ditch (③,④)	Ditch ①	Ditch ②	Ditch ③	Ditch ④	Sum	Ditch ①	Ditch ②	Ditch ③	Ditch ④	Sum	
5	20	14	1.29				87.67	42.7	36.1	35.9	22.5	137.2	224.8
10	10	7	2.06				70.14	42.7	36.9	37.1	22.5	139.1	209.2
15	7	5	2.51				60.29	42.7	36.9	37.1	22.5	139.1	199.6

Table 3. Injection rate and infiltration rate by vertical well interval.

Division		Injection (ton/day)			Infiltration (ton/day)		
		Vertical Well Interval			Vertical Well Interval		
		5 m	10 m	15 m	5 m	10 m	15 m
Ditch	①	42.7	42.7	42.7	34.2	34.9	35.1
	②	36.1	36.9	36.9	32.2	33.2	33.2
	③	35.9	37.1	37.1	32.4	33.4	33.5
	④	22.5	22.5	22.5	20.1	21.2	21.2
	subtotal	137.2	139.2	139.2	118.8	122.7	123.0
Vertical Well	①	25.8	20.6	17.6	20.1	19.0	10.6
	②	25.8	20.6	17.6	18.8	18.1	10.0
	③	18.0	14.4	12.6	13.1	12.5	7.3
	④	18.0	14.4	12.6	13.1	13.0	7.4
	subtotal	87.6	70.0	60.4	65.2	62.6	35.3
Total		224.8	209.2	199.6	184.0	185.2	158.3

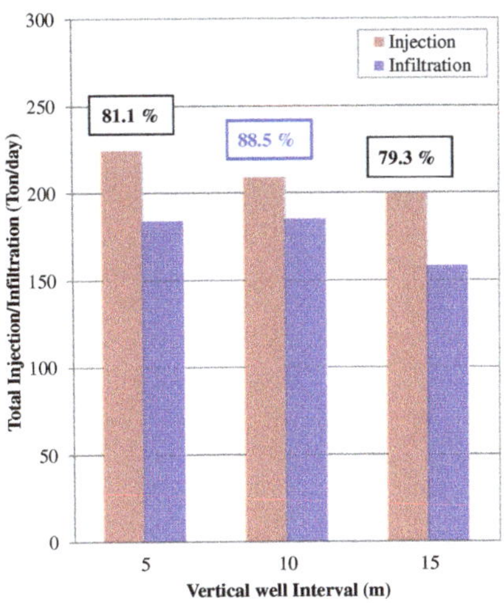

Figure 7. Efficiency (total injection rate/infiltration rate) by vertical well interval.

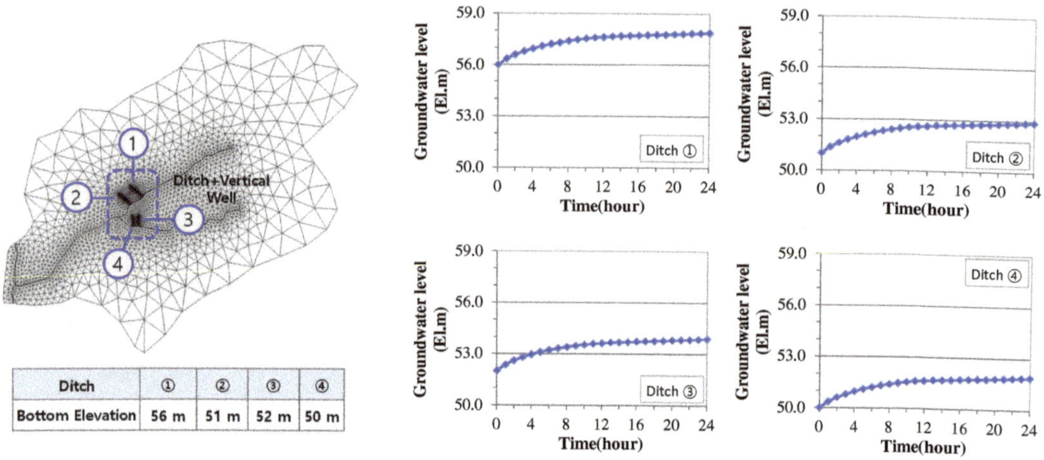

Ditch	①	②	③	④
Bottom Elevation	56 m	51 m	52 m	50 m

Figure 8. Changes of groundwater level in ditch ①, ②, ③, and ④.

3.3. *Evaluation of Optimal Opening Ratio of Horizontal Well*

The intake method for artificially recharged groundwater is divided into vertical type and horizontal radial type. The vertical water intake method takes water from the intake well configured only in the vertical direction, and maintenance and economic feasibility are low. On the other hand, the horizontal radial water intake method installs a well in the aquifer vertically and connects several horizontal wells to the side wall to take in a large amount of water. Although this method requires specialized technology, it is easy to maintain, and thus economic feasibility is high. Therefore, in this section, in order to derive the optimal opening ratio, which is the main design factor of the horizontal radial water intake method, the opening ratio was divided into 10, 20, and 30% and compared with the water level changes at the three downstream observation wells. The simulated conditions were set in such a way that after artificial recharge at 210 ton/day for 30 days, water intake was performed through a horizontal well at 500 ton/day for 10 days.

In this conceptual model, a horizontal well mesh is configured inside the model area reflecting the characteristics of the aquifer, and the inflow of groundwater into the horizontal well from the aquifer can be computed by entering the dimensions of the horizontal well (diameter, length, and opening ratio). The governing equation for this can be described as follows:

$$q = A_o \times k \times (h - H) \tag{1}$$

where q is the rate of groundwater inflow into the horizontal well [L^3t^{-1}], A_o is the effective pore area [L^2], k is the hydraulic conductivity of the horizontal well screen [LT^{-1}], h is the groundwater level in the aquifer [L], and H indicates the water level in the horizontal well [L]. The effective pore area of Equation (1) is defined as follows (Delleur, 2007) [14]:

$$A_o = \left[2 \times 3.14 \times \left(\frac{D}{2}\right) \times L\right] \times O_a \times 50\% \tag{2}$$

where D is the horizontal well diameter [L], L is the horizontal well length [L], and O_α is the horizontal well opening ratio [%].

On the other hand, the loss due to friction in the flow in the horizontal well can be expressed by the Darcy–Weisbach equation, as in Equation (3).

$$\frac{\Delta h_f}{L} = \frac{fV^2}{2gD} \tag{3}$$

where Δh_f is the friction head loss in the horizontal well [L], f is the friction factor, V is the fluid velocity in the horizontal well [LT^{-1}], and g is gravitational acceleration (9.81 m/s^2). Additionally, f is calculated as a function of the Reynold's number and can be expressed in the form of Haaland's equation, as follows:

$$f = \left[1.8\log_{10}\left(\frac{6.9}{R_e}\right) + \left(\frac{\varepsilon}{3.7D}\right)^{1.1}\right]^{-2} \tag{4}$$

where R_e is the Reynolds number and ε is the horizontal roughness coefficient [L].

Therefore, the hydraulic conductivity of the horizontal well screen can be obtained by combining Equations (3) and (4) (Birch, 2004):

$$k = \frac{2gD}{V}\left[1.8\log_{10}\left(\frac{6.9\nu}{VD} + \left(\frac{\varepsilon}{3.7D}\right)^{1.11}\right)\right]^2 \tag{5}$$

where ν is the kinematic viscosity of water [L^2T^{-1}].

As a result of the simulation, the groundwater level fluctuations at the three observation wells dropped 0.10, 0.19, and 0.20 m at 10, 20, and 30% of the opening ratio, respectively (Figure 9). The results reproduced the general behavior of the groundwater level rising due to artificial recharge and falling due to water intake In addition, as the water intake capacity increased as the opening ratio increased, the groundwater level drop increased. However, when the opening ratio was more than 20%, the fluctuation range of the groundwater level was small. Consequently, it was predicted that there would be no significant change in the amount of water intake at an opening ratio of more than 20%, and the appropriate opening ratio of the horizontal well was evaluated to be 20%.

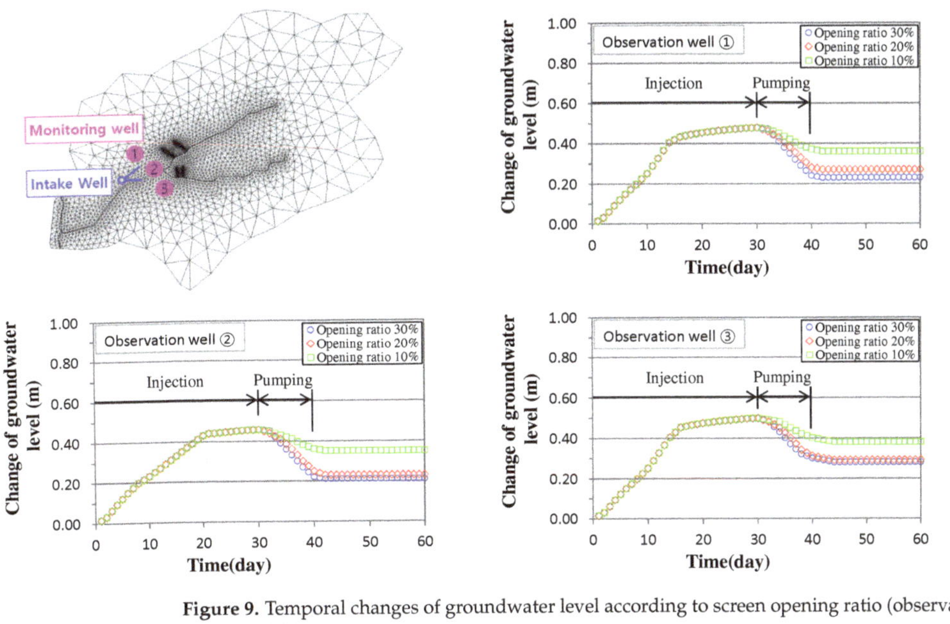

Figure 9. Temporal changes of groundwater level according to screen opening ratio (observation well ①, ② and ③).

3.4. Model Calibration

To verify the numerical model, the measured value of the groundwater level change of the observation well following the injection of the artificial recharge facility constructed on the field and the numerical modeling results were compared and calibrated. In the actual study area, five observation wells for injection experiments and groundwater level monitoring were installed at intervals of 15 m only on Line No.1 of the four ditch-vertical well lines, which were 5 m apart from the line. The structure of the ditch-vertical well of Line No.1 was a mixed type of ditch of 50 mm media size and a vertical well of 10 m intervals installed in consideration of the design factors derived through the optimal evaluation of the artificial recharge facility described above (see Figure 10). In this way, after reflecting on the current status of Line No. 1 installed on the site in the model, the injection conditions of the site and modeling were set equal to injecting 200 tons into the ditch for 54 h and injecting 45 tons into the vertical well, and then the groundwater level changes were compared. The rise of groundwater level at the site was monitored as 49.0 cm in Well No.1, 42.5 cm in Well No.2, 38.9 cm in Well No.3, 40.3 cm in Well No.4, and 44.6 cm in Well No.5, resulting in an average increase of 43.1 cm. On the other hand, the groundwater level fluctuation predicted by the modeling was simulated as an average increase of 44.3 cm, with a rise of 45.8 cm in Well No.1, a rise of 45.2 cm in Well No.2, a rise of 44.0 cm in Well No.3, a rise of 44.3 cm in Well No.4, and a rise of 42.2 cm in Well No.5 (see Table 4). Therefore, the root mean square error (RMSE) of the actual groundwater level fluctuation measurement and the modeling result was 3.6%, indicating that the prediction accuracy of the numerical simulation was good.

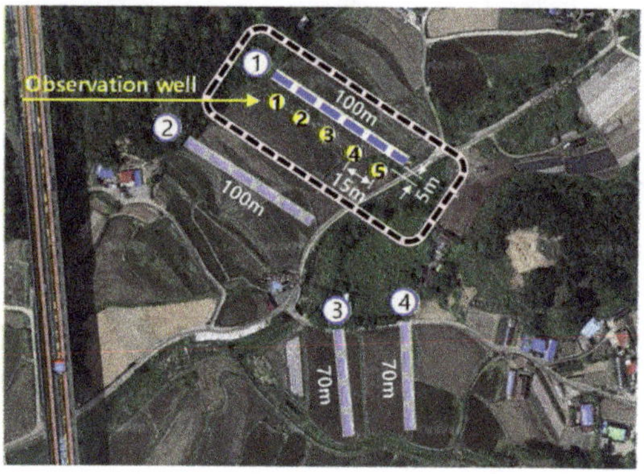

Figure 10. Location of field injection test and observation wells.

Table 4. Comparison of groundwater level rise by measurement and modeling.

Observation Number / Groundwater Elevation (cm)	1	2	3	4	5	Average
Measurement	49.0	42.5	38.9	40.3	44.6	43.1
Modeling	45.8	45.2	44.0	44.3	42.2	44.3

3.5. Performnance Evaluation of Artifical Recharge–Water Intake System

The conceptual model built for performance evaluation of the artificial recharge–water intake system was used to predict the fluctuations in the downstream according to the injection and water intake conditions, reflecting the optimal design factor values of artificial recharge—water intake facilities derived through 3D detailed numerical modeling. First, the spatial distribution of the groundwater level in the modeling area was simulated through a steady-state flow analysis, and a transient simulation was then performed. The change in the groundwater level in the downstream region was compared to the modeling conditions, and the average value of the three observation wells was used. In order to find a suitable artificial injection condition for this area, the injection conditions were configured in three cases of 100, 150, and 200 ton/day through four mixed-type artificial recharge facilities, and the total simulation time and injection period were set to 1 month, and the computational time step was 1 h. As a result, the artificial injection was simulated to rise by 0.29, 0.40, and 0.48 m at 100, 150, and 210 ton/day, respectively (Figure 11a). The rise of groundwater level was similar in the case of artificial injection rates of more than 210 ton/day, so it was estimated that the injection of 210 ton/day for 1 month was the most appropriate for the artificial recharge performance of this study area. In order to evaluate the water intake capacity of the water intake system after the artificial injection is completed, as a result of comparing the change in groundwater level by configuring the water intake conditions in three cases of 300, 400, and 500 ton/day and setting the withdrawal period to 1 month, it was simulated as a fall of 0.16, 0.18, and 0.24 m at 300, 400, and 500 ton/day, respectively (Figure 11b). Thus, the drop in groundwater level at a water withdrawal of 500 ton/day does not affect the storage capacity and initial groundwater level in this study area, and the water supply efficiency within a short period is the highest, so it is estimated to be an appropriate water withdrawal amount.

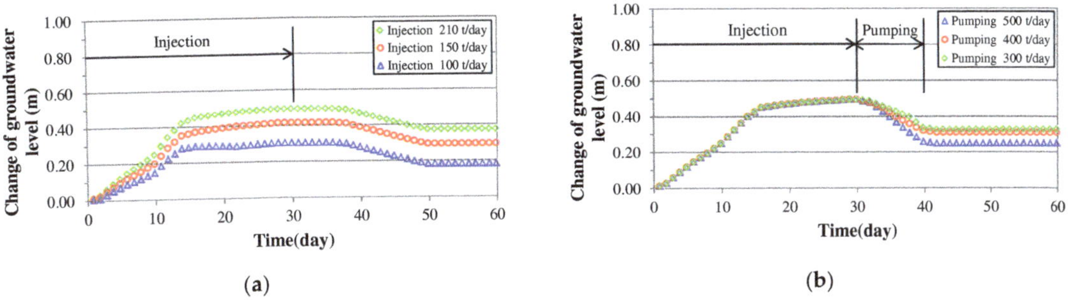

Figure 11. Time-series groundwater level behavior by amount of recharge and withdrawal. (**a**) water injection (**b**) evaluation the water intake capacity.

3.6. Optimal Operation Scenario for the Artificial Recharge–Water Intake System

In order to derive an optimal operation scenario for a sustainable water supply in the event of a drought in this study area, we referred to the agricultural water demand analysis data and the water budget analysis results [15]. In the target area, a shortage of about 30,000 tons of supply compared to demand occurs during the period from April to June, when the dry season and the farming season overlap. Therefore, an operating system was implemented that artificially recharges in March when the demand is relatively low, and water intake was carried out from April to June, and this was reviewed through detailed 3D numerical modeling. The operating stages and scenarios of the artificial recharge–water intake system for the modeling conditions were composed as follows (Table 5):

- Stage 1: Injection rate of 210 ton/day for 30 days (1 March~31 March)
- Stage 2: Injection rate of 400 ton/day for 60 days and withdrawal rate of 500 ton/day for 60 days (1 April~31 May)
- Stage 3: Shut down of recharge and water intake operation (1 June~)

Table 5. Optimal operation scenario for artificial recharge and water intake facilities.

Step	Injection Period	Injection Rate (m^3/day)	Total Injection Rate (ton)	Pumping Rate (m^3/day)	Total Pumping Rate (ton)
①	1 March~31 March	210	6300	0	0
②	1 April~31 May	400	30,300	500	30,000
③	1 June~	Shut down			

As a result of simulating the groundwater level change at the three observation wells in this operation stage, as a total of 6300 tons were injected for 30 days at a 210 ton/day injection rate in the first operation stage, the groundwater level rose by about 0.48 m. In the second stage of operation, following the first stage, a total of 24,000 tons were injected at 400 ton/day for 60 days, resulting in a cumulative injection amount of 30,300 tons. At the same time, as a total of 30,000 tons were withdrawn for 60 days at a 500 ton/day withdrawal rate, the groundwater level dropped by about 0.08 m from 0.48 m to 0.40 m. In the third stage of operation, about 30,000 tons of the required water supply for this study area was withdrawn in the second stage of operation, so all operations for recharge and water intake were stopped, and it was simulated that groundwater level gradually decreased and then stabilized (see Figure 12).

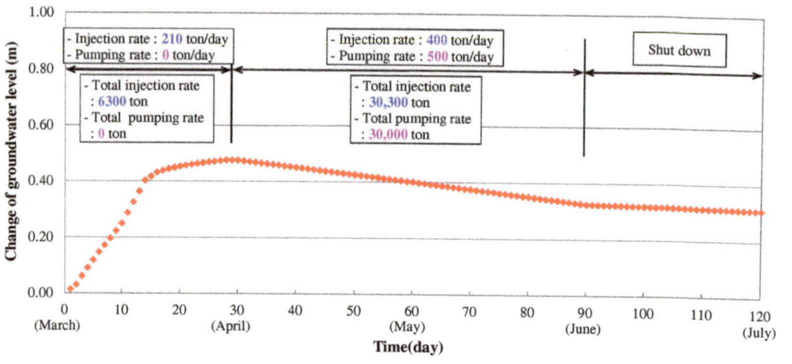

Figure 12. Time-series groundwater level behavior for artificial recharge and water intake.

4. Conclusions

Using 3D detailed numerical modeling, the optimal design factors for an artificial recharge and water intake facility were derived for drought areas where the actual artificial recharge–water intake system construction was completed. The optimal operating scenario for supplying agricultural water to the research area was then suggested through performance evaluation. The results obtained from simulation are summarized as follows:

(1) In order to derive the optimal media size within the ditch of the artificial recharge facility, the media size was divided into 10, 30, 50, 70, and 90 mm. As a result of modeling by setting the fill time of the groundwater level in the ditch, 50 mm was determined as the optimal media size in consideration of economic feasibility because the fill time and infiltration rate were similar at more than 50 mm.

(2) The vertical well intervals were divided into 5, 10, and 15 m to derive the appropriate vertical well interval for the artificial recharge facility, and the efficiency for each interval was estimated after setting the target groundwater level rise at less than 0.1 m below the ground surface in consideration of the depth of crop growth. As a result, efficiency was 81.9, 88.5, and 79.3% at intervals of 5, 10, and 15 m, respectively, which showed that the highest efficiency and the proper spacing of the vertical wells was determined to be 10 m in consideration of economic feasibility and maintenance according to the actual vertical well construction.

(3) In order to derive the proper opening ratio of the horizontal well, which is the main design factor of the water intake facility, the opening ratio was divided into 10, 20, and 30%, and comparative evaluation was performed based on the average value of groundwater level fluctuations at three observation wells. As a results, the groundwater level dropped 0.10, 0.19, and 0.20 m at 10, 20, and 30% of the opening ratio, respectively. At an opening ratio of more than 20%, the fluctuation of the groundwater level was small, and it was predicted that there would be no significant change in the water intake capacity, and the appropriate opening ratio of the horizontal well was determined to be 20%.

(4) A conceptual model was constructed by reflecting the optimal design factor values derived for the performance evaluation of the artificial recharge–water intake system, and the groundwater level fluctuations were predicted according to the in-

jection conditions. After dividing the artificial injection conditions into 100, 150, and 210 ton/day, comparative evaluation was performed through the average value of groundwater level fluctuations at three observation wells. As a result, it was simulated that the groundwater level rose by 0.29, 0.40, and 0.48 m, respectively, and the increase in groundwater level was similar in the case of artificial injection of more than 210 ton/day. Therefore, it was estimated that it was most appropriate to inject 210 ton/day for 1 month for the artificial recharge performance of this study area. In addition, in order to evaluate the water intake capacity of the water intake system after the artificial injection is completed, the water intake conditions were divided into 300, 400, and 500 ton/day, and the average value of the change in groundwater level at the three observation wells was used for comparative evaluation. As a result, it was simulated that the groundwater level dropped by 0.16, 0.18, and 0.24 m, respectively, and the drop in the groundwater level at 500 ton/day water intake does not affect the groundwater storage capacity and the initial groundwater level in this study area. Moreover, because the water supply efficiency within a short period was the highest, it was estimated that water intake at 500 ton/day was the most appropriate.

(5) Based on the optimal design factor values and performance evaluation results of the artificial recharge–water intake facility, the operation scenario of the artificial recharge–water intake system for sustainable water supply was configured by referring to the agricultural water demand analysis data and water budget analysis results in this study area. As a result of simulation, a stabilized groundwater level was confirmed through numerical modeling.

The results of this study can be utilized for the efficient operation and management of artificial recharge–water intake systems installed in drought areas in the future. It is expected that the 3D detailed numerical modeling method used in this study can be very usefully utilized to quantitatively evaluate the performance and effect of artificial recharge–water intake systems.

Author Contributions: Conceptualization, methodology, writing—review and editing, and supervision, J.-Y.L.; software, validation, formal analysis, investigation, data curation, writing—original draft preparation, and visualization, T.-Y.W. All authors have read and agreed to the published version of the manuscript.

Funding: This research was supported by the Korea Environment Industry and Technology Institute (KEITI) through the Demand Responsive Water Supply Service Program, funded by the Korean Ministry of the Environment (MOE), grant number No. 2018002650001.

Institutional Review Board Statement: Not applicable.

Informed Consent Statement: Not applicable.

Data Availability Statement: Not applicable.

Conflicts of Interest: The authors declare no conflict of interest.

References

1. Kim, Y.C.; Kim, Y.J. A review on the state of the art in the management of aquifer recharge. *J. Geol. Soc. Kor.* **2010**, *46*, 521–533.
2. Kim, Y.C.; Seo, J.A.; Ko, K.S. Trend and Barrier in the Patents of Artificial Recharge for Securing Goundwater. *J. Soil Groundw. Env.* **2012**, *17*, 59–75. [CrossRef]
3. Moon, S.H.; Ha, K.C.; Kim, Y.C.; Koh, D.C.; Yoon, H.S. Examination for Efficiency of Groundwater Artificial Recharge in Alluvial Aquifer Near Nakdong River of Changweon Area, Korea. *J. Econ. Env. Geol.* **2014**, *47*, 611–623. [CrossRef]
4. Bouwer, H. Artificial Recharge of Groundwater: Hydrogeology and Engineering. *Hydrogeol. J.* **2002**, *10*, 121–142. [CrossRef]
5. Lee, Y.D.; Shin, D.M.; Kim, B.J.; Kim, G.B. Selecting Aquifer Artificial Recharge Methods based on Characteristics of the Target Aquifer. *J. Eng. Geol.* **2019**, *29*, 483–494.
6. Oh, S.H.; Kim, Y.C.; Koo, M.H. Modeling Artificial Groundwater Recharge In The Hancheon Drainage Area, Jeju island, Korea. *J. Soil Groundw. Env.* **2011**, *16*, 34–45. [CrossRef]
7. Lee, H.J.; Koo, M.H.; Kim, Y.C. Determining Optimal Location of an Artificial Recharge Well using an Optimization-coupled Groundwater Flow Model. *J. Soil Groundw. Env.* **2014**, *19*, 66–81. [CrossRef]

8. Wazir, A. Numerical Simulations of Transient Groundwater Flow to Ditch Drains in Homogeneous Anisotropic Soil Using MODFLOW. *J. Sci. Eng. Res.* **2015**, *6*, 161–168.
9. Hao, Q.; Shao, J.; Cui, Y.; Xie, Z. Applicability of Artificial Recharge of Groundwater in the Yongding River Alluvial Fan in Beijing through Numerical Simulation. *J. Earth Sci.* **2014**, *25*, 575–586. [CrossRef]
10. Lin, D.G.; Chang, K.C.; Ku, C.Y.; Chou, J.C. Three-Dimensional Numerical Investigation on the Efficiency of Subsurface Drainage for Large-Scale Landslides. *Appl. Sci.* **2020**, *10*, 3346. [CrossRef]
11. Cha, J.H.; Lee, J.Y.; Kim, W.S. Development of 3-D Flow Model for Porous Media with Scenario-based Ground Excavation. *J. Kor. Soc. Dis. Sec.* **2017**, *10*, 19–27.
12. Todd, D.K.; Mays, L.W. *Groundwater Hydrology*, 3rd ed.; Wiley: New Jersey, NJ, USA, 2005; pp. 38–41.
13. Choi, M.R.; Kim, G.B. Optimum interval of artificial groundwater recharge wells, considering injection rate and economic feasibility. *Geo. J.* **2022**, *in press*.
14. Delleur, J.W. *The Handbook of Groundwater Engineering*, 2nd ed.; CRC Press: Boca Raton, FL, USA, 2007; pp. 11–18.
15. Kim, G.B.; Hwang, C.I.; Choi, M.R. Assessment of the need and potential for groundwater artificial recharge based on the water supply, water demand, and aquifer properties in a water shortage region of South Korea. *Env. Earth Sci.* **2021**, *80*, 115. [CrossRef]

water

MDPI

Article

Long-Term Evolution of Rainfall and Its Consequences on Water Resources: Application to the Watershed of the Kara River (Northern Togo)

Mozimwè Ani [1,2], Jessy Jaunat [1,*], Béatrice Marin [1], Marie Barel [1] and Kissao Gnandi [2]

1 GEGENAA-EA 3795, University of Reims Champagne-Ardenne, 2 Esplanade Roland Garros, 51100 Reims, France; paulinani11192@gmail.com (M.A.); beatrice.marin@univ-reims.fr (B.M.); m.armandine.barel@gmail.com (M.B.)

2 Département de Géologie, University of Lomé, Bvd Gnassingbé Eyadema, Lomé 01 BP 1515, Togo; kgnandi@yahoo.fr

* Correspondence: jessy.jaunat@univ-reims.fr

Abstract: The Kara River watershed (KRW), northern Togo, is facing climate-change impacts that have never been clearly characterized. Six decades of rainfall data (1961–2020) from six measuring stations ideally distributed across the watershed were used in this study. The flow records from two stations situated in contrasting locations on the KRW were also used. Statistical tests were conducted to assess the spatial and temporal variability of the rainfall and to detect tendencies within these meteorological series. The water balance method and calculation of the dry-off coefficient and of the groundwater volume drained by rivers allowed evaluating the impact of climatic evolution on surface flow and on groundwater volumes during the six decades studied. The results showed contrasting spatiotemporal variability of rainfall (and of aquifer recharge) over the watershed with a decreasing tendency upstream and an increasing one downstream. At the same time, the water volume drained by the aquifer to sustain the river's base flow decreased from −22% to −36% depending on the measuring station. These results constitute a decision-making tool for Togolese water resource managers and are of primary importance for characterizing the fate of water resources worldwide in regions subject to severe droughts.

Keywords: climate change; climatic break; aquifer recharge; drought; western Africa

Citation: Ani, M.; Jaunat, J.; Marin, B.; Barel, M.; Gnandi, K. Long-Term Evolution of Rainfall and Its Consequences on Water Resources: Application to the Watershed of the Kara River (Northern Togo). *Water* **2022**, *14*, 1976. https://doi.org/10.3390/w14121976

Academic Editors: Sang Yong Chung, Gyoo-Bum Kim and Venkatramanan Senapathi

Received: 30 April 2022
Accepted: 11 June 2022
Published: 20 June 2022

Publisher's Note: MDPI stays neutral with regard to jurisdictional claims in published maps and institutional affiliations.

1. Introduction

Groundwater is an important water supply source and contributes to surface freshwater base flows [1]. Therefore, it is worth understanding the impact of climate change on these water resources. Climate change can impact a watershed's hydrological processes and lead to water supply difficulties for the populations that rely on the resource [2]. In combination with certain factors such as insufficient access to drinking water and sanitation, droughts enhance the vulnerability of many African populations [3]. According to Döll et al. [4], the most vulnerable areas are mainly located in Africa and in South America.

Indeed, between the end of the 1960s and the beginning of the 1970s, the Sahelian regions of western Africa suffered a drought that marked a change in the climate regime [5]. This drought had long-term effects, particularly on regions located in the Gulf of Guinea [6–8]. Thus, rainfall and hydrometric deficits were confirmed with a modification of the rainfall–flow relationship in this geographic zone [9,10]. All of these authors used tests from the ICCARE program (Identification and Consequences of Climate Variability in Non-Sahelian Western Africa) to understand the changes that have come about in these climatic regimes. These tests include the Pettitt test [11], Lee and Heghinian test [12] and Hubert's segmentation [13]. The analysis of rainfall tendencies is an essential element for characterizing the consequences of climate change on water resources. The concept of tendency within a hydrometeorological series corresponds to the evolution of its parameters throughout time, independent of seasonal fluctuations [14]. Break tests coupled with tendency tests have often been reported in the literature for the characterization of tendencies within chronological series of climatic data [15–19]. Analyzing past tendencies became an efficient tool for planning, conception and management of water resources, providing valuable information for predictive approaches to evaluate future changes in hydroclimatic parameters [20,21].

During the last years, numerous research studies have used parametric and non-parametric tests to define tendencies within chronological series [22–31] The most commonly used are the Mann–Kendall [32,33] and Sen [34] tests. More recently, Sen [35] developed a new methodological approach, the innovative trend analysis (ITA), to detect tendencies in chronological data series. In contrast to the other methods mentioned above, the advantages of this method are: (1) the dataset does not require any hypothesis to be verified and (2) it is based only on the comparison of the two halves of the original temporal series [36]. This method has been used, for example, in Turkey [37], Nepal [38] and Ethiopia [28]. For instance, Girma et al. [39] applied the ITA method to the upper basin of the Huai River in China to show that the basin's precipitation is characterized by a high variation coefficient during the summer season, and that the annual rainfall had a statistically significant increasing tendency at the Fuyang station during the period 1960–2016. Additionally, by applying this same method to 16 measuring stations in northeast Algeria, Besma et al. [40] predicted that, in the future, some stations will measure drought periods while others will be confronted with significant flooding risks.

The evolution of climate variability and its impact on water resources have been approached by several authors from western Africa [41–44]. Because of limited storage volume and short residence time, bedrock aquifers are particularly dependent on annual rainfall that ensures their recharge. According to Döll et al. [4], the recharge volume could fall by 10% in the most vulnerable zones by 2050. In Togo, Badjana et al. [45] have shown evidence of decreasing precipitation in the extreme northern part of the country during the period 1960 to 2010. More recently, studies conducted in northern Togo [46] have pointed out variability in precipitation and temperature series with significant impacts on cereal production. On the contrary, the examination of precipitation and temperature at the watershed of the Mono River, which is located both in Benin and Togo, revealed a tendency of increasing rainfall and temperature by 2050 [17]. This illustrates the spatial variability of rainfall indices [19]. Thus, the watershed's scale seems the one most adapted to comprehend the hydroclimatic function and tendency within chronological rainfall series. It is in this framework that this study was initiated, to assess the rainfall tendency in the Kara River watershed (KRW) and to highlight its impact on water resources. This research study aims at (i) analyzing statistically the chronological rainfall series and flow data series of the Kara River, (ii) determining rainfall tendencies and analyzing their spatial variability, (iii) analyzing temporal variability of aquifer recharge and, finally, (iv) analyzing the temporal variation of the water volume mobilized by the basin's aquifers to sustain base flow.

2. Study Area

The KRW is a subwatershed of the Oti watershed representing 9.5% of its surface. It is located in the Kara region, one of the five regions of Togo (Figure 1). It extends within the 9°25′ and 10°10′ parallels of northern latitude and the 0°15′ and 1°30′ meridians of eastern longitude. It partially covers the prefectures of Kozah, Binah, Assoli, Guérin-Kouka, Bassar, Doufelgou, Kéran and a very small part of the Oti prefecture (savannah region). It expands to Benin where the Kara River has its source. In this study, only the Togolese part, which has a surface area of approximately 5000 km^2, was taken into account. It is essentially drained by the Kara River and its tributaries and has a dendritic network.

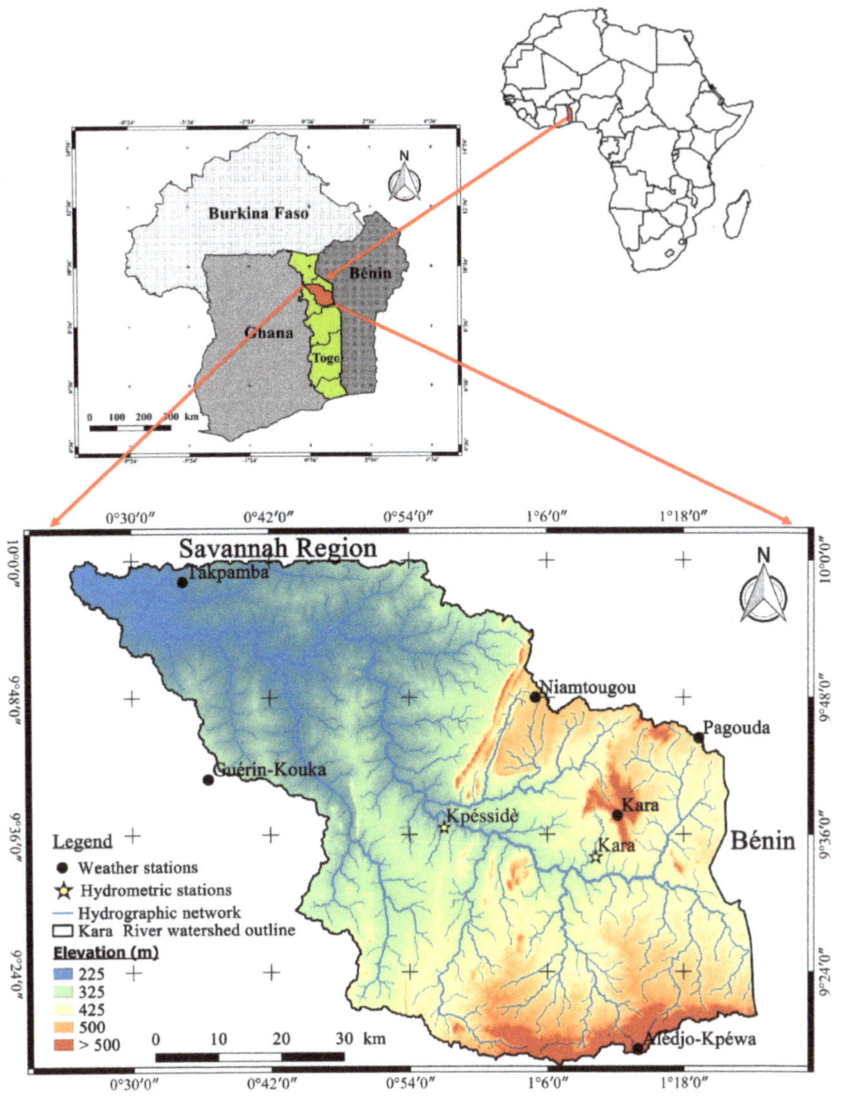

Figure 1. Geographical location, relief of the study zone and location of the different measuring stations.

The climate of the Kara River watershed is characterized by two distinct seasons: a dry season lasting from November until March and a wet season lasting from April until October. The average annual rainfall recorded for the last six decades is 1310 mm ± 96.9 mm. The rainiest months are July, August and September (Figure 2).

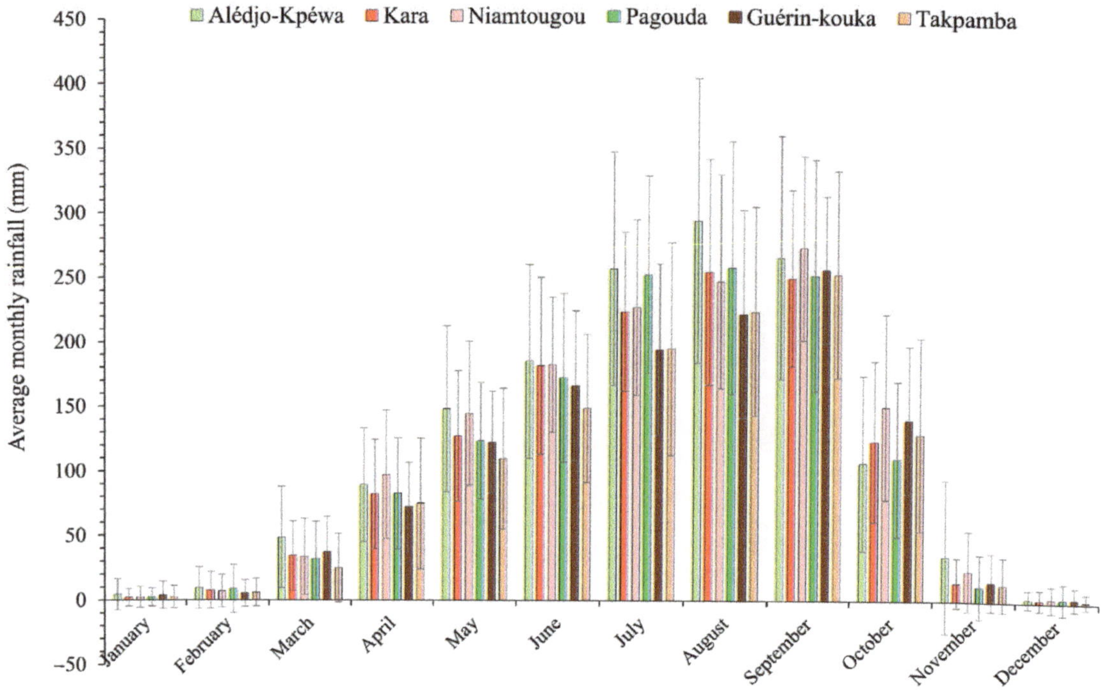

Figure 2. Average level of monthly rainfall at the studied stations between 1961 and 2020 (the location of the stations is presented in Figure 1).

From a geological point of view (Figure 3), the following structural entities occur from east to west [47,48]: (i) the granulites of the suture zone; (ii) the structural unit of Kara–Niamtougou, essentially composed of orthogneiss; (iii) the structural unit of Atacora, essentially composed of quartzites and sericite schists; (vi) the structural unit of Buem, composed of sandstone quartzites in the form of veins; (v) greenish sandstones associated with shales belonging to the great group of Pendjari of the Volta watershed, crossed by hematitic and alluvial formations.

On the hydrogeological level, two main aquifer reservoirs can be identified from surface to depth: an alterite reservoir and a fractured/fissured reservoir. The whole fractured/alterite level acts similar to a unique aquifer system of bilayer type [49]. The wells tapping the water of the alterite reservoir within the first fifty meters of depth are traditional sinks that become dry during the low-water period. The water of the reservoir with a fractured base is exploited by boreholes extending from 40 to 100 m in depth; these sinks provide water year-round (Figure 4).

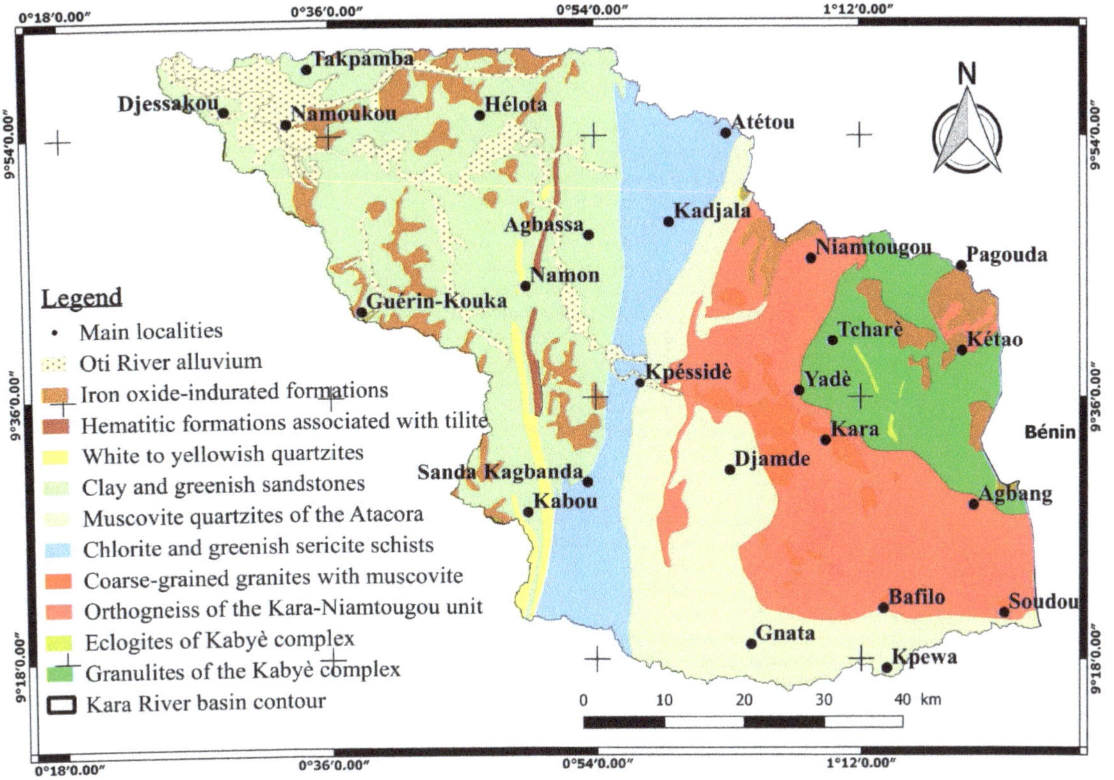

Figure 3. Simplified geological map of the Kara River watershed (modified after General Directorate of Mines and Geology, 1986).

Figure 4. Traditional sinks exploiting the water of the alterite reservoir at Niamtougou (northwest of the study zone) (**a**) and borehole equipped with a human-operated pump, exploiting the water of the fractured reservoir at Bébéda (center of the study zone) (**b**).

3. Material and Methods

3.1. Data Used

Hydroclimatic (precipitation and temperature) and hydrometric (flow) data were used for this study (Table 1 and Figure 1). The hydroclimatic data were provided by the National Directorate of Meteorology of Togo. Hydrometric data were provided by the Directorate of Water Resources of Togo and concerns only the Kpéssidè and Kara stations (which are located southeast of the city of Kara, on a tributary of the Kara River; Figure 1). Unfortunately, this hydrometric measurement network is no longer active since the early 1990s.

Table 1. Characteristics of the hydrometric and hydrological stations. (Meteo. = meteorological; Hydro. = hydrometric; Long. = longitude; Lat. = latitude; Obs. = observation; Var. Coef. = variation coefficient).

Station Type	Localities	Geographical Coordinates		Measured Data	Obs. Period	Time Step Used	Average Rainfall/Flow (mm)/(m^3/h)	Var. Coef. (%)
		Long. E	Lat. N					
Meteo.	Kara	1.203	9.628	Temperature + Precipitation	1961–2020	Monthly	1303 ± 202	16
	Niamtougou	1.083	9.773	Temperature + Precipitation	1961–2020	Monthly	1386 ± 211	15
	Pagouda	1.323	9.741	Temperature + Precipitation	1961–2020	Monthly	1308 ± 243	19
	Alédjo-Kpéwa	1.233	9.287	Temperature + Precipitation	1961–2020	Monthly	1445 ± 255	18
	Guérin-Kouka	0.611	9.608	Temperature + Precipitation	1961–2020	Monthly	1238 ± 184	15
	Takpamba	0.5746	9.969	Temperature + Precipitation	1961–2020	Monthly	1177 ± 234	20
Hydro.	Kpéssidè	0.9555	9.617	Flow	1962–1992	Daily	29 ± 10.9	38
	Kara	1.2833	9.533	Flow	1954–1990	Daily	20 ± 9.4	47

The precipitation records of the meteorological station of Guérin-Kouka showed significant gaps for certain months and/or even years. Unfortunately, this station is the only one located in the southwest part of the studied watershed. The missing data were completed to facilitate the analysis and ease the interpretation of the results. According to [50], missing or erroneous data of a station can be estimated based on the values of nearby stations if they are subject to the same weather conditions and located in the same geographic zone. To highlight the geographical heterogeneities of precipitation, the monthly and annual precipitation data of this station were thereby completed by those of Takpamba. These two stations are approximately 30 km apart, following a north–south axis, and are located in a similar hydroclimatic context.

3.2. Spatiotemporal Variability of Rainfall

3.2.1. Standardized Precipitation Index (SPI)

The standardized precipitation index (SPI) or Nicholson index [51] enables the detection of precipitation variations within a long series of observations. It allows for the calculation of the drought level or the exceeding amount of rainfall in a given region [14,52]. It is defined by Equation (1):

$$SPI = \frac{(Xi - Xm)}{\sigma} \quad (1)$$

where Xi is the precipitation for the year, Xm is the average precipitation of the study period and σ is the standard deviation of the annual rainfall that occurred during the study period. Negative values indicate a rainfall deficit during the period and positive values define excessive rainfall amounts.

3.2.2. Hanning's Low-Pass Filter

Hanning's low-pass filter of order 2, also called the weighted–reduced–centered moving average, enables the elimination of seasonal variations in a chronological series. This method, applied to long precipitation series, is explained elsewhere [8,53,54].

Reduced–centered indices of rainfall amount obtained with this method are useful for distinguishing periods with a deficit of rainfall from periods with an excess of rainfall. Thus, it is possible to identify normal, wet and dry periods within a chronological series. A period is said to be normal when a fluctuation between positive and negative values of the computed index is observed. A period is indicated to be wet when the annual rainfall is superior to the average rainfall (positive index). A period is dry when the annual rainfall is inferior to the average rainfall (negative index).

3.2.3. Geographical Precipitation Analysis

To assess the geographical repartition of rainfall over the watershed, an interpolation based on the kriging method was performed using Surfer software (Version 16). Hence, decennial rainfall was mapped to characterize the evolution and the repartition of rainfall during the six decades studied (1961–1970, 1971–1980, 1981–1990, 1991–2000, 2001–2010 and 2011–2020).

3.2.4. Break in Precipitation

According to Lubès-Niel et al. [55], a break is defined by a change in the probability law of the random variables, of which successive realizations define the studied chronological series. To highlight climatic breaks, several methods were previously used: the Pettitt test [11], the Bayesian method of Lee and Heghinian [12], and Hubert's segmentation method [13]. The use of these statistical methods is justified by their robustness. Based on the literature [6,7,55], we choose to use these three methods to detect statistical breaks in the precipitation levels during the period considered. These three statistical methods were applied to our data by utilizing Khronostat software (Version 1.01). which was developed by the mixed research unit for Hydrosciences of Montpellier [56].

3.2.5. Long-Term Trends of Rainfall

The Mann–Kendall non-parametric test [32,33] explores the tendencies within a data series without specifying whether the tendency is linear or not [57]. The Mann–Kendall tendency test is a simple and robust test that accepts missing data but hardly detects annual trends [14]. This test was used in other studies to evaluate the importance of a trend within chronological series of climatic data [20,22,24–27,58–60]. A positive value for the Mann–Kendall slope expresses an upward trend and a negative value implies a downward trend.

The Sen test [34] is used when the trend is thought to be linear [57,61]. A positive value for the Sen slope expresses an upward trend and a negative value implies a downward trend [57].

Finally, the method of innovative trend analysis (ITA) was recently developed by Şen [35,36]. It enables a visual inspection and an identification of categorical tendencies within chronological data [37]. The most commonly used methodology is the Mann–Kendall (MK) method, but it requires some basic assumptions, which may not be valid in natural hydroclimatological series. The ITA method used in this paper does not require any assumptions and is based on the comparison of the two ascending ordered halves from the original time series. The steps of this method follow:

- The temporal data series is subdivided into two parts (the first and the second half of the record) and the rainfall levels are ranged in increasing or decreasing order.
- The arranged rainfall data are then placed in a dispersion diagram (rainfall levels of one half of the series versus rainfall levels of the other half of the series) with equal vertical and horizontal scales. The curves thus obtained are then compared with the line curve: y = x.

There is no significant trend if the dispersion of the points occurs along the straight line of slope value equal to 1. If the dispersion of the points is below or above the unitary line curve, then a significant tendency either upward or downward exists. Based on random variables, five types of trends can be discriminated: a trendless times series, a monotonic positive trend, a monotonic negative trend, a non-monotonic positive trend and a non-monotonic negative trend [37].

3.2.6. Water Balance

To estimate the aquifer's recharge rates, the water balance method was adopted. Numerous methods describe ways to evaluate the parameters for the water balance (e.g., the Penman method [62], the Thornthwaite method [63] and the Turc method [64]). Considering the available data, the Thorntwaite method was selected because it only requires rainfall, temperature and the station's latitude data. As reported in the literature, this method is regularly used to estimate water balance parameters (e.g., [65–67]). The parameters computed via this method are the potential evapotranspiration (PET), the actual evapotranspiration (ETa) and the effective precipitation (EP). In this study, we used a value for the useful water reserves (UWR) of 100 mm, in agreement with work to estimate the recharge of the base aquifers in the plateau regions of Togo [68], carried out in a similar environment.

Runoff was estimated based on the empirical formula from Tixéront and Berkaloff, as reported in [69]. This method uses the evapotranspiration value computed with the Thornthwaite method by using the following equation:

$$R = \frac{P^3}{3 \times PET^2} \quad (2)$$

with R: runoff (mm), P: precipitation (mm) and PET: potential evapotranspiration (mm).

3.2.7. Recession Coefficient and Water Volume Mobilized by the Aquifers

The analysis of the rivers' drying constitutes a fundamental aspect for comprehending the watershed's function during the low-water period. In absence of precipitation, the flow of certain rivers is reduced to only groundwater contributions from the aquifer [70]. The analysis of recession episodes thus enables determination of the contributions coming from the aquifers and the drainage rhythm of the underground reservoirs.

Different mathematical methods are suggested for calculating the recession coefficient and the water volume mobilized by the aquifers: [71–73]. All of these methods are inspired by the Maillet model; in this study, we used dichotomous resolution, which is widely mentioned in literature (e.g., [73–76]).

In this method, the expression of the dry-off coefficient is given by Equation (3):

$$\frac{e^{-kt}}{k} + \frac{V}{Q_0} + \frac{1}{k} = 0 \quad (3)$$

with Q_0: initial flow ($m^3 \cdot s^{-1}$), k: recession coefficient (day^{-1}) and t: time (s).

The water volume mobilized by all of the watershed's aquifers is then given by the expression:

$$V_{mobilised} = \int_0^{+\infty} Q_0 \cdot e^{-kt} dt = \frac{Q_0 \times 86{,}400}{k} \quad (4)$$

4. Results and Discussion

4.1. Spatiotemporal Variation of Precipitation

The SPI values and Hanning's low-pass filter of order 2 values applied to the chronological precipitation series of six stations of the KRW are presented in Table 2. Figure 5 shows an alternation of wet, normal and dry periods of which the amplitude varies according to the station. The period from 1961 until the end of 1970 was humid for the Kara, Niamtougou, Pagouda and Guérin-Kouka stations, and this continued until the beginning of the 1980s for the Alédjo-Kpéwa station. Takpamba's station stands out with a precipitation deficit that persisted until the beginning of the 1990s. A return to normal precipitation was observed after the 1990s and until 2011 at the Kara, Niamtougou and Pagouda stations. During the last decade, a decrease in precipitation was significant at most of the stations apart from Guérin-Kouka and Takpamba, which recorded an increase in precipitation with positive SPI values reaching +2.5. The return of a rise in rainfall at these stations was also noticed by some authors [52,77].

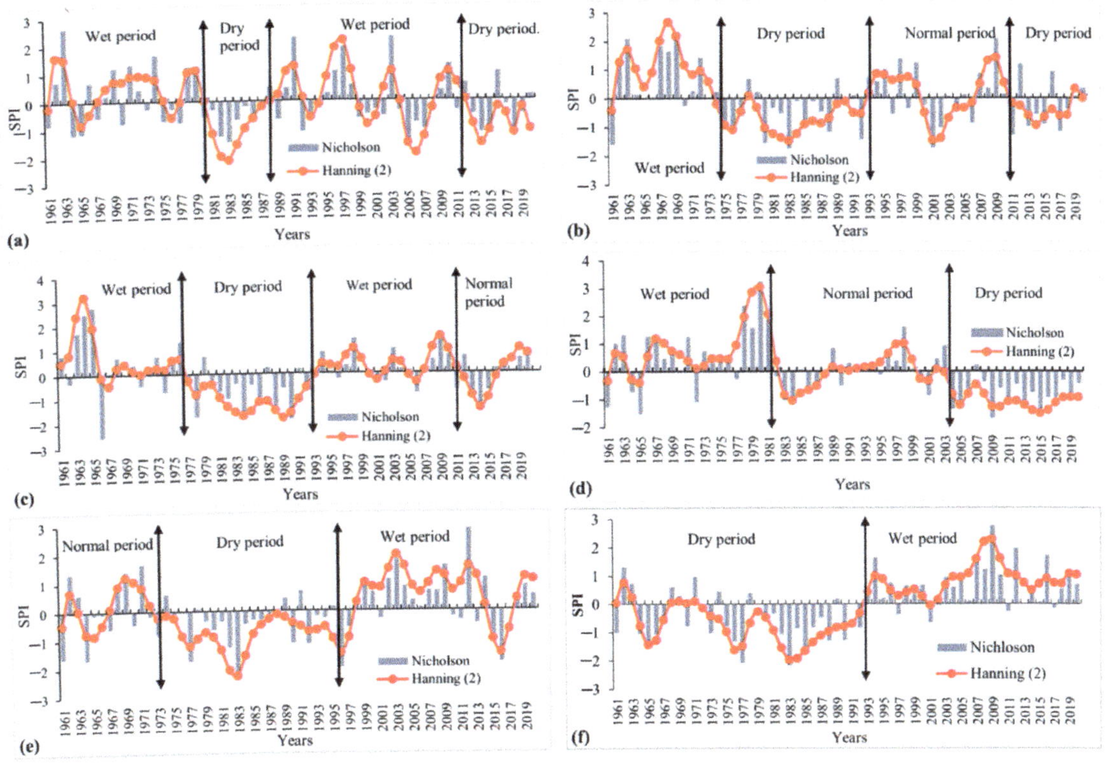

Figure 5. Nicholson's index and Hanning's low-pass filter of 2 applied to the KRW stations: (**a**) Kara; (**b**) Niamtougou; (**c**) Pagouda; (**d**) Alédjo-Kpéwa; (**e**) Guérin-Kouka and (**f**) Takpamba.

Table 2. Alternation of different phases and breaks detected within the precipitation series recorded at the measuring stations.

Stations	Periodic Trend within the Series			Breaks		
	Wet Period	Normal Period	Dry Period	Pettitt	Lee and Heghinian	Hubert
Kara	1961–1980	1989–2012	1981–1988 2013–2020	-	1963	-
Niamtougou	1961–1973	1993–2020	1974–1992	-	1972	1972
Pagouda	1961–1976 1994–2011	2012–2020	1977–1993	-	1965	1965
Alédjo-Kpéwa	1961–1982	1983–1999	2000–2020	2003	2003	1977; 1981; 2003
Guérin-Kouka	1998–2020	1961–1972	1973–1997	1997	1997	1997
Takpamba	1993–2020	-	1961–1992	1992	1992	1992

Two tendencies emerged when analyzing the SPI values. The stations located at higher altitudes, upstream from the KRW (Kara, Niamtougou, Pagouda and Alédjo-Kpéwa) are subject to a tendency of decreasing precipitation while those located at lower altitudes in the downstream portion (Guérin-Kouka and Takpamba) experience a tendency of increasing precipitation, even if the average annual rainfall is lower. The decrease in precipitation at the stations located upstream was also noted by [77,78]. Despite the long drought that marked the precipitation regime of the savannah region in extreme northern Togo (Mango and Dapaong station) since the 1970s, [79] showed that the region has been affected by increasingly more frequent exceptional rainfall events since the 1990s. Analysis of extreme climatic events in the Oti watershed in northern Togo suggest rising rates of total precipitation of between 1% and 16% with a rise of the rainfall level of about 13.89 mm [80]. The proximity of stations in the savannah region (Guérin-Kouka and Takpamba) is in agreement with the previous results. The periods 1998–2020 and 1993–2020, respectively, correspond to the wet periods at the Guérin-Kouka and Takpamba stations, as defined by [79] in the savannah region. This climate variability which induced a decrease in precipitation is due to the disorganization of seasons, which is potentially useful for agriculture [79]. Noting a significant effect on the yields of grain cultures, ref. [46] suggested measures to adapt to the change, such as the use of seeds that are tolerant to drought, low-cost irrigation practices, diversification of cultures and the application of agroecological practices for sustainable agricultural production in northern Togo.

The results produced by the spatial interpolation of precipitation over the watershed are presented in Figure 6. The repartition of isohyet curves for the six decades show contrasting precipitation temporal variability.

The period 1961–1970 (Figure 6a) was humid with maximum values at the Niamtougou and Alédjo-Kpéwa stations and the emergence of 1500 mm isohyets. Minimum values are observed in the downstream part of the watershed with 1200 and 1100 mm isohyets. This contrast can be explained by the influence of the watershed's geomorphology. According to Badjana et al. [45], the Dahomeyides mountain range, extending through the watershed, interrupts the trade winds coming from the southwest and promotes the formation of clouds.

The decade 1971–1980 (Figure 6b) experienced a decrease in precipitation at the Takpamba, Niamtougou (disappearance of the 1500 mm isohyet), Kara and Pagouda stations, but the Alédjo-Kpéwa station, located in the southern part of the watershed, recorded rainfall up to 1600 mm.

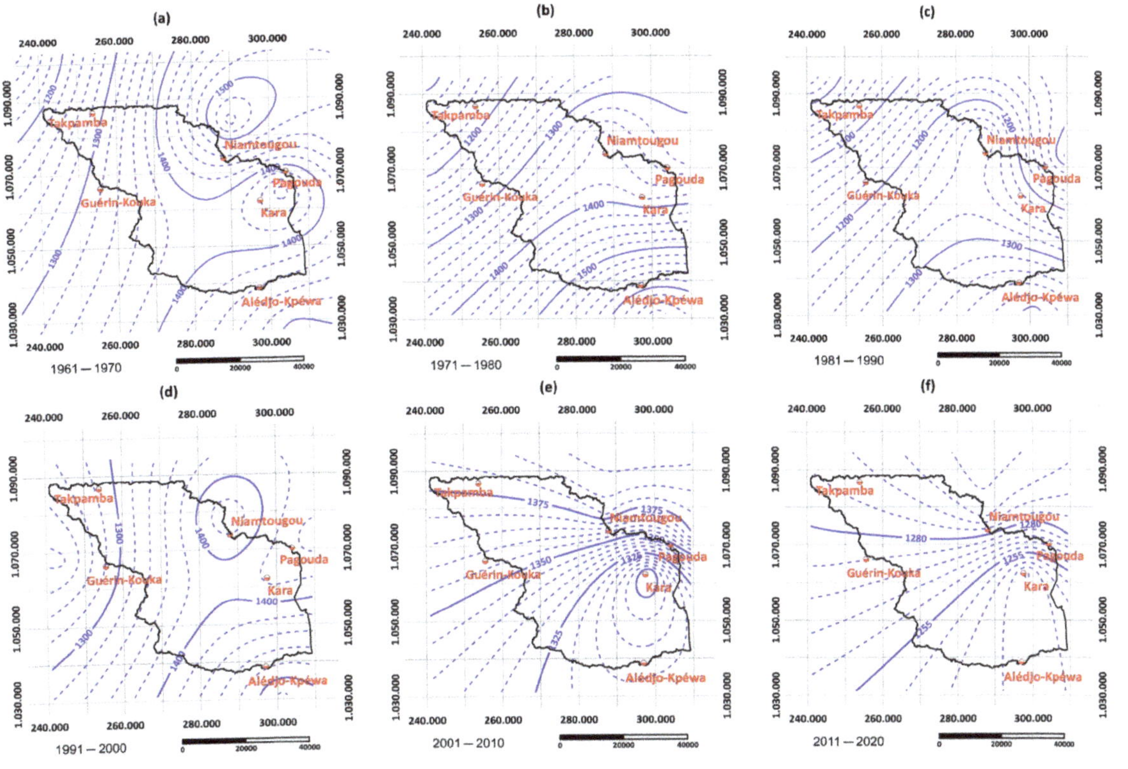

Figure 6. Spatial and temporal variation of decadal precipitation over the Kara River watershed during 1961 to 2020: (**a**) 1961–1970; (**b**) 1971–1980; (**c**) 1981–1990; (**d**) 1991–2000; (**e**) 2001–2010; (**f**) 2011–2020. The isohyets are represented by blue lines (the difference in precipitation levels between two isohyets varies according to the map).

A general decrease in precipitation was observed during the decade 1981–1990 (Figure 6c) at all stations. We notice the appearance of the 1100 mm isohyet and the migration of the 1200 and 1300 mm isohyets to the northeast and southwest in the upstream part of the watershed.

The return to normal precipitation was observed during the decade 1991–2000 (Figure 6d). The consequence of this is the migration of the 1300 mm isohyet to the northwest, where the Guérin-Kouka and Takpamba stations are located, and the reappearance of the 1400 mm isohyet.

A slight decrease in precipitation was observed during the decade 2001–2010 (Figure 6e) at the Alédjo-Kpéwa, Kara, Niamtougou and Pagouda stations. This slight decrease was not observed at the Takpamba and Guérin-Kouka stations.

During the decade 2011–2020 (Figure 6f), a general decrease in precipitation in the watershed occurred but remained moderate in the north while it was accentuated in the south and center of the basin.

The spatial distribution in precipitation over the watershed thereby confirms the results obtained by studying the temporal variations, with the disappearance of the 1500 and 1400 mm isohyets at the watershed's scale during the two last decades. This spatiotemporal variation of precipitation was due to the drought that occurred in 1970 in the Sahelian countries, which continues in the southern countries [6]. The appearance of the 1300 mm isohyet in the upstream portion of the basin is the consequence of a return to normal precipitation during the decade 1991–2000 (Figure 6d). During the last two decades, [81] observed a slight phase of increased rainfall in northern Ghana, the magnitude of which is inferior to the phase of strong precipitation during the 1950s and 1960s. This situation can be described as an improvement even if the rainfall levels are not comparable to those of the 1950s and 1960s.

4.2. Detection of Climatic Breaks

Several statistical break tests were applied to the different chronological precipitation series of each of the stations (Table 2). The Pettitt test shows no break at the Kara, Niamtougou and Pagouda stations but indicates a break for 2003, 1997 and 1992 at the Alédjo-Kpéwa, Guérin-Kouka and Takpamba stations, respectively. The Bayesian method of Lee and Heghinian shows breaks for 1963, 1965, 1972, 1997, 1992 and 2003 at the Kara, Niamtougou, Pagouda, Alédjo-Kpéwa, Guérin-Kouka and Takpamba stations, respectively. The Hubert segmentation shows no break at the Kara, Guérin-Kouka and Takpamba stations. Using this method, only the Alédjo-Kpéwa station recorded three breaks, in 1977, 1981 and 2003. The Kara and Pagouda stations recorded early breaks in 1963 and 1965 in opposition to Alédjo-Kpéwa station, which recorded a late break in 2003.

The various breaks that were calculated with the different tests are due to the statistical characteristics of the series which, according to [55], are normality, stationarity and non-autocorrelation:

- The more normal the series, the less probable the break; the non-observation of a break means normal behavior of the series and good autocorrelation.
- The methods for computing breaks do not produce the same response elements when considering the stationary nature of the series [55]. The Pettitt test is only valid for abrupt breaks and is invalid for continuous breaks [14].

As a general rule, the existence of a break in a series of hydrometeorological data indicates a sudden modification of the average in the series. Consequently, it is equivalent to a sudden temporal variability, hence the need to use various methods of break detection to characterize this modification within series and better discern climate variability at the considered stations.

The breaks occurred during 1963 to 2003. These break dates agree with values determined in the west African subregion [6,82–85]. The rainfall rates after these breaks varied from one station to another. At the Kara, Niamtougou, Pagouda and Alédjo-Kpéwa stations, the observed rates of precipitation decrease are about −12.5%, −13.2%, −23%, and −13.8%, respectively. At the Guérin-Kouka and Takpamba stations, precipitation increase of 12% and 27.3% were calculated. These results confirm the tendencies observed with the Nicholson method and Hanning's low-pass filter of order 2 used earlier.

4.3. Long-Term Tendencies of Precipitation

The results of the Mann–Kendall and the Sen tests applied to the interannual precipitation records of the six stations considered are presented in Table 3. The tendency is non-significant at the Kara, Niamtougou, Pagouda, and Guérin-Kouka stations as the corresponding p-value is superior to the level of significance $\alpha = 0.05$. These results support those of [47] who analyzed the temporal evolution of precipitation and temperature and their impact on cereal products in northern Togo. Significant statistical tendencies are observed at the Alédjo-Kpéwa and Takpamba stations (p-value < 0.05). The value of the S parameter is less than zero for the Alédjo-Kpéwa station, indicating a monotonous decreasing tendency within the series. On the contrary, the S parameter is more than zero at the Takpamba station, thus indicating a monotonous increasing tendency at this station. The amplitude of the reduction in the precipitation levels is −6.59 mm at the Alédjo-Kpéwa station, located in the more elevated upstream portion of the watershed. The level is rising by +6.83 mm at the Takpamba station, located in downstream portion of the watershed at a low altitude.

Table 3. Results of the Mann–Kendall and Sen tests.

Parameters	Kara	Niamtougou	Pagouda	Alédjo-Kpéwa	Guérin-Kouka	Takpamba
Rate of Kendall	−0.052	−0.13	−0.022	−0.30	0.11	0.31
S	−92	−222	−38	−536	266	556
p-value	0.56	0.16	0.81	0.6×10^{-3}	0.092	0.2×10^{-3}
Alpha	0.05	0.05	0.05	0.05	0.05	0.05
Tendency	No	No	No	Yes	No	Yes
Slope of Sen	−1.04	−2.3	−0.30	−6.59	2.24	6.83

Figure 7 illustrates the tendencies of the series computed with the ITA method. For the Kara station, we observe a non-monotonous positive tendency. A non-monotonous negative tendency is noted at the Niamtougou and Pagouda stations. The Alédjo-Kpéwa station records a monotonous negative tendency and, finally, the Guérin-Kouka and Takpamba stations show a monotonous positive tendency. A good interpretation of the annual rainfall requires their classification into three categories [35]: weak, average and strong. For example, for the Pagouda station (Figure 7c), weak rainfall tended toward increase in the second half of the series (1991–2003) as compared to the first half of the series (1961–1990), average rainfall did not present any tendency and strong rainfall had a tendency to decrease during the first half (1961–1990).

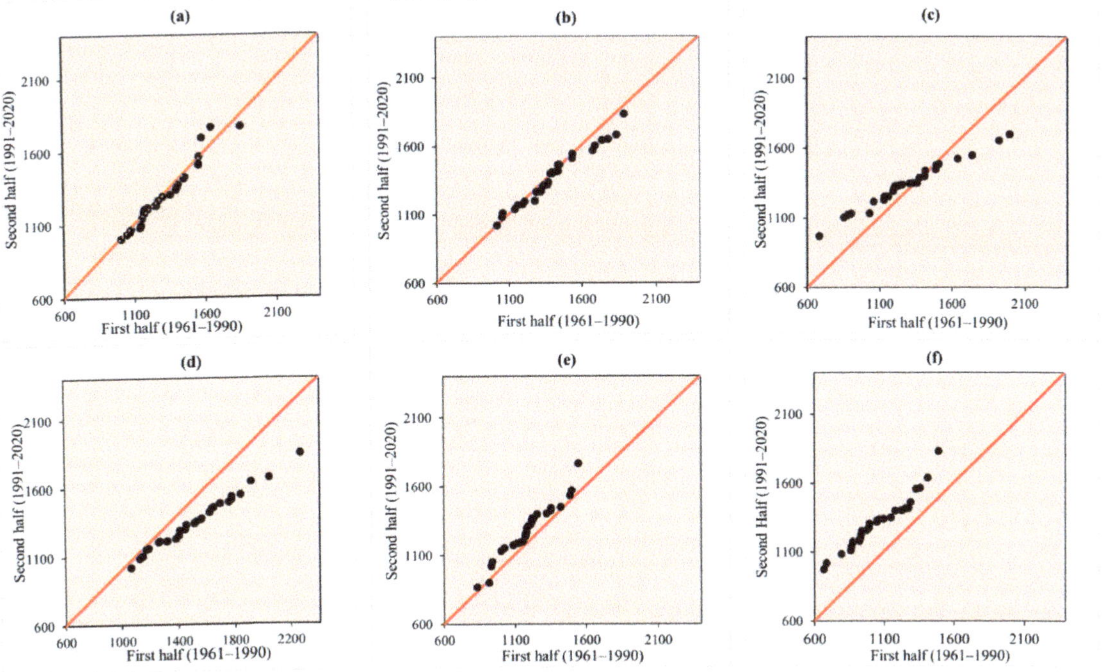

Figure 7. Results of the ITA method applied to the six stations in the Kara River watershed: (**a**) Kara; (**b**) Niamtougou; (**c**) Pagouda; (**d**) Alédjo–Kpéwa; (**e**) Guérin–Kouka; (**f**) Takpamba.

The results of the ITA method and the Mann–Kendall method are partially comparable. The tendencies found with the Mann–Kendall method have been confirmed by the ITA method, in particular at the Alédjo-Kpéwa and Takpamba stations. Concomitantly, in opposition to the Mann–Kendall method, the ITA method points out tendencies within the Kara, Niamtougou and Pagouda station series. According to Şen [35], the segmentation into three categories (weak, average and strong) provides detailed information about the structure of the internal tendency of the series considered. Other authors, who used the ITA method in other regions, confirmed that the method is more adapted to detect tendencies within the series [28,37,38]. The results obtained with the ITA method are important because they enable better comprehension of the hydroclimatic context and provide elements for water resource management [28].

The increased rainfall at the Guérin-Kouka and Takpamba stations during the 1961 to 2020 period (Table 3) indicates a return to normal precipitation at the watershed's scale. This confirms the results found earlier. The same observations were made by [86], who analyzed the climatic tendency in the Volta watershed during the period 1981–2010. These authors indeed observed a positive tendency in the precipitation of the Sahelian zone of the western part of Africa, which would indicate the establishment of a wet period. It could persist in time. The different results show that the tendency of increasing rainfall in the downstream part of the watershed of Kara is rather regionalized.

4.4. Consequences of the Climatic Variability on the Watershed's Water Resources

The climatic context is the principal factor of the spatiotemporal variability of groundwater recharge. Precipitation is the climatic factor that affects it the most [87]. Table 4 summarizes the evolution of the water balance parameters at the studied stations before and after the breaks. The tendency observed regarding the spatiotemporal variability of precipitation is also observed for aquifer recharge. Indeed, an increase in the recharge is observed at the Guérin-Kouka (+2%) and Takpamba (+16%) stations, and a decrease at the Kara (−15%), Niamtougou (−7%), Pagouda (−2%) and Alédjo-Kpéwa (−73%) stations. These results are similar to those obtained in studies carried out in analogous conditions. In the square degree of Grand Lahou in southwest Côte d'Ivoire, [88] observed a decrease in aquifer recharge at the Gagnoa (−80.5%) and Sassandra (−66%) stations. Results reported in [89] also showed the influence of precipitation changes on the recharge of aquifers in the Bandama watershed (northern Côte d'Ivoire), where the authors noted a 42.63% decrease at the Sirasso weather station and 72% at the Dikodougou weather station.

Table 4. Water balance parameters before and after the dates of break at the six KRW stations.

Water Balance Parameters	Kara		Niamtougou		Pagouda		Alédjo-Kpéwa		Guérin-Kouka		Takpamba	
	Break = 1963		Break = 1972		Break = 1965		Break = 1977		Break = 1997		Break = 1992	
	before	after	before	before	before	after	before	after	before	after	before	after
Rainfall ($mm \cdot y^{-1}$)	1480	1294	1553	1607	1607	1277	1607	1385	1191	1334	1049	1335
PET ($mm \cdot y^{-1}$)	1689	1862	1681	1764	1764	1751	1764	1563	1757	2007	2144	2532
ETa ($mm \cdot y^{-1}$)	944	952	973	975	975	909	975	972	904	1031	869	1100
Excedent (R-ETa) ($mm \cdot y^{-1}$)	536	342	580	632	632	368	632	413	287	303	180	235
Runoff ($mm \cdot y^{-1}$)	379	208	442	445	445	226	445	363	182	196	84	124
Effective Infiltration ($mm \cdot y^{-1}$)	157	134	138	187	187	142	187	50	105	107	96	111

To specify the impact of climatic fluctuations on the flow of the Kara River, two chronological series of hydrometric data of the Kara River (Kara and Kpésside stations) were subjected to statistical break detection tests. The break appeared in 1974 (Pettitt test) at the Kara station (Figure 8a).

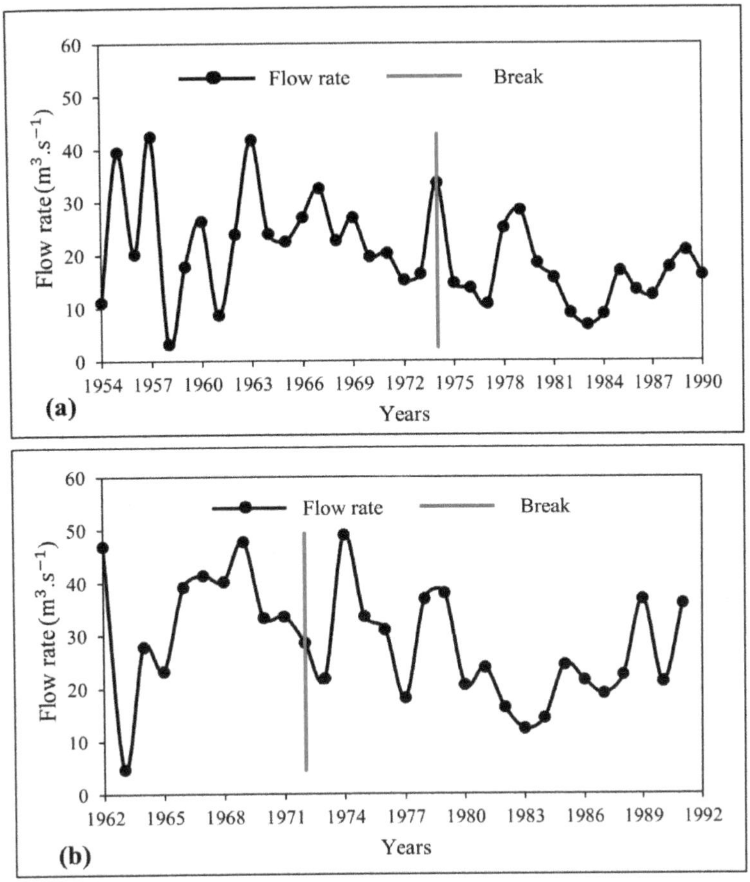

Figure 8. Interannual variation of flow at the Kara (**a**) and Kpéssidè (**b**) hydrological stations. The red line corresponds to years for which a statistical break has been computed.

The average flow of the Kara River dropped from 23.63 $m^3 \cdot s^{-1}$ before the break to 15.6 $m^3 \cdot s^{-1}$ after the break, meaning a 34% decrease in flow during the period 1954–1990. At the Kpéside station (Figure 8b), only the Bayesian method of Lee and Heghinian detected a break in 1972. The flow of the Kara River dropped from 43.20 $m^3 \cdot s^{-1}$ before the break to 28.11 $m^3 \cdot s^{-1}$ after the break, representing a rate of decrease of 35%. Another report [90] also observed a deficit of 43% in the Mono-Couffo watershed, located between Togo and Benin, during the period 1961 to 2000. According to these authors, this diminution was mainly due to a precipitation deficit but also to strong evapotranspiration and anthropogenic effects.

The flow regime of the Kara River clearly indicates a deficit due to the decrease in the precipitation regime observed in the 1970s over Africa's western region and, more particularly, the KRW. The dry-off coefficients computed for the Kara River and the water volumes mobilized by the aquifers are compiled in Table 5.

Table 5. Results of the calculation of the recession coefficients and the water volumes mobilized by the aquifers before and after the break dates.

Stations	Period	Recession Coefficient (day^{-1})	Rate of Increase (%)	Water Volume Mobilized (km^3)	Rate of Decrease (%)	Recession Duration (day)
Kara	1954–1974	2.8×10^{-2}	25	0.32	22	36
	1975–1990	3.5×10^{-2}		0.25		29
Kpéssidè	1962–1972	3.1×10^{-2}	10	0.47	36	32
	1973–1991	3.4×10^{-2}		0.3		29

The evolution of the recession coefficients at the Kara hydrometric station during the different periods (Figure 9a) shows that the recession coefficient varies between 2.0×10^{-2} and 3.6×10^{-2} day^{-1} before the year of break with an average of 2.8×10^{-2} day^{-1} (inverse of 36 days). After the break, the values are between 2.1×10^{-2} and 4.8×10^{-2} day^{-1} with an average of 3.5×10^{-2} day^{-1} (inverse of 29 days). At the Kpéssidè station (Figure 9b), the recession coefficient values are between 1.9×10^{-2} and 4.3×10^{-2} day^{-1} with an average of 3.1×10^{-2} day^{-1} (inverse of 32 days) before the break and between 2.6×10^{-2} and 4.1×10^{-2} day^{-1} with an average of 3.4×10^{-2} day^{-1} (inverse of 29 days) after the break.

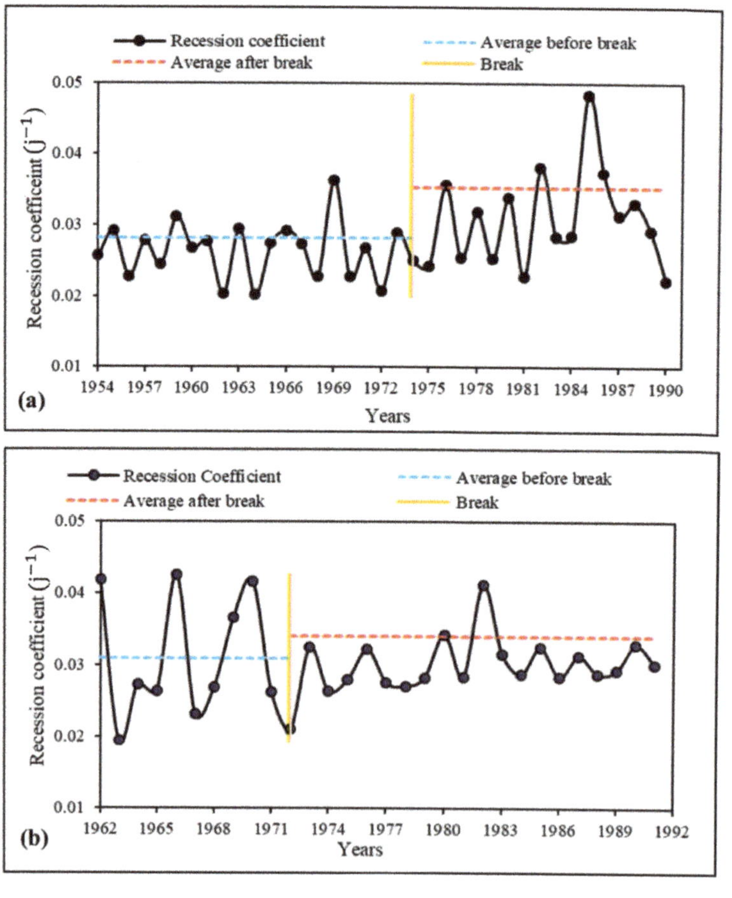

Figure 9. Variation of the dry-off coefficient at the Kara (**a**) and Kpéssidè (**b**) stations.

At the Kara station (Figure 10a), the average annual water volumes mobilized by the aquifers vary between 0.09 and 0.56 $km^3 \cdot yr^{-1}$ before the break with an annual average of 0.32 $km^3 \cdot yr^{-1}$. After the break, these values vary from 0.08 to 0.43 $km^3 \cdot yr^{-1}$ with an annual average of 0.25 $km^3 \cdot yr^{-1}$. The water volume mobilized by the aquifers thus fell by 22% after the climatic break. At the Kpéssidè station (Figure 10b), the water volume mobilized fluctuates between 0.04 and 0.9 $km^3 \cdot yr^{-1}$ before the break with an annual average of 0.47 $km^3 \cdot yr^{-1}$. After the break, the volume varies between 0.03 and 0.52 $km^3 \cdot yr^{-1}$ with an annual average of 0.3 $km^3 \cdot yr^{-1}$. The decline rate is thus about 36%.

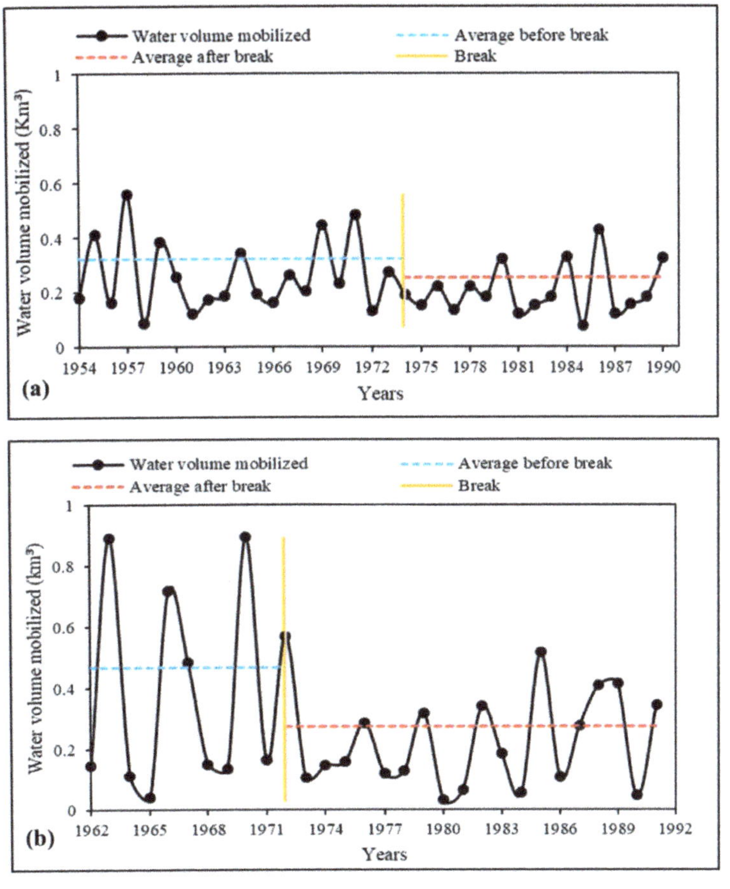

Figure 10. Variation of the water volume mobilized by the aquifers at the Kara (**a**) and Kpéssidè (**b**) stations.

The results show a 25% and 10% increase in the recession coefficient after the break dates at the Kara and Kpéssidè stations, respectively, and a shortening of the dry-off time by 8 days at Kara and 3 days at Kpéssidè (Table 5). Inversely, a diminution of the water volumes mobilized by the aquifers of the KRW was observed. The decreases are 22% and 36% at Kara and Kpéssidè, respectively. The recession coefficient is proportional to the drainage rate of the aquifer and increases in times of drought [91]. In this study, the increase in the recession coefficient indicates rapid drainage of the aquifers.

The values of the recession coefficient agree with those found in certain regions of western Africa [65,74,75,90,91]. These authors observed an increase in the recession coefficient with a decrease in the volume drained by the aquifers, which is partially due to the decrease in precipitation observed at the end of the 1960s and the beginning of the 1970s. The impoverishment of the base flow is due to a reduction in the water volume within the aquifers which, in principle, supplies the rivers during the dry-off phase [75]. Using this methodological approach, other authors [73–75,89,91] also showed the same decrease in water volumes mobilized by the aquifers at other localities.

5. Conclusions

Hydroclimatic functioning within the Kara River watershed in northern Togo was studied, based on classical statistics described in the literature. A mixed spatiotemporal variability of rainfall levels was highlighted with an alternation of wet, normal and dry periods. The spatial analysis revealed a return to normal precipitation in the watershed during the decade 1991–2000. Overall, the effect of the last drought during the 1970s was felt longer for the stations located in the upstream portion of the watershed (Kara, Niamtougou, Alédjo-Kpéwa and Pagouda) in contrast to the downstream stations (Guérin-Kouka and Takpamba), where an increase in rainfall amount was observed sooner. This tendency was confirmed by the Mann–Kendall test, the Sen test and the ITA test. The last method was adapted more to the detection of changes in tendency within hydrometeorological series than the other two. The recharge rate of the aquifers and the flow regime of the surface waters mimicked the climatic tendency prevailing in the watershed. The recharge decreased in the upstream part and increased in the downstream part of the watershed. Then, during dry periods, the aquifer contribution to the Kara River base flow was smaller and the recession coefficients had increased.

Given the climatic variability prevailing in the watershed of the Kara River, special attention has to be drawn to the sustainable management of the water resource and related activities such as agriculture and livestock production, which are performed by more than half of the population living within the area studied. Measures and adaptation methods could be considered for the agricultural sector, as a large part of the food consumed in sub-Saharan Africa is produced by small farmers. These results demonstrate the importance, in Togo and elsewhere worldwide, for developing and managing long-term climate and hydrometric monitoring networks. Indeed, the hydrometric network used for this work is no longer active, while the climatic conditions are likely to evolve in the coming years given the context of climate change that we are facing. The spatially heterogeneous reactivity of the hydrosystem studied shows that some sectors are more vulnerable than others to drought episodes. For these areas, as a priority, it is necessary to anticipate possible future climate crises.

Since precipitation is the only source for the supply of water to the aquifers, a study based on isotope hydrology could provide more precision on the origin of the rainwater, help to define the main recharge zones and enable a better understanding of the recharge modes for aquifers in the KRW. While knowledge of aquifer recharge and the quantitative approach proposed within this study are essential for management, a qualitative characterization of the surface water and groundwater resources would complete this dataset. Complementary work is thus ongoing—on one hand to ensure good water quality for human consumption and, on the other hand, to specify the flowpaths within the aquifer.

Author Contributions: Conceptualization, methodology, formal analysis, investigation, data curation, writing—original draft preparation, writing—review and editing, visualization, and resources, M.A.; conceptualization, validation, writing—review and editing, visualization, supervision, and project administration, J.J.; validation, writing—review and editing, visualization, supervision, and project administration, B.M.; writing—review and editing, M.B.; conceptualization, methodology, resources, supervision, project administration, and funding acquisition, K.G. All authors have read and agreed to the published version of the manuscript.

Funding: Mozimwè Ani receives a scholarship from French Government to complete part of his Ph.D. thesis in France (grant number 971551G). The APC was funded by GEGENAA research unit and the University or Reims ChampagneArdenne.

Institutional Review Board Statement: Not applicable.

Informed Consent Statement: Not applicable.

Data Availability Statement: The data presented in this study are available on request from the National Directorate of Meteorology of Togo and from Directorate of Water Resources of Togo.

Acknowledgments: The authors thank the National Directorate of Meteorology of Togo and the Directorate of Water Resources of Togo for access to their data. We also thank Campus France and the French Embassy in Togo for the management of Mozimwè Ani's scholarship. The anonymous reviewers are also thanked for their valuable comments that helped to improve the quality of the manuscript.

Conflicts of Interest: The authors declare no conflict of interest.

References

1. Hughes, A.; Mansour, M.; Ward, R.; Kieboom, N.; Allen, S.; Seccombe, D.; Charlton, M.; Prudhomme, C. The impact of climate change on groundwater recharge: National-scale assessment for the British mainland. *J. Hydrol.* **2021**, *598*, 126336. [CrossRef]
2. Raposo, J.R.; Dafonte, J.; Molinero, J. Assessing the impact of future climate change on groundwater recharge in Galicia-Costa, Spain. *Hydrogeol. J.* **2013**, *21*, 459–479. [CrossRef]
3. Urama, K.C.; Ozor, N. Impacts of climate change on water resources in Africa: The Role of Adaptation. Climate change and water ressources in Africa. *Afr. Technol. Policy Stud. Netw.* **2010**, *29*, 1–29.
4. Döll, P. Vulnerability to the impact of climate change on renewable groundwater resources: A global-scale assessment. *Environ. Res. Lett.* **2009**, *4*, 035006. [CrossRef]
5. Nicholson, S.E. Recent rainfull fluctuation in Africa and their relationship to past conditions over the continent. *Holocene* **1994**, *4*, 121–131. [CrossRef]
6. Paturel, J.E.; Servat, E.; Kouame, B.; Boyer, J.F.; Lubes, H.; Masson, J.M. Manifestations de la sécheresse en Afrique de l'Ouest non sahélienne. Cas de la Côte d'Ivoire, du Togo et du Bénin. *Sécheresse* **1995**, *6*, 95–102.
7. Servat, E.; Paturel, J.-E.; Lubès-Niel, H. La sécheresse gagne l'Afrique tropicale. *La Rech.* **1996**, *290*, 24–25.
8. Servat, E.; Paturel, J.E.; Kouame, B.; Travaglio, M.; Ouédraogo, M.; Boyer, O.; Lubes-Niel, H.; Fritsch, J.-M.; Masson, J.-M.; Marieu, B. Identification, Caractérisation et Conséquences D'une Variabilité Hydrologique en Afrique de l'Ouest et Centrale. In *Water Ressources Variability in Africa during the XXth Century*; International Association of Hydrological Sciences: Abidjan, Côte d'Ivoire, 1998; pp. 323–337.
9. Ouédraogo, M.; Servat, E.; Lubes-Niel, H.; Masson, J.M. Caractérisation d'une modification éventuelle de la relation pluie-débit autour des années 1970 en Afrique de l'Ouest et centrale non sahélienne. In *Water Ressources Variability in Africa during the XXth Century*; International Association of Hydrological Sciences: Abidjan, Côte d'Ivoire, 1998; Volume 252, pp. 315–332.
10. Paturel, J.E.; Servat, E.; Delattre, M.O.; Lubes-Niel, H. Analyse de séries pluviométriques de longue durée en Afrique de l'Ouest et Centrale non sahélienne dans un contexte de variabilité climatique. *Hydrol. Sci. J.* **1998**, *43*, 937–946. [CrossRef]
11. Pettitt, A.N. A Non-Parametric Approach to the Change-Point Problem. *Appl. Stat.* **1979**, *28*, 126–135. [CrossRef]
12. Bruneau, P.; Rassam, J.-C. Application of à Bayesian model to detect shifts in the mean of a series. *Hydrol. Sci. J.* **1983**, *28*, 341–354. [CrossRef]
13. Hubert, P.; Servat, E.; Bendjoudi, H.; Lubes-Niel, H. La Procédure de Segmentation, Dix ans Après. In *Water Ressources Variabilty in Africa during the XXth Century*; IAHS: Abidjan, Côte d'Ivoire, 1998; pp. 267–273.
14. Bouba, L.; Sauvagnargues, S.; Gonne, B.; Ayral, P.-A.; Ombolo, A. Trends in rainfall and flood hazard in the Far North region of Cameroon. *Geo-Eco-Trop* **2017**, *41*, 339–358.
15. Zarenistanak, M.; Dhorde, A.G.; Kripalani, R.H. Trend analysis and change point detection of annual and seasonal precipitation and temperature series over southwest Iran. *J. Earth Syst. Sci.* **2014**, *123*, 281–295. [CrossRef]
16. Khosravi, H.; Sajedi-Hosseni, F.; Nasrollahi, M.; Gharechaei, H.R. Trend analysis and detection of precipitation fluctuations in arid and semi-arid regions. *Desert* **2017**, *22*, 77–84.
17. Lawin, E.A.; Houngué, N.R.; Biaou, C.A.; Badou, D.F. Statistical Analysis of Recent and Future Rainfall and Temperature Variability in the Mono River Watershed (Benin, Togo). *Climate* **2019**, *7*, 8. [CrossRef]
18. Nouaceur, Z.; Murarescu, O. Rainfall variability and trend analysis in west africa (Senegal, Mauritanie, Burkina). *Water* **2020**, *12*, 1754. [CrossRef]
19. Getahun, Y.S.; Li, M.-H.; Pun, I.-F. Trend and change-point detection analyses of rainfall and temperature over the Awash River basin of Ethiopia. *Heliyon* **2021**, *7*, e08024. [CrossRef]
20. Yue, S.; Wang, C. The Mann-Kendall Test Modified by Effective Sample Size to Detect Trend in Serially Correlated Hydrological Series. *Water Resour. Manag.* **2004**, *18*, 201–218. [CrossRef]
21. Kumar, S. Non-Parametric and Parametric Analysis of Runoff in Satluj River Basin, Himachal Pradesh, India. *Int. Lett. Chem. Phys. Astron.* **2015**, *54*, 15–36. [CrossRef]
22. Some'e, B.; Ezani, A.; Tabari, H. Spatiotemporal trends of aridity index in arid and semi-arid regions of Iran. *Theor. Appl. Climatol.* **2013**, *111*, 149–160. [CrossRef]

23. Khatiwada, K.R.; Panthi, J.; Shrestha, M.L.; Nepal, S. Hydro-Climatic Variability in the Karnali River Basin of Nepal Himalaya. *Climate* **2016**, *4*, 17. [CrossRef]
24. Koudahe, K.; Kayode, A.J.; Samson, A.O.; Adebola, A.A.; Djaman, K. Trend Analysis in Standardized Precipitation Index and Standardized Anomaly Index in the Context of Climate Change in Southern Togo. *Atmos. Clim. Sci.* **2017**, *7*, 401–423. [CrossRef]
25. Hu, M.; Sayama, T.; Try, S.; Takara, K.; Tanaka, K. Trend Analysis of Hydroclimatic Variables in the Kamo River Basin, Japan. *Water* **2019**, *11*, 1782. [CrossRef]
26. Daba, M.; Ayele, G.T.; You, S. Long-Term Homogeneity and Trends of Hydroclimatic Variables in Upper Awash River Basin, Ethiopia. *Adv. Meteorol.* **2020**, *2020*, 1–21. [CrossRef]
27. Braimah, M.; Asante, V.A.; Ahiataku, M.; Ansah, S.O.; Otu-Larbi, F.; Yahaya, B.; Ayabila, J.B. Seasonal Rainfall Variability over Southern Ghana. *Preprints* **2021**, 202108. [CrossRef]
28. Harka, A.E.; Jilo, N.B.; Behulu, F. Spatial-temporal rainfall trend and variability assessment in the Upper Wabe Shebelle River Basin, Ethiopia: Application of innovative trend analysis method. *J. Hydrol. Reg. Stud.* **2021**, *37*, 100915. [CrossRef]
29. Navatha, N.; Sreenivas, G.; Umareddy, R. Rainfall and Temperature Trends in Jagtial District of Telangana State. *Int. J. Environ. Clim. Change* **2021**, *11*, 47–59. [CrossRef]
30. Sarkar, A.; Saha, S.; Sarkar, D.; Mondal, P. Variability and trend analysis of the rainfall of the past 119 (1901-2019) years using statistical techniques: A case study of Uttar Dinajpur, India. *J. Clim. Change* **2021**, *7*, 49–61. [CrossRef]
31. Tirkey, N.; Parhi, P.K.; Lohani, A.K.; Chandniha, S.K. Analysis of precipitation variability over Satluj Basin, Himachal Pradesh, India: 1901-2013. *J. Water Clim. Change* **2021**, *12*, 127–135. [CrossRef]
32. Mann, H.B. Nonparametric Tests Against Trend. *Econometrica* **1945**, *13*, 245. [CrossRef]
33. Kendall, M.G. *Rank Correlation Methods*; Griffin: London, UK, 1975; ISBN 9780852641996.
34. Sen, K.P. Estimates of the regression coefficient based on Kendall's Tau. *J. Am. Assoc.* **1968**, *63*, 1379–1389. [CrossRef]
35. Şen, Z. Innovative Trend Analysis Methodology. *J. Hydrol. Eng.* **2012**, *17*, 1042–1046. [CrossRef]
36. Şen, Z. Innovative trend significance test and applications. *Theor. Appl. Climatol.* **2015**, *127*, 939–947. [CrossRef]
37. Güçlü, Y.S. Multiple Şen-innovative trend analyses and partial Mann-Kendall test. *J. Hydrol.* **2018**, *566*, 685–704. [CrossRef]
38. Thapa, S.; Li, B.; Fu, D.-L.; Shi, X.; Tang, B.; Qi, H.; Wang, K. Trend analysis of climatic variables and their relation to snow cover and water availability in the Central Himalayas: A case study of Langtang Basin, Nepal. *Theor. Appl. Climatol.* **2020**, *140*, 3–4. [CrossRef]
39. Girma, A.; Qin, T.; Wang, H.; Yan, D.; Gedefaw, M.; Abiyu, A.; Batsuren, D. Study on recent trends of climate variability using innovative trend analysis: The Case of the upper Huai River Basin. *Pol. J. Environ. Stud.* **2020**, *29*, 2199–2210. [CrossRef]
40. Besma, B.; Şen, Z.; Hamouda, B. Climate change impact on rainfall in north-eastern Algeria using innovative trend analyses (ITA). *Arab. J. Geosci.* **2021**, *14*, 511. [CrossRef]
41. Al-Gamal, S.A.; Sokona, Y.; Dodo, A.-K. Climatic changes and groundwater resources in Africa. *Int. J. Clim. Change Strateg. Manag.* **2009**, *1*, 133–145. [CrossRef]
42. Roudier, P.; Ducharne, A.; Feyen, L. Climate change impacts on runoff in West Africa: A review. *Hydrol. Earth Syst. Sci.* **2014**, *18*, 2789–2801. [CrossRef]
43. Tirogo, J.; Jost, A.; Biaou, A.; Valdes-Lao, D.; Koussoubé, Y.; Ribstein, P. Climate variability and groundwater response: A case study in Burkina Faso (West Africa). *Water* **2016**, *8*, 171. [CrossRef]
44. Agoh, C.F.; Lekadou, T.T.; Saley, M.B.; Gala, B.T.J.; Danumah, J.H.; Coffi, P.-M.J.; Koffi, Z.E.B.; Goula, B.T.A. Impact of Climate Variability on Water Resources: The Case of Marc Delorme-Cnra Station, Southeast of Ivory Coast. *J. Water Resour. Prot.* **2021**, *13*, 726–749. [CrossRef]
45. Badjana, M.H.; Wala, K.; Batawila, K.; Akpagana, K.; Edjaména, K.S. Analyse de la variabilité temporelle et spatiale des séries climatiques du nord du Togo entre 1960 et 2010. *Eur. Sci. J.* **2014**, *10*, 257–275.
46. Affaton, P. Le Bassin des Volta (Afrique de l'Ouest): Une Marge Passive, D'âge Protérozoique Supérieur, Tectonisée au Panafricain (600 (+ ou−) 50 Ma). Ph.D. Thesis, Université d'Aix-Marseille III, Marseille, France, 1987; 496p.
47. Gadédjisso-Tossou, A.; Adjegan, K.I.; Kablan, A.K.M. Rainfall and Temperature Trend Analysis by Mann–Kendall Test and Significance for Rainfed Cereal Yields in Northern Togo. *Science* **2021**, *3*, 17. [CrossRef]
48. Tairou, M.S. La Tectonique Tangentielle Panafricaine au Nord-Togo. Doctoral Dissertation, Université de Lomé, Lomé, Togo, 2006.
49. Lachassagne, P.; Wyns, R.; Dewandel, B. The fracture permeability of Hard Rock Aquifers is due neither to tectonics, nor to unloading, but to weathering processes: Weathering and permeability of Hard Rock Aquifers. *Terra Nova* **2011**, *23*, 145–161. [CrossRef]
50. Musy, A.; Higy, C. *Hydrologie: Une Science de la Nature*; Presses Polytechniques et Universitaires Romandes: Vaud, Switzerland, 2004; p. 314.
51. Nicholson, S.E.; Kim, J.; Hoopingarner, J. *Atlas of African Rainfall and Its Interannual Variability*; Department of Meteorology, Florida State University Tallahassee: Floride, FL, USA, 1988; p. 237.
52. Oguntunde, P.G.; Abiodun, B.J.; Lischeid, G. Impacts of climate change on hydro-meteorological drought over the Volta Basin, West Africa. *Glob. Planet. Change* **2017**, *155*, 121–132. [CrossRef]
53. Assani, A.A. Analyse de la variabilité temporelle des précipitations (1916–1996) à Lubumbashi (Congo-Kinshasa) en relation avec certains indicateurs de la circulation atmosphérique (oscillation australe) et océanique (El Niño/La Niña). *Sci. Changements Planétaires/Sécheresse* **1999**, *10*, 245–252.

54. Lamb, P.J. Persistence of subsaharan drought. *Nature* **1982**, *299*, 46–47. [CrossRef]
55. Lubes-Niel, H.; Masson, J.M.; Paturel, J.E.; Servat, E. Climatic variability and statistics. A similation approch for estimating power and robustness of tests of stationarity. *J. Water Sci.* **1998**, *11*, 383–408. [CrossRef]
56. *Khronostat*, version 1.0; Logiciel D'analyse Statistique de Séries Chronologiques, Développé Par l'Orstom UR 2-21; l'Ecole des Mines de Paris et l'UMR GBE de l'Université de Montpellier II: Montpellier, France, 1998. Available online: http://www.hydrosciences.org/index.php/2020/09/04/khronostat/ (accessed on 4 September 2020).
57. Machiwal, D.; Jha, M.K. *Hydrological Times Series Analysis: Theory and Practice*; Springer Science & Business Media: Berlin/Heidelberg, Germany, 2012; p. 315. [CrossRef]
58. Xiong, L.; Guo, S. Trend test and change-point detection for the annual discharge series of the Yangtze River at the Yichang hydrological station. *Hydrol. Sci. J.* **2004**, *49*, 99–112. [CrossRef]
59. Oguntunde, P.G.; Friesen, J.; van de Giesen, N.; Savenije, H.H.G. Hydroclimatology of the Volta River Basin in West Africa: Trends and variability from 1901 to 2002. *Phys. Chem. Earth* **2006**, *31*, 1180–1188. [CrossRef]
60. Djaman, K.; Koudahe, K.; Ganyo, K.K. Trend Analysis in Annual and Monthly Pan Evaporation and Pan Coefficient in the Context of Climate Change in Togo. *J. Geosci. Environ. Prot.* **2017**, *5*, 41–56. [CrossRef]
61. Eshetu, M. Hydro-climatic Variability and Trend Analysis of Modjo River Watershed, Awash River Basin of Ethiopia. *J. Environ. Earth Sci.* **2021**, *11*, 38–48. [CrossRef]
62. Penman, H.L. Natural Evaporation from Open Water, Bare Soil and Grass. *Proc. R. Soc. Lond.* **1948**, *193*, 120–145. [CrossRef]
63. Thornthwaite, C.W. An Approach toward a Rational Classification of Climate. *Geogr. Rev.* **1948**, *38*, 55–94. [CrossRef]
64. Turc, L. *Evaluation des Besoins en Eau Irrigation, Evaporation Potentielle*; Annales Agronomiques: Paris, France, 1961; pp. 13–49.
65. Assémian, A.E.; Kouame, K.F.; Djagoua, V.E.; Affian, K.; Jourda, J.P.J.; Adja, M.; Lasm, T.; Biemi, J. Impact of climatic variabilty on water ressources in a humid tropical climate. *J. Water Sci.* **2013**, *26*, 247–261. [CrossRef]
66. Santoni, S.; Huneau, F.; Garel, E.; Celle-Jeanton, H. Multiple recharge processes to heterogeneous Mediterranean coastal aquifers and implications on recharge rates evolution in time. *J. Hydrol.* **2018**, *559*, 669–683. [CrossRef]
67. Kawoun, A.G.; Ahamide, B.; Chabi, A.; Ayena, A.; Adanddedji, F.; Vissin, E. Rainfall Variability and Impacts on Surface Water in the lower Ouémé Valley in South-East Bénin. *Int. J. Progress. Sci. Technol.* **2020**, *23*, 52–65.
68. Akpataku, K.V. Apports de L'hydrogéochimie et de L'hydrologie Isotopique à la Compréhension du Fonctionnement des Aquifères en Zones de Socle Dans la Région des Plateaux au Togo. Ph.D. Thesis, Université de Lomé, Lomé, Togo, 2018; 259p.
69. Faye, C.; Mendy, A. Climatic variability and hydrological impacts in case of the Gambiawatershed (Senegal). *Environ. Water Sci. Public Health Tertitorial Intellignece J.* **2018**, *2*, 54–66.
70. Lang, C.; Gille, E. A recession analysis method for low flow forecasting. *Norois* **2006**, *201*, 31–43. [CrossRef]
71. Maillet, E. La vidange des réserve de réservés. *Ann. Des Ponts Et Chaussées Mémoires Et Doc.* **1906**, *21*, 218.
72. Castany, G. Courbes de tarissement et calcul des réserves régulatrices. *Gen. Assem. Berkeley Int. Assoc. Sci. Hydrol.* **1964**, *63*, 319–328.
73. Savané, I.; Coulibaly, K.M.; Gioan, P. Etude comparative de trois méthodes de calcul du coefficient de tarissement des cours d'eau. *Secheresse* **2003**, *14*, 37–42.
74. Goula, B.T.A.; Savane, I.; Konan, B.; Fadika, V.; Kouadio, G.B. Impact de la variabilité climatique sur les ressources hydriques des bassins de N'Zo et N'Zi en Côte d'Ivoire (Afrique tropicale humide). *VertigO* **2006**, *7*, 1–12. [CrossRef]
75. Kouassi, A.M.; Kouamé, K.F.; Saley, M.B.; Biémi, J. Application du modèle de maillet à l'étude des impacts des changements climatiques sur les ressources en eau en Afrique de l'Ouest: Cas du bassin versant du N'zi-Bandama (Côte d'Ivoire). *J. Asian Sci. Res.* **2013**, *3*, 214–228.
76. Berhail, S. The impact of climate change on groundwater resources in northwestern Algeria. *Arab. J. Geosci.* **2019**, *12*, 770. [CrossRef]
77. Badjana, H.M.; Renard, B.; Helmschrot, J.; Edjamé, K.S.; Afouda, A.; Wala, K. Bayesian trend analysis in annual rainfall total, duration and maximum in the Kara River basin (West Africa). *J. Hydrol. Reg. Stud.* **2017**, *13*, 255–273. [CrossRef]
78. Pilabina, S.; Yabi, I.; Kola, E. Changements Climatiques et Sensibilité des Ressources en eau Dans le Bassin Versant de la Kara au Nord du Togo. Conference Paper. In Proceedings of the XXXII ème Colloque International de l'AIC: The Climatic Change Variability and Climatic Risks, Thessaloniki, Grèce, 29–31 May 2019; pp. 107–112.
79. Adéwi, E.; Badameli, K.M.S.; Dubreuil, V. Evolution des saisons des pluies potentiellement utiles au Togo de 1950 à 2000. *Climatologie* **2010**, *7*, 89–107. [CrossRef]
80. Adjoussi, P. Analyse des extremes climatiques dans le bassin de l'Oti au nord Togo. *J. De La Rech. Sci. De L'université De Lomé* **2021**, *23*, 115–133.
81. Owusu, K. Rainfall changes in the savannah zone of northern Ghana 1961–2010. *Weather* **2018**, *73*, 46–50. [CrossRef]
82. Doumouya, I.; Kamagaté, B.; Bamba, A.; Ouedraogo, M.; Ouattara, I.; Savané, I.; Goula, B.T.A.; Biémi, J. Impact de la variabilité climatique sur les ressources en eau et végétation du bassin versant du Bandama en milieu inetrtropical (Côte d'Ivoire). *Rev. Ivoir. Des Sci. Et Technol.* **2009**, *14*, 203–215.
83. Irié, G.R.; Soro, G.E.; Goula, B.T.A. Recent variations in space-time of rainfall settings and their impact on flow river Marahoué (Ivoiry Coast). *Larhyss J.* **2016**, *25*, 241–258.

84. Assémian, E.A.; Kanga, K.H.M.; Kouassi, K.; Yao, K.J.-J.; Koffi, B.E. Analyse des variations de la recharge des eaux souterraines du socle de Dimbokro, Zone tropicale humide du Centre-Est de la Côte d'Ivoire, face au changement climatique. *Rev. Ivoir. Des Sci. Et Technol.* **2018**, *31*, 84–107.
85. Gnamba, F.M.; Kpan, O.J.G.; Yapi, A.F.; Oga, M.S.Y. Rainfall variability from Kotiala region in the North of Côte d'Ivoire (West Africa). *Eur. Sci. J.* **2020**, *16*, 169–188. [CrossRef]
86. Darko, D.; Adjei, K.A.; Odai, S.N.; Obuobie, E.; Asmah, R.; Trolle, D. Recent climate trends for the Volta Basin in West Africa. *Weather* **2019**, *74*, S13–S22. [CrossRef]
87. Amanambu, A.C.; Obarein, O.; Mossa, J.; Li, L.; Ayeni, S.S.; Balogun, O.; Oyebamiji, A.; Ochege, U.F. Groundwater System and Climate Change: Present Status and Future Considerations. *J. Hydrol.* **2020**, *589*, 125163. [CrossRef]
88. Soro, T.D.; Soro, N.; Oga, Y.M.-S.; Lasm, T.; Soro, G.; Ahoussi, K.E.; Biémi, J. La variabilité climatique et son impact sur les ressources en eau dans le degré carré de Grand-Lahou (Sud-Ouest de la Côte d'Ivoire). *Physio-Géo. Géographie Phys. Et Environ.* **2011**, *5*, 55–73. [CrossRef]
89. Kamagaté, A.; Koffi, Y.B.; Kouassi, A.M.; Kouakou, B.D.; Seydou, D. Impact des évolutions climatiques sur les ressources en eau des pétits bassins en Afrique Sub-Saharienne: Application au bassin versant du Bandama à Tortiya (Nord Côte d'Ivoire). *Eur. Sci. J.* **2019**, *15*, 84–105. [CrossRef]
90. Amoussou, E.; Camberlin, P.; Mahé, G. Impact de la variabilité climatique et du barrage Nangbéto sur l'hydrologie du système Mono-Couffo (Afrique de l'Ouest). *Hydrol. Sci. J.* **2012**, *57*, 805–817. [CrossRef]
91. Kouakou, E.; Koné, B.; N'Go, A.; Cissé, G.; Speranza, C.I.; Savané, I. Groundwater sensitivity to climate variability in the white Bandama basin, Ivory Coast. *SpringerPlus* **2014**, *3*, 226. [CrossRef]

water

MDPI

Article

An Assessment of the Effectiveness of Riverbank Filtration in a Sewage Plant Effluent-Impacted River Using a Full-Scale Horizontal Well

Jin-Hyung Noh [1], Soo-Hyun So [1], Ji-Won Park [1], Sang-Yeob Kim [1], Kyung-Guen Song [2], Jaewon Choi [3], Gyoo-Bum Kim [4], Heejong Son [5], Heeyoung Kim [5] and Sung-Kyu Maeng [1,*]

1 Department of Civil and Environmental Engineering, Sejong University, Seoul 05006, Korea; nowdaji@hanmail.net (J.-H.N.); sue_5732@naver.com (S.-H.S.); jija88@naver.com (J.-W.P.); sangyeob.kim29@gmail.com (S.-Y.K.)
2 Water Cycle Research Center, Korea Institute of Science and Technology (KIST), Seoul 02792, Korea; kgsong@kist.re.kr
3 K-Water Institute, Water Quality & Safety Research Center, Daejeon 34350, Korea; choijw@kwater.or.kr
4 Department of Construction Safety and Disaster Prevention, Daejeon University, Daejeon 34520, Korea; geowater@dju.kr
5 Busan Water Quality Institute, Busan Water Authority, Busan 50804, Korea; menuturk@hanmail.net (H.S.); hykim86@korea.kr (H.K.)
* Correspondence: smaeng@sejong.ac.kr

Abstract: From 2014 to 2020, a full-scale horizontal well was operated to investigate the performance of full-scale riverbank filtration (RBF) in the Nakdong River in Korea, which is significantly impacted by the effluents from sewage treatment plants. In this study, an individual lateral full-scale horizontal collector well was investigated for the first time in Korea, and its performance was determined based on the turbidity and levels of iron, total nitrogen, dissolved organic matter, and four selected trace organic contaminants (TrOCs) (tebuconazole, hexaconazole, iprobenfos, and isoprothiolane) in the RBF and Nakdong River. The turbidity of the river was high with an average of 10.8 NTU, while that of the riverbank filtrate was 0.5 NTU or less on average. The average dissolved organic carbon (DOC) concentrations were 2.5 mg/L in the river water and 1.4 mg/L in the riverbank filtrate, which indicated a 44% reduction in DOC content during the RBF. Out of the 10 laterals, 8 laterals exhibited similar levels of iron, manganese, total nitrogen, DOC, and total hardness, electrical conductivity, and turbidity. The characteristics of the remaining two laterals were different. Because the groundwater inflow was relatively low (<10%), the laterals were contaminated by agricultural land use before the installation of the RBF. This is the first study to report changes in water quality according to individual laterals in a river affected by wastewater effluents. The filtration unit exhibited more than 90% removal rates for tebuconazole and hexaconazole. However, the removal rate for iprobenfos was approximately 77%, while that for isoprothiolane was 46%. The four selected TrOCs in this study were not detected in the groundwater. We found that some organic micropollutants were effectively removed by the RBF.

Keywords: riverbank filtration; dissolved organic matter; Nakdong river; trace organic contaminants

Citation: Noh, J.-H.; So, S.-H.; Park, J.-W.; Kim, S.-Y.; Song, K.-G.; Choi, J.; Kim, G.-B.; Son, H.; Kim, H.; Maeng, S.-K. An Assessment of the Effectiveness of Riverbank Filtration in a Sewage Plant Effluent-Impacted River Using a Full-Scale Horizontal Well. *Water* **2022**, *14*, 1873. https://doi.org/10.3390/w14121873

Academic Editor: Andrea G. Capodaglio

Received: 30 April 2022
Accepted: 7 June 2022
Published: 10 June 2022

1. Introduction

Countries that rely on surface water for drinking have difficulty securing high-quality water resources due to excess nutrients, climate change, industrialization, and urbanization [1]. Among the trace organic contaminants (TrOCs) that have recently become an issue, refractory organic substances are not effectively removed from sewage treatment plants, and these substances can affect aquatic environments [2]. TrOCs exposed to aquatic environments can influence tap water quality directly or indirectly; therefore, further research is required [3]. Advanced treatment methods, such as advanced oxidation, granular activated carbon, and membrane filtration, are effective ways to remove TrOCs for drinking water production. Some advanced drinking water treatment processes are not sustainable, owing to their high energy consumption and unwanted transformation products. No single water treatment process can eliminate all contaminants, thus multibarrier water treatment is essential [4,5]. Water bodies affected by industrial wastewater effluent or sewage treatment plants have a high risk of contamination by TrOCs and chemical accidents. Therefore, eco-friendly, chemical-free, and sustainable water treatment techniques, such as riverbank filtration (RBF), are effective methods for multibarrier water treatment.

RBF has been used in Europe for more than 100 years, particularly in the Rhine, Donau, and Elbe rivers [6]. Some wells in Germany have been operating for more than a century [7]. Countries including Germany, Austria, Hungary, and the Netherlands have been successful in improving drinking water sources via RBF, artificial recharge, or both. RBF has been used for many years to improve drinking water quality and provide protection against water accidents, such as chemical spills. The RBF is an intake system for water that has traveled in an aquifer for a substantial amount of time, depending on the location of the production wells. It consists of river water flowing through the aquifer with some groundwater and can remove contaminants during soil passage [8]. By pumping surface water through vertical wells or horizontal collection wells in aquifers around rivers or lakes, it induces river and lake water into alluvial aquifer formations to improve the water quality. By maintaining a steady water temperature in the winter, the RBF prevents freezing in the supply pipe network. Furthermore, the use of coagulants and the cost of sludge treatment can be reduced by reducing the amount of suspended matter [9]. RBF provides high-quality filtrate via sorption, biodegradation, and groundwater mixing as the raw water passes through the aquifers. In an RBF, a portion of dissolved organic matter (DOM) is removed by the complex biofilm layer present on the upper part of the filtration layer (i.e., schmutzdecke), and biodegradation plays key role in improving the water quality [10,11]. RBF is a natural water treatment technology for the reduction in TrOCs levels, such as pharmaceutically active compounds [12]. However, a disadvantage of RBF is that the concentration of iron in the riverbank filtrate is usually high, and the produced sludge requires additional treatment at water treatment facilities, in case of further treatment applied. Moreover, local farmers in South Korea are concerned about the possible changes in groundwater levels and quality that could be caused by RBF. The contaminant removal efficiency of RBF is affected by various factors, such as the raw water quality, the soil characteristics of the aquifer, the distance between the river and intake well, and the dilution of the groundwater [13,14]. The RBF are functional in the United States and Europe; however, academic research and field studies in other countries are limited. Consequently, the process for importing these RBF is not well-established; hence, studies in other countries are required.

Natural organic matter (NOM) in aquatic environments combines with heavy metals and hydrophobic compounds, increasing the mobility of contaminants in aquatic environments and decreasing bioavailability [15]. It is also essential to understand the behavior of DOM in RBF because natural organic matter is a known precursor of disinfection by-products [14]. To understand the behavior of natural organic matter in an RBF, it is necessary to investigate the characteristics of the DOM in the riverbank filtrate. In many previous studies on RBF, the behavior of DOM has been investigated using column studies, but limited field studies have been done. Moreover, only a few reports have been published on the effects of TrOCs in filtrates of full-scale RBF in Asian countries.

Various water pollution sources threaten several cities downstream of the Nakdong River (Busan, South Korea). Few studies have investigated RBF, especially in areas highly affected by sewage treatment plant effluents, such as the Nakdong River. According to the Ministry of Environment, 1117 public sewage treatment plants and 58 public industrial wastewater treatment plants have been operating in the Nakdong River watershed as of 2019 [16]. Many industrial wastewater treatment facilities discharge effluents containing substantial amount of pollutants, such as TrOCs, into the Nakdong River. Currently, three drinking water treatment plants are using RBF as a pretreatment technology to provide water from the Nakdong River. The city of Changwon, which is on the Nakdong River in South Korea, has been providing 80,000 m^3/day of drinking water since 2006, using RBF with vertical and horizontal collector wells. This system was the first RBF site to be installed to supply drinking water in South Korea. Moreover, using the RBF of the Nakdong River, it is currently providing 127,000 m^3/day to Gimhae and 20,000 m^3/day to Haman-gun. However, limited research has been conducted to improve the water quality of RBF facilities.

In this study, we investigated the performance of an RBF affected by wastewater effluent and examined the characteristics of DOM using a full-scale horizontal collector well. This study also provides insights for water utilities interested in implementing RBF. This is the first time in Korea that water quality parameters have been compared to determine the significance of lateral location in a horizontal collector.

2. Materials and Methods

2.1. Full-Scale Horizontal Collector Well

This field study on a horizontal collector well will be used to design the remaining 9 horizontal collector wells that supply drinking water to the city of Busan. Before the installation of all 10 horizontal collector wells, which provide 280,000 m^3/day, a horizontal well was constructed and used to determine the performance of RBF. This study was conducted from October 2014 to September 2020, focusing on RBF via a horizontal collector well installed in Changnyeong-gun (Kyeongsangnam-do, South Korea), located downstream of the Nakdong River. Several observation wells had been installed and monitored before and after the collection well operation. The horizontal collector and monitoring wells selected for this study are shown in Figure 1. Since the operation of the horizontal collector well began in May 2015, the quality of the river water, riverbank filtrate, and 8 monitoring wells (OMW 1–8) were analyzed. All 8 monitoring wells were constructed to a depth of approximately 40 m and located in the horizontal and vertical directions of the stream flow (Figure 1). The RBF used in this study consisted of 10 laterals, which were installed at elevations of −20.5, −22, and −23 m from the surface for 3 laterals in the upper layer, 3 laterals in the middle layer, and 4 laterals in the lower layers. The diameter of the collecting pipe was 300 mm, and the total flow rate of the facility was 25,000 m^3/day.

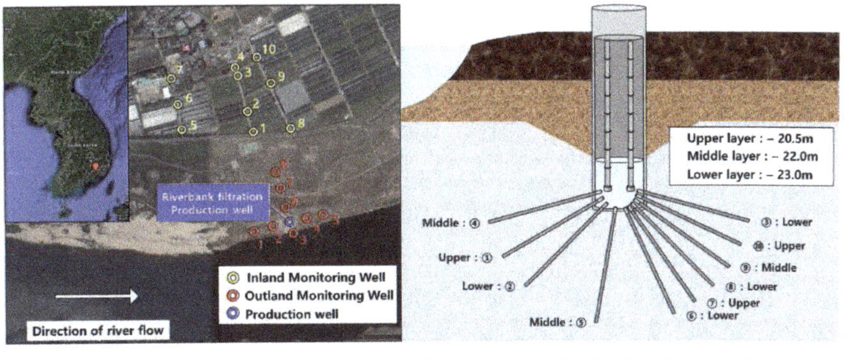

Figure 1. A horizontal collector well with 10 inland monitoring wells (IMW) and 8 outland monitoring wells (OMW) (Nakdong River, Korea).

2.2. Nitrogen and Phosphorus Content

Standard methods [17] were used to analyze the total nitrogen and ammonia (NH_4-N) content in the water. Chloride ions were analyzed using ion chromatography (ICS-900, Thermo Fisher Scientific Inc., Waltham, MA, USA).

2.3. Characteristics of Dissolved Organic Matter (DOM)

DOC level and ultraviolet absorbance at 254 nm (UV254) of the water samples were characterized; each sample was filtered using a 0.45 μm filter (Whatman, Kent, UK). The specific UV Absorbance (SUVA) value was obtained by dividing the UV254 value with the DOC value and multiplying the obtained value by 100. For the characterization of the DOM, the fluorescence excitation–emission matrix (F-EEM) was measured using an RF-5301 spectrofluorometer (Shimadzu, Japan). Four peak components were selected: T1 (tryptophan-like, ex/em: 220–240 nm/em: 330–360 nm), T2 (tyrosine-like, ex/em: 270–280 nm/em: 330–360 nm), A (fulvic-like, ex/em: 230–260 nm/400–450 nm), and C (humic-like, ex/em: 300–340 nm/400–450 nm) [5]. The organic matter was characterized according to molecular weight, and liquid chromatography-organic carbon detection (LC-OCD) analysis was performed using Model 8 (DOC Labor, Karlsruhe, Germany); the DOM fractions were separated into biopolymers (proteins and polysaccharides, 20 kDa or higher), humic substances (350–1000 Da), building blocks (300–500 Da), low-molecular-weight (LMW) (350 or lower) neutrals, and acids according to the retention time based on separation and molecular size [18].

2.4. Trace Organic Contaminants (TrOCs)

The organic micropollutants were analyzed using an online sample concentration method, which involved column switching as a sample pretreatment method. An Equan Max (Thermo Fisher Scientific Inc., Waltham, MA, USA) UPLC model was used for the analysis, which was equipped with a Hypersil Gold aQ (20 mm, 12 μm) concentration column and a Hypersil Gold C18 (50 mm, 1.9 μm) analytical column. The sample injection volume was set to 1 mL, and all the samples used for calibration and analysis were filtered using a 0.2 μm filter (Whatman, Kent, UK). Mass spectrometry was performed using the Orbitrap Exactive mass spectrometer (Thermo Fisher Scientific Inc., Bremen, Germany) model that used the high-resolution full scan method. The resolution and mass accuracy of the equipment were measured under the conditions of 50,000 and 5 ppm, respectively. After analyzing the equipment, a Quan Browser (version 2.1) was used for the qualitative and quantitative analyses.

3. Results and Discussion

3.1. Groundwater and Riverbank Filtration

To understand the contaminant removal efficiency of an RBF, it is necessary to determine the effects of groundwater dilution. In this study, the effects of groundwater were estimated using the levels of chloride and magnesium ions, electrical conductivity, and total dissolved solids of the ground water, and the mixing ratio of the river water and groundwater were calculated based on these values (Figure 2). The average groundwater inflow rates calculated based on the levels of chloride and magnesium ions were approximately 12% and 13%, respectively. Therefore, the ratio of river water in the riverbank filtrate was estimated to be 88%, which was relatively high. Additionally, the groundwater ratio in the riverbank filtrate was 12%, as per electrical conductivity, and 20%, as per the total dissolved solids, indicating that the groundwater ratios as per different parameters were within 20%. The average groundwater ratio in riverbank filtrate was approximately 14.2%, and the river water ratio in the riverbank filtrate was approximately 85.8%. From the previous study on RBF, the groundwater ratio in the riverbank filtrate was approximately 20–40% [19]. The reason behind the high ratio of river water in the riverbank filtrate was that the horizontal collector well was located close to the river and the 10 lateral wells were located below the river.

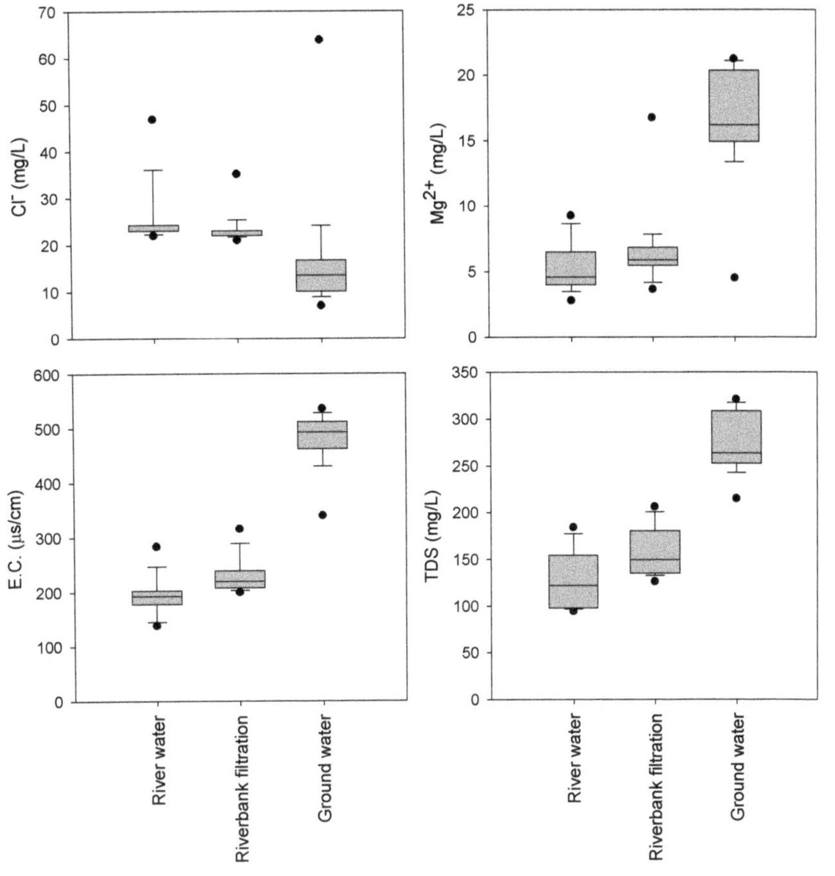

Figure 2. Concentration of Cl^- and Mg^{2+}, electrical conductivity (E.C.), and total dissolved solids (TDS) of the river water, riverbank filtrate, and groundwater (OMW 8G) (n = 20–32).

To confirm the effect of the RBF on the original groundwater quality, the change in the water quality of the groundwater observation well near the RBF facility was assessed (Figure 3). The concentration range of chloride ions was approximately 6–26 mg/L, which varied over time; however, a trend in the concentration was not observed before and after the operation of the RBF facility (Figure 3a). The electrical conductivity was not affected by the RBF, and no significant difference in the concentrations of magnesium and total dissolved solids in the groundwater of the collecting well during the operation period were observed (Figure 3b). Therefore, groundwater accounted for approximately 14% of the riverbank filtrate, and its effect on the original groundwater was not significant. The effect of riverbank filtrate on the quality of groundwater must be observed over a long period.

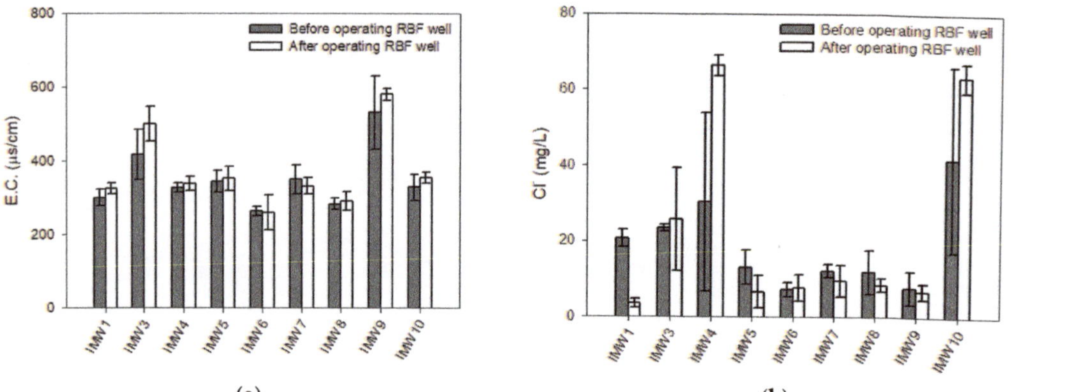

Figure 3. Concentration of chloride ions (Cl^-) **(a)** and electrical conductivity (E.C.). **(b)** in inland monitoring wells (IMWs) before and after the operation of a horizontal collector well.

3.2. Performance of a Horizontal Collector (Temperature, Turbidity, and Concentrations of Fe, Mn, and DOC)

While the temperature of the riverbank filtrate varied from 10 °C to 25 °C, the temperature of the river water ranged from 10 °C to 30 °C. The temperature changes in the riverbank filtrate were smaller because the aquifer temperature differed from that of the river. A small change in water temperature compared with that of river water was also confirmed in the monitoring wells, where no significant change in water temperature was observed. For the well OMW 8G, located 160 m from the horizontal well, which represented the groundwater, the temperature change was between 5 °C and 15 °C. The water temperatures of the monitoring wells parallel to the river were between 15 °C and 28 °C. Moreover, no significant difference in water temperature was observed even in the monitoring wells installed with sand and gravel layers. Therefore, the effect of temperature according to the location was greater than that of depth. In winter, the temperature of the riverbank filtrate did not change significantly, which could act as an advantage because the coagulants in the drinking water treatment plant can be easily controlled, and pipe freezing can be prevented.

The turbidity of the river water was higher than that of the riverbank filtrate, with average turbidity of river water and riverbank filtrate being 10.8 and 0.5 NTU, respectively. Additionally, even when the river water turbidity reached 100 NTU or higher, the turbidity of the riverbank filtrate remained fairly stable (e.g., 1 NTU or less). The change in turbidity in the horizontal collector well is presented in detail in Figure 4 and Section 3.3. The average concentration of iron in the river water was approximately 0.5 mg/L, and the concentration of manganese was exceptionally low, with an average of approximately 0.1 mg/L. However, the average iron concentration of the riverbank filtrate was 9.3 mg/L, and the average manganese concentration was 1.4 mg/L, which significantly increased during soil passage. An increase in iron and manganese concentrations in the riverbank filtrate was observed due to reducing conditions resulting from the biodegradation of organic matter, which requires electron acceptors (i.e., oxygen) [20]. Moreover, the iron content in the soil of this site was relatively high compared with that in other regions (data not shown). The concentrations of DOC in the river water and riverbank filtrate were 1.4 mg/L and 2.5 mg/L, respectively, with an organic matter reduction rate of 44%. The reduction in the DOM content due to RBF is known to occur predominantly by biodegradation and adsorption. Dissolved oxygen (DO) levels during RBF decreased significantly from the river, 9.9 mg/L to RBF > 1 mg/L, indicating that biodegradation played an important role in pollutant removal. The DOC of the OMW 3 and OMW 6 wells was similar to that of the riverbank filtrate. OMW 7 and OMW 8G had average DOC concentrations from 4 to 6 mg/L, which were higher than that of the river water (2.8 mg/L). The monitoring wells OMW 7 and 8G represented the groundwater instead of the river water. Because OMG 8G well was located near agricultural land, the risk of increase in DOC level existed.

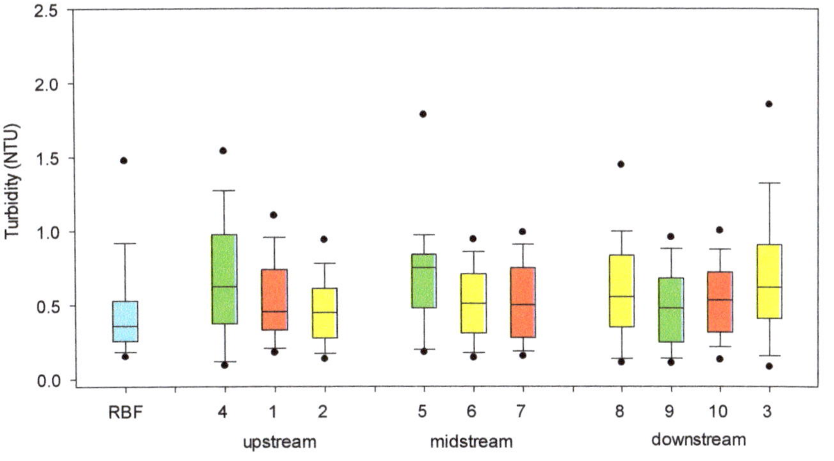

Figure 4. Turbidity changes in a collector well and 10 laterals.

3.3. Performance of Laterals (Temperature, Turbidity, and Concentrations of Fe, Mn, and DOC)

We collected samples with a 1 L bailer from 10 laterals below the water table in monitoring wells to observe how the water quality differed according to the location of each lateral in the horizontal collector well. The temperature changes were not significant, and the turbidity results are shown in Figure 4, which includes the production well and 10 laterals. The turbidity of the river water was high at approximately 11 NTU. The riverbank filtrate had an average turbidity of only 0.5 NTU, which was significantly lower than that of river water, and was within the range of 1 NTU or less for each lateral.

Figure 5 presents the concentrations of iron, manganese, and total nitrogen, total hardness, electrical conductivity, and DOC for each lateral. The iron and manganese concentrations in the individual laterals varied based on their location. The iron concentration was high in lateral #4, and the manganese concentrations were high in laterals #9, #10, and #3. In the riverbank filtrate, an increase in the concentration of iron and manganese was observed, which could be due to reducing conditions; however, the concentrations increased in the lateral direction, where the concentration was measured only in a specific location; this observation could be due to the geological characteristics of the area and not the river water. Lateral #4, which had high levels of iron, exhibited DOC and nitrogen levels different from those in the river water and other laterals. In addition to the effects of upstream and middle streams, the effect of fertilizers accumulated on the land could affect the riverbank filtrate from lateral #4 because that specific location was used as an agricultural land.

In general, when the hardness of the riverbank filtrate is high, the inflow rate of groundwater is expected to be high; however, in this study, the hardness of the riverbank filtrate was not significantly different from that of river water. Owing to the geological characteristics of the target area and the location of the horizontal collector well, the riverbank filtrate consisted of a significant proportion of surface water. The total nitrogen levels in laterals #3 and #4 were slightly higher than those of iron, manganese, and hardness detected in other laterals. The total nitrogen levels in laterals #5, #6, and #7, located close to the river, were lower than those in other pipes, and the ammonia nitrogen levels were more than twice those found in other laterals. The total nitrogen levels in laterals #5 and #6 were also lower than those in other laterals by approximately 60–88%.

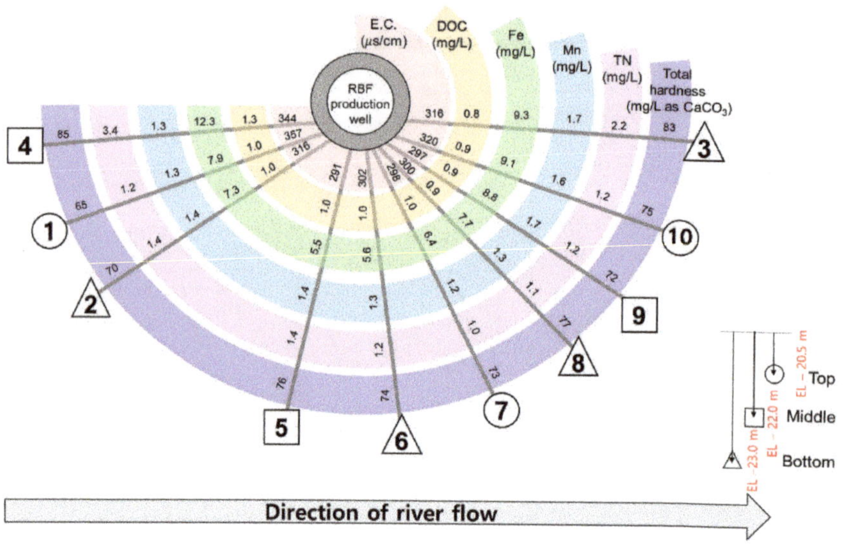

Figure 5. The water quality parameters of lateral collecting pipes (n = 41–192).

After RBF, it was confirmed that the average DOC concentration was reduced by 43.6%. All laterals, except for lateral #4, showed similar DOC removal rates. The total nitrogen level detected in all the laterals except for #3 and #4 appeared to be similar. This could be the result of agricultural activity in the regions close to laterals #3 and #4. The water quality of the laterals connected to the horizontal collector well in this study varied slightly according to the location of the laterals (Figure 5). To improve the water quality of the riverbank filtrate, reduction in the flow rates of laterals #3 and #4 was considered. As a result of this study, the change in the water quality characteristics of the laterals in the RBF facility was examined with respect to the depth of the laterals. It is expected that these results will be useful for water utilities where RBF facilities are installed.

3.4. Characteristics of Dissolved Organic Matter (DOM) (LC-OCD and F-EEM)

LC-OCD was used to analyze the characteristics of organic matter according to its molecular weight in the riverbank filtrate (Figure 6). In the groundwater (OMW 8G), humic substances comprised 64% of the DOM, whereas biopolymers comprised 1% compared to those in the river water. In the groundwater, humic substances made up the majority of the DOM, whereas biopolymers were relatively less abundant in comparison to their abundance in the river water. The biodegradable organic matter content in the groundwater was low. The biopolymers in the river water in this study were measured with an OCD detector, as less amount was detected with a UVD detector. During RBF, biopolymers (97%), humic substances (70%), and building blocks (30%) were removed from the river water, indicating that organic substances with relatively large molecular weights were removed preferentially. Furthermore, because biopolymers such as proteins and polysaccharides can be removed using biological filtration methods [21,22], the hydrophilic biopolymers in the river water can be also removed using biological filtration in RBF, and similar findings have been reported in previous studies on RBF [10,20,23].

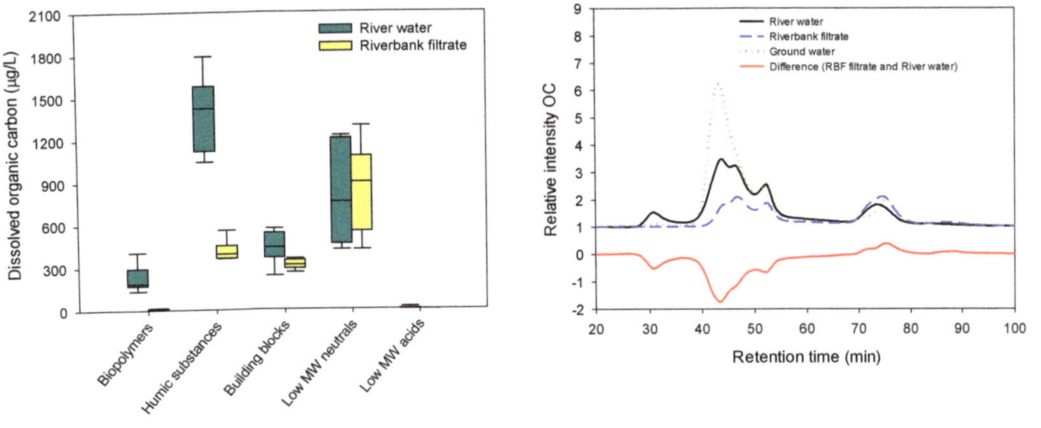

(**a**) (**b**)

Figure 6. Changes in dissolved organic matter fractions determined by LC−OCD (biopolymers, humic substances, building blocks, low molecular weight (MW) neutrals, and low MW acids) (**a**) (n = 9), and (**b**) LC−OCD chromatograms of the river water, riverbank filtrate, and groundwater (OMW 8G) (n = 7).

The fluorescence characteristics of the DOM were used to compare the results of the F-EEM analyses (Figure 7). The fluorescence intensity in the river water was characterized by four peak regions, which were divided into T1 and T2 regions, with tryptophan- and protein-like characteristics, respectively, and A and C regions, with humic-like characteristics [5,24,25]. The fluorescence intensity of the river water was high in T1 (1182 ± 546), T2 (748 ± 377), A (639 ± 150), and C (418 ± 128), whereas those of the groundwater were high in A (2784 ± 576), C (1760 ± 387), T1 (965 ± 201), and T2 (593 ± 133). Compared with the groundwater, the river water contained more humic-like substances, and protein-like substances were also prevalent. According to the F-EEM analysis of the riverbank filtrate, its fluorescence intensity was weaker than that of the river water and groundwater, and fluorescence was detected near T1 and T2, suggesting that it was affected by the river water rather than the groundwater with relatively high A and C fluorescence. Similar reductions in the fluorescence intensity were observed for the four peak regions by approximately 36 to 42%, and the result was consistent with the DOC removal rate of approximately 42%. Although the results of the F-EEM analyses were difficult to compare quantitatively with DOC removal rate, it was possible to understand the characteristics of DOM in the RBF. This finding was also consistent with the LC-OCD results, which identified biopolymers in the river water and high levels of humic substances in the groundwater.

3.5. Trace Organic Contaminants (TrOCs) (Tebuconazole, Hexaconazole, Iprobenfos, and Isoprothiolane)

The removal rates of the RBF for the selected TrOCs (tebuconazole, hexaconazole, iprobenfos, and isoprothiolane) were investigated. The removal rates were 90% for tebuconazole and hexaconazole, 77% for iprobenfos, and 46% for isoprothiolane. Because none of the substances were detected in the groundwater, and the dilution rate of the groundwater was not significant, a significant portion of the selected TrOCs was removed from the riverbank filtrate. The physicochemical properties of the TrOCs (i.e., pesticides) selected in this study indicate that they were hydrophobic and were removed by soil adsorption (Table 1). When filtered through an RBF, even if these TrOCs are easy to remove, their removal can increase the effectiveness of activated carbon in drinking water treatment plants. Consequently, RBF removes pollutants through biodegradation and adsorption, which makes it an effective method for treating water to remove TrOCs.

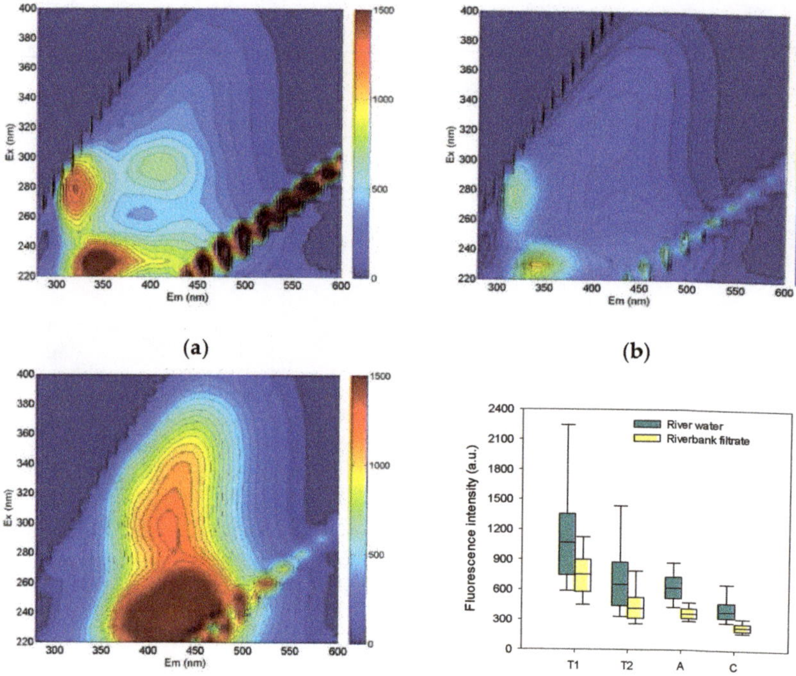

Figure 7. Fluorescence excitation–emission matrix (F-EEM) spectroscopy contour plots of river water (**a**), riverbank filtrate (**b**), groundwater (OMW 8G) (**c**), and changes in the fluorescence intensities in selected regions (T1, T2. A, C peaks) in the river water and riverbank filtrate (**d**) ($n = 12$).

Table 1. Physicochemical properties of selected trace organic contaminants.

Compound	MW (g/mol)	CAS# [1]	pK a [2]	log Kow [3] (pH = 8)	logD [4] (pH = 7.4)	Classification @ pH = 8 [5]	Use
Hexaconazole	314.21	79983-71-4	2.3	3.66	3.63	Hydrophobic–Ionic	Fungicide
Iprobenfos	288.34	26087-47-8	−8.2	3.57	3.21	Hydrophobic–Neutral	Fungicide
Isoprothiolane	290.40	50512-35-1	−7	2.79	3.44	Hydrophobic–Neutral	Fungicide
Tebuconazole	307.82	107534-96-3	2.3	3.89	3.74	Hydrophobic–Ionic	Fungicide

[1] Chemical abstracts service registry numbers. [2] pKa calculated from SPARC. Available online: http://www.archemcalc.com/sparc.html (accessed on 4 May 2021). [3] log Kow value reported for neutral molecule form; based upon U.S. environmental protection agency, 2005, log Kow calculation, EPI Suite KOWWIN Program. Available online: https://www.epa.gov/tsca-screening-tools/epi-suitetm-estimation-program-interface (accessed on 9 June 2022). [4] Chemspider search and share chemistry. Available online: http://www.chemspider.com (accessed on 10 April 2021). [5] For acidic pharmaceuticals: hydrophobic, log D > 1; hydrophilic, log D < 1 at pH 7.4; for neutral pharmaceuticals: hydrophobic, log Kow > 2; hydrophilic, log Kow < 2.

4. Conclusions

LC-OCD and F-EEM analyses of water from deep observation wells confirmed that the groundwater contained humic-like DOM of soil origin. Water quality analyses of the 10 laterals installed in the collector well of RBF revealed that the water quality trends of 8 laterals were similar. The water quality of laterals #3 and # 4 was poor compared with that of the other laterals. It is necessary to examine various possibilities regarding the water quality of laterals #3 and #4, such as whether the aquifer characteristics differ from those of other areas or whether agricultural activity was carried out in the area. To improve the quality of the riverbank filtrate, it is necessary to reduce the flow rates of lateral #3 and

#4. Based on these results, it can be concluded that controlling the lateral valve has the potential to change the water quality of the final riverbank filtrate. This study demonstrates that, in the future, in the horizontal collection well of the riverbank filtrate, it is necessary to conduct a detailed geological investigation into the water quality changes according to the location of the collection pipe and the target area.

Author Contributions: Conceptualization, S.-K.M. and K.-G.S.; methodology, S.-H.S.; formal analysis, J.C., H.S. and H.K.; investigation, J.-H.N. and J.-W.P.; resources, G.-B.K.; writing—original draft preparation, J.-H.N. and S.-K.M.; writing—review and editing, S.-K.M. and S.-Y.K.; funding acquisition, G.-B.K. All authors have read and agreed to the published version of the manuscript.

Funding: This research was funded by the Korea Ministry of the Environment (MOE) and the Korea Environmental Industry and Technology Institute (KEITI) as "Demand Responsive Water Supply Service Program (#146523)".

Institutional Review Board Statement: Not applicable.

Informed Consent Statement: Not applicable.

Data Availability Statement: Not applicable.

Conflicts of Interest: The authors declare no conflict of interest.

References

1. Agudelo-Vera, C.; Avvedimento, S.; Boxall, J.; Creaco, E.; de Kater, H.; Di Nardo, A.; Djukic, A.; Douterelo, I.; Fish, K.E.; Rey, P.L.I.; et al. Drinking Water Temperature around the Globe: Understanding, Policies, Challenges and Opportunities. *Water* **2020**, *12*, 1049. [CrossRef]
2. Rogowska, J.; Cieszynska-Semenowicz, M.; Ratajczyk, W.; Wolska, L. Micropollutants in treated wastewater. *Ambio* **2020**, *49*, 487–503. [CrossRef] [PubMed]
3. Kim, K.Y.; Ekpe, O.D.; Lee, H.J.; Oh, J.E. Perfluoroalkyl substances and pharmaceuticals removal in full-scale drinking water treatment plants. *J. Hazard Mater.* **2020**, *400*, 123235. [CrossRef] [PubMed]
4. Sudhakaran, S.; Maeng, S.K.; Amy, G. Hybridization of natural systems with advanced treatment processes for organic micropollutant removals: New concepts in multi-barrier treatment. *Chemosphere* **2013**, *92*, 731–737. [CrossRef]
5. Noh, J.H.; So, S.H.; Park, J.W.; Maeng, S.K. Influence of algal organic matter on the attenuation of selected trace organic contaminants and dissolved organic matter in managed aquifer recharge: Column studies. *Environ. Sci. Water Res. Technol.* **2020**, *6*, 2789–2799. [CrossRef]
6. Shamrukh, M.; Abdel-Wahab, A. Riverbank filtration for sustainable water supply: Application to a large-scale facility on the Nile River. *Clean Technol. Environ. Policy* **2008**, *10*, 351–358. [CrossRef]
7. Gillefalk, M.; Massmann, G.; Nützmann, G.; Hilt, S. Potential Impacts of Induced Bank Filtration on Surface Water Quality: A Conceptual Framework for Future Research. *Water* **2018**, *10*, 1240. [CrossRef]
8. Maeng, S.K.; Sharma, S.K.; Lekkerkerker-Teunissen, K.; Amy, G.L. Occurrence and fate of bulk organic matter and pharmaceutically active compounds in managed aquifer recharge: A review. *Water Res.* **2011**, *45*, 3015–3033. [CrossRef]
9. Jaramillo, M. Riverbank Filtration: An Efficient and Economical Drinking-Water Treatment Technology. *Dyna* **2012**, *79*, 148–157.
10. Maeng, S.K.; Sharma, S.K.; Abel, C.D.; Magic-Knezev, A.; Amy, G.L. Role of biodegradation in the removal of pharmaceutically active compounds with different bulk organic matter characteristics through managed aquifer recharge: Batch and column studies. *Water Res.* **2011**, *45*, 4722–4736. [CrossRef]
11. Hoppe-Jones, C.; Dickenson, E.R.; Drewes, J.E. The role of microbial adaptation and biodegradable dissolved organic carbon on the attenuation of trace organic chemicals during groundwater recharge. *Sci. Total Environ.* **2012**, *437*, 137–144. [CrossRef] [PubMed]
12. Maeng, S.K.; Ameda, E.; Sharma, S.K.; Grutzmacher, G.; Amy, G.L. Organic micropollutant removal from wastewater effluent-impacted drinking water sources during bank filtration and artificial recharge. *Water Res.* **2010**, *44*, 4003–4014. [CrossRef] [PubMed]
13. Ray, C.; Grischek, T.; Schubert, J.; Wang, J.Z.; Speth, T.F. A Perspective of Riverbank Filtration. *J. Am. Water Works Assoc.* **2002**, *94*, 149–160. [CrossRef]
14. Partinoudi, V.; Collins, M.R. Assessing RBF reduction/removal mechanisms for microbial and organic DBP precursors. *J. Am. Water Works Assoc.* **2007**, *99*, 61–71. [CrossRef]
15. Hur, J.; Schlautman, M.A. Molecular weight fractionation of humic substances by adsorption onto minerals. *J. Colloid Interface Sci.* **2003**, *264*, 313–321. [CrossRef]
16. Ministry of Environment. *Occurrence of Industrial Wastewater in Korea*; Ministry of Environment: Sejongsi, Korea, 2021.

17. American Public Health Association (APHA); the American Water Works Association (AWWA); the Water Environment Federation (WEF). *Standard Methods for the Examination of Water and Wastewater*, 21st ed.; American Public Health Organisation: Washington, DC, USA, 2005.
18. Huber, S.A.; Balz, A.; Abert, M.; Pronk, W. Characterisation of aquatic humic and non-humic matter with size-exclusion chromatography—Organic carbon detection—Organic nitrogen detection (LC-OCD-OND). *Water Res.* **2011**, *45*, 879–885. [CrossRef]
19. Hoppe-Jones, C.; Oldham, G.; Drewes, J.E. Attenuation of total organic carbon and unregulated trace organic chemicals in U.S. riverbank filtration systems. *Water Res.* **2010**, *44*, 4643–4659. [CrossRef]
20. Maeng, S.K.; Lee, K.H. Riverbank Filtration for the Water Supply on the Nakdong River, South Korea. *Water* **2019**, *11*, 129. [CrossRef]
21. Baghoth, S.A.; Maeng, S.K.; Rodríguez, S.G.S.; Ronteltap, M.; Sharma, S.; Kennedy, M.; Amy, G.L. An urban water cycle perspective of natural organic matter (NOM): NOM in drinking water, wastewater effluent, storm water, and seawater. *Water Supply* **2008**, *8*, 701–707. [CrossRef]
22. Pharand, L.; Van Dyke, M.I.; Anderson, W.B.; Yohannes, Y.; Huck, P.M. Full-Scale Ozone Biofiltration: Seasonally Related Effects on NOM Removal. *J. Am. Water Works Assoc.* **2015**, *107*, E425–E435. [CrossRef]
23. Grünheid, S.; Amy, G.; Jekel, M. Removal of bulk dissolved organic carbon (DOC) and trace organic compounds by bank filtration and artificial recharge. *Water Res.* **2005**, *39*, 3219–3228. [CrossRef]
24. Coble, P.G. Marine optical biogeochemistry: The chemistry of ocean color. *Chem. Rev.* **2007**, *107*, 402–418. [CrossRef]
25. Fellman, J.B.; Hood, E.; Spencer, R.G.M. Fluorescence spectroscopy opens new windows into dissolved organic matter dynamics in freshwater ecosystems: A review. *Limnol. Oceanogr.* **2010**, *55*, 2452–2462. [CrossRef]

water

MDPI

Article

Effects of Hybrid-Type Artificial Groundwater Recharge and Underground Barrier in a Small Basin

Myoung-Rak Choi [1] and Gyoo-Bum Kim [2,*]

[1] Department of Disaster Prevention, Graduate School of Daejeon University, Daejeon 34520, Korea; audfkr18@naver.com
[2] Department of Construction Safety and Disaster Prevention, Daejeon University, Daejeon 34520, Korea
* Correspondence: geowater@dju.kr; Tel.: +82-42-280-4574

Abstract: Climate change is exacerbating water shortages in upstream basins in the Korean peninsula that lack agricultural water supply systems. The basin investigated in this study requires an extra 208 $m^3 \cdot d^{-1}$ of agricultural water during May (the busiest month for agriculture). The purpose of this study was to assess a hybrid-artificial recharge and circulation system, which was composed of a hybrid-recharge source and re-infiltration of pumped water in the field, and to estimate yield capacity by a field injection test and a numerical model. Injecting pretreated stream water for 42 d increased groundwater levels in the recharge basin. Water budget analysis in MODFLOW simulations revealed that injecting water increased groundwater levels as well as stream discharge due to the terrain's gentle slope. To prevent downstream discharge and maintain groundwater levels after injection, we assumed the installation of an underground barrier at the basin outlet in the model, following which changes in groundwater levels and water balance were simulated. Water level was persistently maintained after a ~31-cm water level rise, and 590 $m^3 \cdot d^{-1}$ of water could be supplied from the collector well, which can ease water shortages. Therefore, it is necessary to develop structures to prevent recharged water escape when artificially recharging groundwater in small upstream basins. In upstream areas where reservoirs or water supply conduits are unfeasible, artificial recharge systems could solve water shortages.

Keywords: artificial recharge; groundwater; numerical model; drought; water shortage

Citation: Choi, M.-R.; Kim, G.-B. Effects of Hybrid-Type Artificial Groundwater Recharge and Underground Barrier in a Small Basin. *Water* **2022**, *14*, 1849. https://doi.org/10.3390/w14121849

Academic Editor: Shui-Long Shen

Received: 6 May 2022
Accepted: 3 June 2022
Published: 8 June 2022

1. Introduction

Climate change is causing severe droughts in many countries and, as such, is having major impacts on the economy, human society, and the environment. In recent decades, changes in the duration, frequency, and severity of droughts combined with low precipitation levels have resulted in water issues. The Korean peninsula has been experiencing droughts every four to six years, with extreme droughts occurring between 2014 and 2015.

Many countries use groundwater as their main source of domestic and agricultural water [1,2]. Thus, recent increases in water demand are leading to increased usage of groundwater. Additionally, drought events are becoming more extreme, meaning that the sustainability of groundwater is becoming an important concern.

In recent decades, Korea has utilized groundwater using wells as the primary means of securing water in the event of a drought. The country has 1.64 million wells that provide 2.9 billion m^3 of domestic, industrial, and agricultural water per year [3].

Artificial groundwater recharge can contribute to sustainable groundwater management by increasing groundwater levels. Therefore, this approach can mitigate severe droughts by ensuring a sufficient water supply. Artificial groundwater recharge is currently promoted as a solution to water shortages in many countries as one of many tools used to manage water resources. However, its application is sometimes limited by technical and socioeconomic factors. The benefits of artificial recharge have been clearly demonstrated because aquifers can store large volumes of water. Numerous schemes exist to recharge groundwater; they can be broadly grouped into the following categories: spreading methods, open wells, drilled wells, riverbank filtration, sand dams, and rainwater harvesting. The effectiveness of an artificial recharge system is governed by several factors: hydrology, geography, geology, hydrogeology, source water availability, water quality, management issues, environmental effect, water policy, regulation, socioeconomic factors, and institution.

Many researchers have studied the effectiveness of artificial groundwater recharge using field tests and numerical models [4–11]. Abraham and Mohan studied the effects of different recharge structures, namely, check dams and percolation ponds, in India; they revealed that combining these methods could increase groundwater levels more effectively than using individual structures [12]. Meanwhile, Chitsazan and Movahedian used the MODFLOW code of GMS software to assess the artificial recharge effect in Gotvand Plain, Iran. They found that it positively affected the aquifer [13]. Shi et al. studied changes in groundwater quantity, quality, land subsidence rate, and aquifer thermal energy storage after long-term artificial recharge in Shanghai [14]. Akhter and Hossain predicted that groundwater levels would decrease to approximately −120 m in Dhaka City after 20 years but could be maintained at −70 m following the injection of a quantity of water equal to that of the abstraction [15]. Sashikkumar et al. studied the effect of artificial recharge using an integrated Geographic Information System (GIS) approach combined with MODFLOW numerical groundwater modeling techniques for the watershed of the Kodaganar river basin in Tamil Nadu [16]. They revealed that the groundwater volume could be increased by 15–38% by conducting artificial recharge using several kinds of artificial recharge methods; the major methods included percolation ponds and check dams. Meanwhile, Xu et al. revealed that the efficiency of artificial groundwater recharge was affected by the infiltration basin location and recharge density using the Finite Element subsurface FLOW (FEFLOW) system [17]. Mohammadzadeh-Habili and Khalili investigated how artificial recharge effects could be increased by changing the structure of recharge dams using numerical modeling; they suggested the excavation of vertical sidewall pits inside the reservoirs of recharge dams could improve artificial groundwater recharge [18].

The current study aimed to evaluate the effects of an artificial groundwater recharge facility installed in a small upstream basin in Korea, based on a numerical model and field test data. Furthermore, this study aimed to evaluate the need for installing an underground barrier at the outlet of the basin to ensure a stable water supply in a drought region. Finally, this study evaluated how the artificial recharge system combined with an underground barrier can store enough water in the unconsolidated aquifer by preventing water loss to the outside of the basin.

2. Materials and Methods

2.1. Study Site

This study was conducted in Ungok-maeul, Galsan-myeon, Hongseong-gun, Chungcheongnam-do, which is located in the mid-west region of Korea (Figure 1a). This region suffered from a severe agricultural drought in 2015. The most-upstream basin of this watershed experiences frequent water shortages during the busy farming season in spring, and it mostly relies on groundwater during the dry season due to the lack of surface water. The basin area is ~2.5 km^2, and the stream flows from the mountainous area in the east to the flat area in the west. The strata of the study area consisted of Quaternary alluvial and Precambrian granitic gneiss. The alluvial layer is composed mostly of sand, clay, and silt and is mainly distributed to the west of the study area and along the Singok Stream with a thickness of approximately 10 m (Figure 1b).

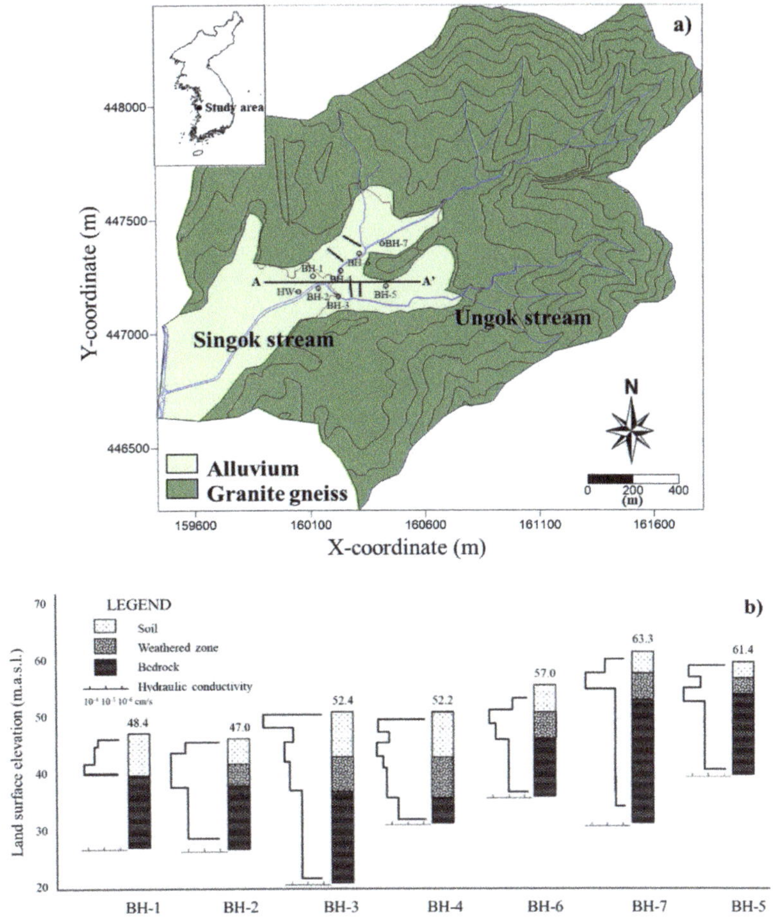

Figure 1. (**a**) Location of the study area and (**b**) geologic cross section along A–A′ [19].

2.2. Water Shortage at the Study Area

In general, drought is caused by abnormal climate (a long period of time during which no rain falls), but it may also appear every several years due to abnormally low rainfall in an area where a water supply system is not effective. Many upstream areas in Korea have a weak water supply system and are easily exposed to drought during irrigational season. Water balance analysis was performed in a previous study, by assuming that there was little rainfall during the dry season (from January to June) [19]. As per the analysis, no water shortages occurred in April, as rice farming mainly started at the end of April. In May, when agricultural activities began in earnest, however, there was a water shortage of ~208 m^3 of water per day, as the existing facilities (wells) could only supply 9423 m^3 of water, which is significantly less than the maximum demand of 15,870 m^3 (512 $m^3 \cdot d^{-1}$; Table 1). June was not considered in this study as water shortages are generally alleviated by rainfall during this month. The upper part of the saturation zone can store approximately 959–1040 m^3 of water per day from April to May; hence, it was estimated that extra water could be supplied by securing a source of artificial groundwater recharge.

Table 1. Analysis of water shortage from January to June in the study area (modified from [19]).

	Period	Main Demand Period and Dry Season (m^3/Month)						Total (m^3/6 Month)
	Month	**1**	**2**	**3**	**4**	**5**	**6**	
Demand analysis	Water demand	527	476	527	1690	9857	37,665	50,742
	(a) Domestic	527	476	527	510	527	510	3077
	(b) Rice farming	0	0	0	1150	8480	33,920	43,550
	(c) Dry field farming	0	0	0	30	850	3235	4115
	Maximum water demand [1]	527	476	527	2010	14,427	51,110	69,077
	(a) Domestic	527	476	527	510	527	510	3077
	(b) Rice farming	0	0	0	1300	11,200	41,000	53,500
	(c) Dry field farming	0	0	0	200	2700	9600	12,500
	Maximum demand with safety factor [2]	580	524	580	2211	15,870	56,221	75,985
Supply analysis	Present water supply capability of wells	572	445	428	5184	9423	8842	24,894
	(a) Domestic wells	572	445	428	410	480	538	2873
	(b) Agricultural wells	-	-	-	4774	8944	8304	22,021
Shortage analysis [3]		−7	−79	−152	2973	−6446	−47,379	−51,090

[1] The maximum water demand is the maximum value of the monthly demand for each purpose from 2007 to 2016; [2] a safety factor of 10% is applied; and [3] shortage = maximum demand with the safety factor—present water supply capability of wells.

2.3. *Hybrid-Type Artificial Recharge and Circulation System*

Considering that the research area is a small upstream basin, a new method was reviewed for artificial groundwater recharge. A hybrid-type artificial recharge and circulation system for groundwater was applied to diversify and reuse sources of groundwater recharge (Figure 2a). Using stream water as an artificial recharge source may have been challenging due to the small size of the stream and its depletion during the dry season. Therefore, the base flow and deep groundwater were considered as further sources of groundwater recharge. The deep bedrock water was located at a depth of several hundred meters and had little hydraulic interaction with the alluvial groundwater. Thus, even small amounts (several to tens of m^3) could be used as an artificial recharge source and added to the unconsolidated layer. The circulation system ensured that following the artificial recharge of groundwater, water withdrawn downstream was supplied to farmland, naturally re-infiltrated, and then withdrawn from the downstream well again. As such, this system aimed to ensure that all water resources throughout the small basin could be stored for a long time without any water escaping (as much as possible).

To maintain the high quality of the injected source water, water quality pretreatment devices were installed in two sections. From upstream to downstream, the pretreatment device consisted of a gravel oxidation section, a vegetation water treatment section, a gravel section, and a gabion section (Figure 2b).

Artificial recharge technology was applied combined with ditches and wells to boost the effect of adding groundwater recharge sources. Four recharge lines were built in the north and south of the study area: two 100 m-long ditch lines including nine injection wells in the north and two 70 m-long ditch lines including seven injection wells in the south (Figure 2c). The widths and depths of the ditches were approximately 3 and 3.5 m, respectively, and the size of the gravel section was 20–75 mm. Smaller gravel layers were used at the deeper bottom. The depths and diameters of the injection wells were 10 and 100 mm, respectively.

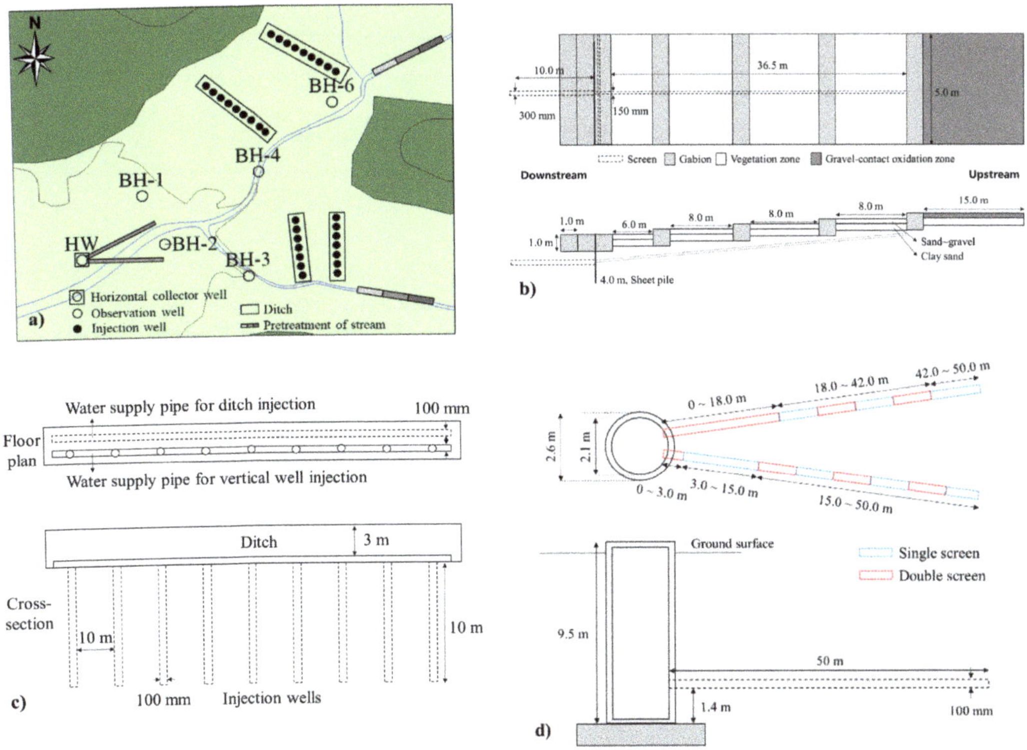

Figure 2. (**a**) Location of hybrid-type artificial recharge system in the study area. (**b**) pretreatment system of stream water. (**c**) the combined artificial recharge structure of ditch and multiple wells and (**d**) collector well: a caisson with two horizontal wells.

A collector well with two horizontal pipes was installed at the downstream recharge section to ensure a secure water supply. The horizontal pipes were 50 m long, with a diameter of 100 mm. A collector well was also installed 20 m from the river; it had a caisson with a diameter of 2.1 m and a depth of 9.5 m (Figure 2d). Water withdrawn from the collector well was transferred to a water tank upstream of the basin through a 100 mm pipe and then supplied to the farmland through natural flow.

2.4. Field Test and Numerical Modeling

Field injection test and numerical modeling using Visual MODFLOW (Version 4.6) were carried out to evaluate the effect of artificial recharge and diagnose the need to install an underground barrier. The accuracy of the model was verified by comparing the results of the field injection test with the results of the numerical model. This model was used to predict the synergistic effect of the groundwater level and the stability of water intake when installing an underground barrier wall, and ultimately evaluate whether the water shortage in the study area can be resolved through artificial recharge.

Artificial recharge was conducted at the four recharge lines for 42 d from 10 January to 21 February 2022. The total injection rate of water added was 150–210 $m^3 \cdot d^{-1}$, which included 90–150 $m^3 \cdot d^{-1}$ from two lines in the north and ~60 $m^3 \cdot d^{-1}$ from two lines in the south (Figure 3a). During this period, the groundwater level and water quality were measured at five groundwater observation wells in the basin, and changes after injection were analyzed (see Figure 2a).

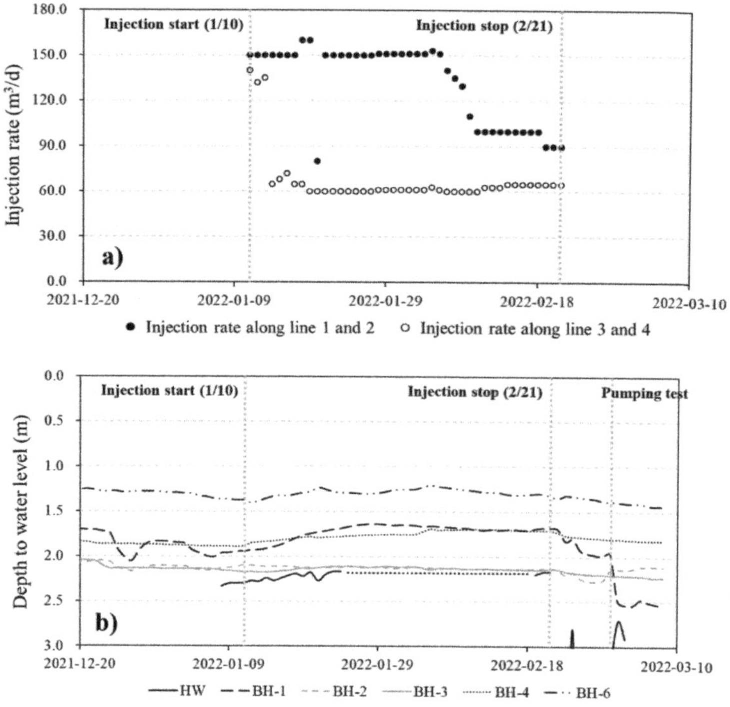

Figure 3. (**a**) Groundwater injection rate along recharge lines and (**b**) groundwater level at the monitoring wells during artificial recharge period.

In addition to the field test, the groundwater flow and changes in groundwater levels and water balance caused by artificial recharge were examined using Visual MODFLOW (flow engine: MODFLOW-NTW). The topography, strata distribution, groundwater level distribution, hydraulic conductivity, river and streamflow characteristics, pumping well distribution, and groundwater usage were investigated through field investigations and indoor experiments; numerical modeling was then performed after establishing a conceptual model.

The study area of the numerical model consisted of 5 m × 5 m grids of 490 columns and 470 rows (Figure 4a). The mountainous areas in the north, east, and south were set as no-flow boundaries, and the north–south flowing river in the west and east–west stream in the study area were defined as river boundaries, of which conductance ranged from 0 $m^2 \cdot d^{-1}$ at the uppermost to 50 $m^2 \cdot d^{-1}$ at the lowermost. Additionally, the branches of the east–west stream were set as a drain package with a drain conductance of 3 $m^2 \cdot d^{-1}$, as the bedrock was exposed at the bottom of the stream (Figure 4b).

During modeling, strata are usually divided into the alluvial layer, weathered zone, and bedrock layer (from top to bottom). The alluvial layer and the weathering zone were considered as one layer as they showed similar hydraulic conductivity in the survey area. In addition, using the excavation data of the observation well, the upper part was composed of an unconsolidated layer, and the lower part was made of bedrock at 12 m below the surface. The distribution of the unconsolidated layer was estimated based on field drilling data (9 wells) in the area around the stream passing through the center of the study area. Hydraulic conductivity was calculated by soil particle analysis results and field pumping tests, respectively. Soil particle analysis was performed on a total of 14 samples, and the pumping test was conducted in one well. The values considered for the unconsolidated and bedrock layers in the modeling were 1.27×10^{-5} to 1.04×10^{-2}

and 2.47 × 10^{-6} to 1.31 × 10^{-5} cm·s^{-1}, respectively (Figure 4c). The flow rate of the stream was calculated using the water level measured at the V-notch, and the groundwater level was automatically measured every hour at nine groundwater observation wells. After calculating the groundwater recharge rate at the points of the groundwater observation network distributed across the country by the "water table fluctuation" method, which calculates the groundwater recharge rate based on the fluctuation range of the groundwater level, an artificial neural network model for the relationship with the characteristic factors of each point was developed. This made it possible to estimate the groundwater recharge rate in the unmeasured area [20]. Subsequently, the average natural groundwater recharge rate in the study area was calculated to be 12.8%, and 130 mm·y^{-1} after applying the 10-year average rainfall.

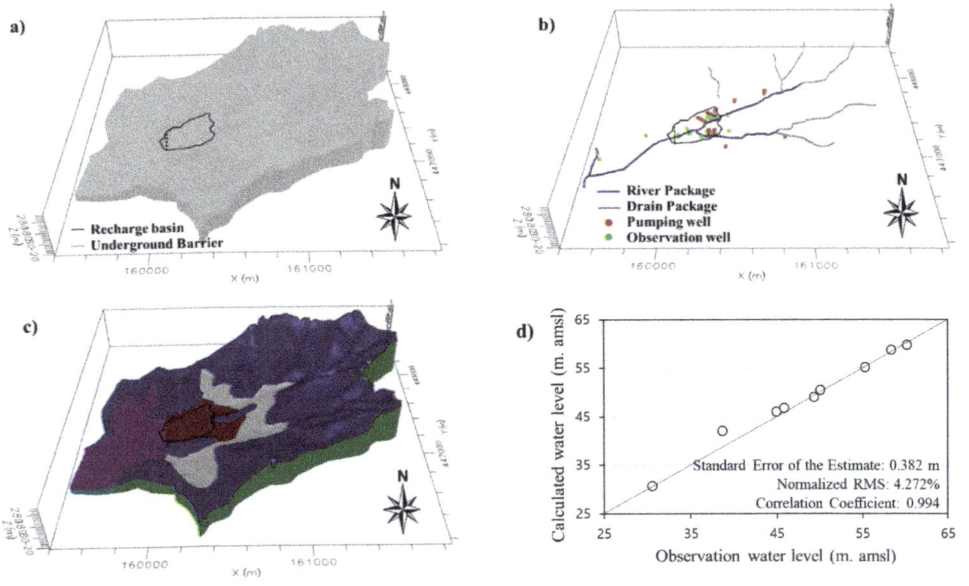

Figure 4. Steady-state model construction and calibration result: (**a**) grid composition of the model, (**b**) stream and wells, (**c**) distribution of hydraulic conductivity and (**d**) comparison between actual and predicted groundwater levels.

A steady-state flow model was performed under the above conditions, and model calibration was conducted by repeatedly measuring the fitness of the actual and predicted groundwater levels using the trial-and-error method.

3. Results

3.1. Artificial Recharge Field Test

Regarding changes in groundwater levels after injection, BH-6, BH-4, BH-3, and BH-2, which were adjacent to the stream, showed several cm of groundwater level rise, while BH-1 (located at the center of the basin) experienced ~30 cm of water level rise (see Figure 3b). In addition, larger rises were observed closer to the injection lines, showing more rapid changes in groundwater levels. The maximum value was recorded 12 d after injection in BH-6, whereas BH-1 showed the largest groundwater level rise approximately 18 d later. These results show that, during the injection period, the groundwater level did not increase rapidly; instead, it remained relatively constant. In the downstream horizontal wells, the change began several days after injection, resulting in a water level rise of ~13 cm. The local anisotropy of the strata seems to have been the reason for BH-1 showing a higher groundwater level rise than other sites. Low hydraulic conductivity at BH-1, which was estimated by the grain size analysis of soil samples, caused this anisotropy. BH-1 had a hydraulic conductivity of 1.03×10^{-4} cm·s^{-1}, which was 5–10 fold smaller than the surrounding areas; this seems to have caused the observed differences in water level rise following water injection.

The fact that the groundwater level remained constant without a significant rise despite the continued injection indicates that some of the injected water was discharged back into the stream. When injections from all lines were stopped, the groundwater levels returned to the original natural water level for approximately 15 d.

As the study site was upstream of a basin where groundwater was discharged down a gentle slope, the increase in groundwater levels during artificial recharge was not significant, and it was difficult to store the injected water for a long time. Nevertheless, the increase in groundwater levels resulting from artificial groundwater recharge could likely satisfy water demand during the busy farming season by increasing the water volume being withdrawn from the collector well, though it could be hard to maintain the water supply during a period of prolonged drought. Therefore, it is necessary to develop a plan to increase the retention period of injected groundwater while maintaining water added from the collector well.

3.2. Numerical Modeling of Artificial Recharge Effects

Groundwater levels were adjusted step-by-step to produce the best fit model by calibrating the geologic layer distribution and hydraulic conductivity, which were relatively unclear factors in a steady-state model of Visual MODFLOW. The steady-state model solver used the Newton formulation, and the solver option used COMPLEX, which is used for a model consisting of one or more non-pressure layers representing complex geological and surface–groundwater interactions [21]. The steady-state flow model achieved a correlation coefficient between the calculated and actual water levels of 0.994, with a standard error of 0.38 m, which was achieved by the calibration process via trial and error to minimize some measure of goodness of fit between two values (see Figure 4d). It was assumed that this error was due to limited data on the hydraulic conductivities and the uncertainty of pumping amount of existing wells.

To estimate the change in groundwater levels after injecting water for 42 d, transient flow model was run by reflecting the actual injection conditions at the site. As a result, observation sites BH-6 and BH-4 showed maximum groundwater level rises of 28.3 and 21.4 cm, respectively (Table 2). Regarding the time series changes after injection, BH-3, BH-4, and BH-6 showed similar groundwater level fluctuations during the injection test in the field, verifying the fitness of the numerical model (Figure 5). At BH-1, where hydraulic conductivity was relatively low, a difference was found between the actual and modeled fluctuations; this occurred because the numerical model did not accurately reflect the partial anisotropy of the hydraulic conductivity at this location. Overall, the modeled time series changes were similar to the actual observations, indicating that the simulation was relatively successful. The model predictions revealed that continuous injection led to a gradual rise in the groundwater levels while stopping injections resulted in a slow decrease in water levels over dozens of days. Based on this numerical model, the changes in water balance in the recharge basin were measured following the artificial recharge.

Table 2. Prediction of groundwater level rise due to artificial recharge and underground barrier by using a numerical model.

Water Level Rise (m)	HW	BH-6	BH-4	BH-3	BH-2	BH-1
Artificial recharge	0.021	0.283	0.214	0.042	0.023	0.059
Artificial recharge with underground barrier	0.033	0.358	0.246	0.044	0.043	0.133
Artificial recharge and re-injection with underground barrier	0.035	0.363	0.253	0.049	0.052	0.144

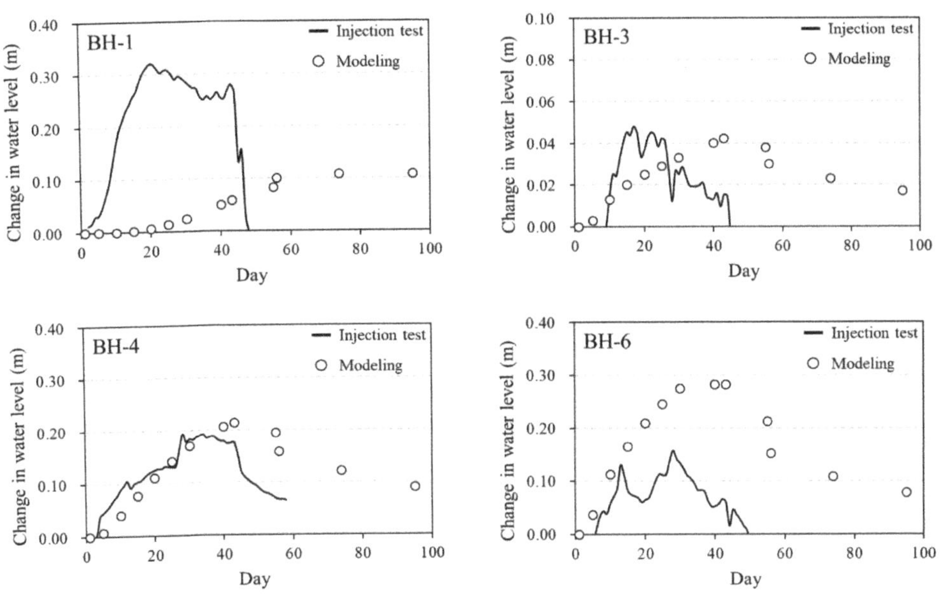

Figure 5. Comparison of actual and predicted groundwater levels after artificial recharge.

Meanwhile, the effects of long-term injection were compared under steady-state conditions (no injection) and without underground barrier. The retention caused by injection in the recharge area increased by 98 $m^3 \cdot d^{-1}$, while discharge into the stream occurred simultaneously at a rate of 80 $m^3 \cdot d^{-1}$ (Figure 6). This implies that the injected water did not stay in the recharge basin for a long time and was released into the stream. This contributed to the observed increase in stream flow, explaining the relatively slow groundwater level rise compared to the amount of water added. Therefore, installation of the horizontal pipe of the collector well through the lower layer of the stream bed would increase the total pumping rate by adding the indirect withdrawal of stream water, as designed above (see Figure 2a).

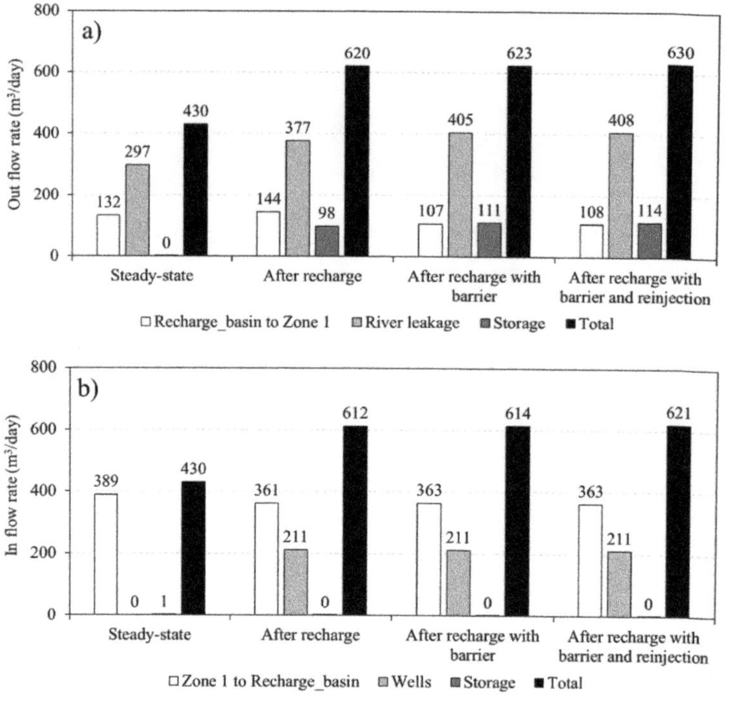

Figure 6. Result of water budget of three cases for the artificial recharge basin by using a numerical model: steady-state condition without artificial recharge, after 45 days recharge without underground barrier, after 45 days recharge with underground barrier, and after 45 days recharge with underground barrier and reinjection: (**a**) outflow rate, (**b**) inflow rate.

3.3. *Effects of Underground Barrier and Water Circulation*

To minimize the loss of injected water downstream and secure a stable water supply to the horizontal wells, the changes in water levels and water balance were simulated after installing a low permeability barrier at the downstream outlet. It was assumed that a 200-m-long low permeability barrier was installed deep in the bedrock and that the hydraulic conductivity of the barrier was 2×10^{-6} cm·s^{-1} (i.e., smaller than that of the bedrock).

If 500 m^3·d^{-1} of water was supplied from the collector well to ~50,000 m^2 of farmland, it was equivalent to supplying 10 mm·d^{-1} of water per 1 m^2. Additionally, 12.8% (1.28 mm·d^{-1}) of this amount was considered to be re-injected through the farmland and it was reflected in the numerical model. Therefore, the changes in groundwater levels and water balance were estimated using the numerical model by assuming that this re-injection was maintained for 30 d, that is, assuming that 38.4 mm of total extra recharge occurred for 30 d after Day 31 of the first injection.

When only the installation of underground barrier was considered, the groundwater levels increased by 35.8 cm in BH-6 and 24.6 cm in BH-4. However, when re-injection in the farmland was considered, the predicted water level rises increased to 36.3 and 25.3 cm, respectively (Table 2). In particular, as significant water level increases near the river contributed to the modeled rise in stream flow, the water supply from the horizontal wells located at the bottom of the river was expected to increase.

The elevated groundwater levels were maintained for a long time or even increased when underground barrier was installed or when re-injection occurred. This indicates that the horizontal wells provided a more stable water supply (Figure 7a,b). A clear change was also found in the water balance; compared to artificial recharge without barrier, adding the barrier and re-injection increased the stream flowrate and groundwater storage by 31 and 16 $m^3 \cdot d^{-1}$, respectively (see Figure 6).

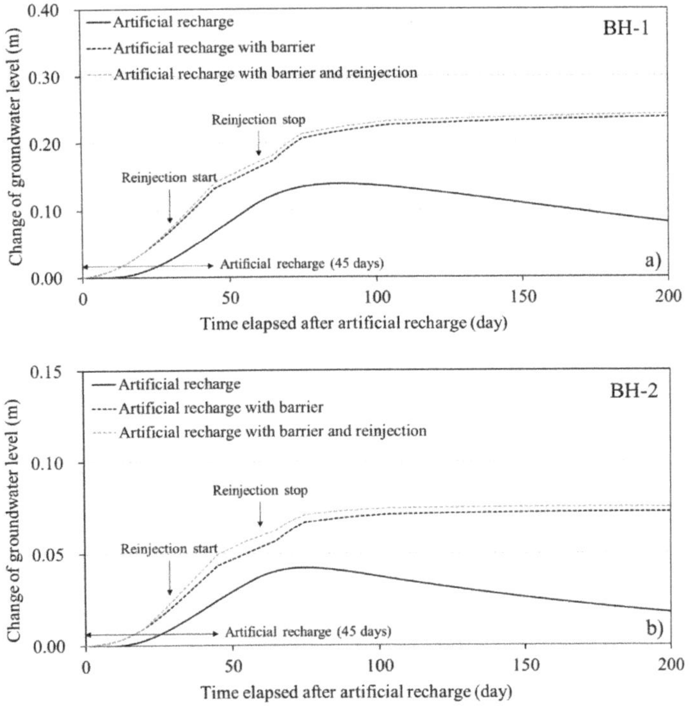

Figure 7. Prediction of groundwater level rise at two monitoring wells after artificial recharge: (**a**) BH-1, (**b**) BH-2.

3.4. Estimation of Groundwater Yield at Horizontal Collector Well

As mentioned above, a collector well with two horizontal pipes was installed at the bottom of the unconsolidated layer to secure as much groundwater as possible, and horizontal pipes were designed to pass through the lower layer of the stream bed. The amount of water withdrawn from the collector well near the river was estimated using the following equation [22]:

$$Q = \frac{2\pi k T_a s_w}{\ln\left(\frac{2b}{R_e}\right)} \tag{1}$$

$$R_e = \exp\left[\ln(2b) - \frac{2\pi \sqrt[3]{\frac{b}{L}}}{3\left(1 - 10^{-0.36(1+n)}\right)\left(1.598 - 0.66\frac{T_a^2}{L^2} + 0.222\frac{T_a^4}{L^4}\right)}\right] \tag{2}$$

In Equations (1) and (2), Q indicates the amount of withdrawn water ($m^3 \cdot d^{-1}$), k is the hydraulic conductivity of the aquifer ($m \cdot d^{-1}$), T_a is the thickness of the aquifer (m), s_w is the drop in the water level (m), b is the distance from the stream channel to the well (m), R_e is the equivalent radius of the collector well (considered as a large vertical well [m]), n is the number of the horizontal pipes, and L is the average length (m) from the center of the well to the end of the horizontal pipe.

To calculate the water intake of the horizontal collector well, we considered a 3-m drop in the water level, which is approximately 50% of the natural head, hydraulic conductivity of 2.53 $m \cdot d^{-1}$, and average value of BH-1 and BH-2 points (Table 3).

Table 3. Result of yield estimation at the horizontal collector well by Babac and Babac (2009) and pumping test.

Condition	k ($m \cdot d^{-1}$)	T_a (m)	b (m)	n	L (m)	R_e (m)	Drawdown (m)	Yield Estimation ($m^3 \cdot d^{-1}$)	Method
Before recharge							2.1	345	Step-drawdown pumping test
	2.53	12	20	2	51	13.72	3	535	Babac and Babac, 2009 [22]
After artificial recharge with underground barrier	2.53	12	20	2	51	13.72	3.3	588	
After artificial recharge with underground barrier and reinjection	2.53	12	20	2	51	13.72	3.31	590	

In this case, the hydraulic conductivity according to a previous study, "hydraulic conductivity of artificial aquifer recharge attenuated with time power function and was smaller than the hydraulic conductivity of pumping. When the quantity of water recharge per unit time was smaller, the attenuation of hydraulic conductivity was slower [23]". The possibility of change in hydraulic conductivity exists in artificial recharge areas depending on the characteristics of alluvial layers, injection cycle, interval, and clogging occurrence. This research facility was installed in December 2021 and is currently under trial operation, so it is necessary to derive the characteristics of this change in the operation process over the next few years.

Under the condition that artificial groundwater recharge cannot be carried out, the water intake of the horizontal well was estimated to be approximately 535 $m^3 \cdot d^{-1}$. However, when artificial recharge and underground walls are applied, they could cause an additional drop in water level by approximately 30 cm (average of water level rises for BH-4 and BH-6), thereby resulting in a total water level drop of 3.3 m; the water intake of the horizontal collector well at this time becomes approximately 588 $m^3 \cdot d^{-1}$. Furthermore, in the case of artificial recharge and re-injection through underground walls and agricultural land, a total water level drop of 3.31 m is possible, and a water intake of 590 $m^3 \cdot d^{-1}$ is secured.

In Table 3, the methods by Babac and Babac (2009) need an assumption that nearby stream level should be constant and water is continuously supplied from the stream. However, the actual stream level in the study area decreases in the dry season and this assumption is not satisfied. The step-drawdown pumping test with 7 steps at the collector well has been done for 700 min before artificial recharge and has shown that the safe yield is approximately 345 $m^3 \cdot d^{-1}$. However, long-term artificial recharge produced an increase in stream discharge as well as groundwater level, thereby satisfying the above assumption.

This volume would likely be sufficient to meet the daily demand of 512 $m^3 \cdot d^{-1}$ in May, and could completely overcome the existing facilities' shortfall (208 $m^3 \cdot d^{-1}$) (see Table 1). Although recharging the supplied water did not lead to a sharp increase in the amount of withdrawn water, it helped to stabilize the water supply by maintaining the increased groundwater level; this effect was also boosted by the installation of the barrier.

4. Discussion

Artificial groundwater recharge is a method of storing water in the subsurface for use later when necessary; this approach can be implemented in dry areas. Korea has an average annual rainfall of ~1300 mm, and hence is not at risk of a water crisis in terms of its total water resources. However, some upstream watersheds that are not properly equipped with water supply systems, such as the present study area, can suffer water shortages during the dry, busy farming season. As water shortages in such areas result in a water shortfall, it is necessary to develop a solution.

As upstream watersheds have rapid discharge regarding both surface water and groundwater, a solution should be introduced to minimize the escape of water outside of the watershed. In the research area, a hybrid-type artificial recharge and circulation system for groundwater, differs from conventional artificial recharge systems, was applied. The applied system increased the total amount of water resources by 1.5 fold compared to before recharge. In addition, it supplied a more stable water supply by releasing water from horizontal wells installed at the lower zone of the stream bed.

The depth at which water is collected from the pretreatment system is essential; here, no stream water was released during the severe dry period, and the base flow, was instead supplied from the bottom of the stream. Therefore, a deeper water collection system was required based on the depth of the alluvial layer at the bottom of the stream. Thus, more water could be collected, leading to a higher water injection volume. As injected water rapidly flowed downstream due to the slope of the ground, it was also important to determine the right location for the collector well. By locating the horizontal pipe where the injected groundwater accumulates (thereby raising the water level) and at the lower zone of the stream bed, more water could be supplied due to the drop in the water level that occurred during pumping.

In general, groundwater dams store groundwater through the construction of underground barriers in flat areas. However, in areas with gentle slopes, such as the study site, it is difficult for an underground barrier to serve as an efficient groundwater dam. Therefore, if a groundwater dam is installed on a slope where groundwater is released relatively quickly, it is hard to secure a sufficient amount of groundwater, as the storage volume upstream of the dam will likely not be significant. Therefore, it will be necessary to minimize water loss downstream by combining artificial groundwater recharge with the construction of underground barrier. It will be also necessary to ensure that the supplied water is re-injected into the recharge basin to achieve a sustainable water supply.

5. Conclusions

In this study, on-site experiments and numerical modeling were employed to explore the effects of an artificial groundwater recharge system using an underground barrier to secure water in a small basin upstream within a watershed that experiences frequent droughts. The water demand in May in the study area was significant (512 $m^3 \cdot d^{-1}$); hence, various techniques could be used to secure water sources, such as constructing a reservoir, obtaining a water supply from outside the watershed, and drilling wells into the bedrock. However, these solutions were not realistic for the study area due to their expensive costs and the insufficient groundwater yield from the bedrock. As such, introducing an artificial recharge system, combined with underground barrier, could solve water shortages within the watershed; this approach can store more water in the alluvial layer and allow more water to re-drain into the river in a small basin. Therefore, the total quantity of water within the basin increases while water loss is minimized, which indicates direct rapid surface runoff out of the basin. This system can be effective upstream where conventional supply methods such as reservoirs cannot be introduced. In the future, if an artificial recharge and water pumping scenario is constructed in consideration of a drought period, whether water can be secured within the basin itself can be assessed.

As water discharged into the river will quickly flow out of the basin (if it is not interrupted), it can be regarded as water loss. Accordingly, a plan should be developed to minimize or utilize water loss. Using the streamflow and base flow in the dry season as a source of artificial recharge and storing them in the aquifer can delay discharge out of the basin. Installing underground barrier at the outlet of the basin to increase retention capacity can also help to stabilize the water supply. In this study, introducing a barrier and artificial recharge in the target area at the same time enabled ~590 $m^3 \cdot d^{-1}$ of water withdrawal from the collector well, which could help to solve existing water shortage problems.

In the Korean peninsula, where faults and joints have developed as a result of several tectonic movements, the sediment layer and weathered zone are both relatively well developed in small upstream basins along these structures. Therefore, once the water retention capacity of the permeable strata has been identified through geological investigations, hybrid-type systems for the artificial recharge and circulation of groundwater could likely be applied in many areas to solve water shortage problems. Water shortages in mid-latitude regions are generally caused by a temporary imbalance between water supply and demand, unlike those in dry desert areas. Therefore, an effective solution would be to not only to build a reservoir, but also implement a system that secures water within the basin through artificial recharge.

Author Contributions: Conceptualization, G.-B.K.; writing—review and editing, M.-R.C. and G.-B.K.; resources and data curation, M.-R.C. and G.-B.K.; visualization, M.-R.C.; supervision, G.-B.K.; funding acquisition, G.-B.K. All authors have read and agreed to the published version of the manuscript.

Funding: This research was funded by the Korea Ministry of the Environment (MOE) and the Korea Environmental Industry & Technology Institute (KEITI) through the "Demand Responsive Water Supply Service Program (#146523)".

Data Availability Statement: Not applicable.

Conflicts of Interest: The authors declare no conflict of interest.

References

1. Shah, T. Climate change and groundwater: India's opportunities for mitigation and adaptation. *Environ. Res. Lett.* **2009**, *4*, 035005. [CrossRef]
2. Margat, J.; Gun, J. *Groundwater around the World: A Geographic Synopsis*; CRC Press/Balkema: Rotterdam, The Netherlands, 2013; p. 341.
3. Ministry of Environment. *2021 Groundwater Annual Report*; Ministry of Environment: Sejong, Korea, 2021; p. 389.
4. Bouwer, H.; Back, J.T.; Oliver, J.M. Predicting infiltration and ground-water mounds for artificial recharge. *J. Hydrol. Eng.* **1999**, *4*, 350–357. [CrossRef]
5. Gale, I.; Neumann, I.; Calow, R.; Moench, M. *The Effectiveness of Artificial Recharge of Groundwater: A Review*; British Geological Survey: Nottingham, UK, 2002; p. 60.
6. Taheri, A.; Zare, M. Groundwater artificial recharge assessment in Kangavar Basin, a semi-arid region in the western part of Iran. *Afr. J. Agric. Res.* **2011**, *6*, 4370–4384.
7. Alrehaili, A.M.; Tahir Hussein, M. Use of remote sensing, GIS and groundwater monitoring to estimate artificial groundwater recharge in Riyadh, Saudi Arabia. *Arab. J. Geosci.* **2012**, *5*, 1367–1377. [CrossRef]
8. Ahirwar, S.; Malik, M.S.; Ahirwar, R.; Shukla, J.P. Identification of suitable sites and structures for artificial groundwater recharge for sustainable groundwater resource development and management. *Groundw. Sustain. Dev.* **2020**, *11*, 100388. [CrossRef]
9. Abraham, M.; Priyadarshini; Manikannan, K. Numerical modeling for groundwater recharge. In *Groundwater Resources Development and Planning in the Semi-Arid Region*; Pande, C.B., Moharir, K.N., Eds.; Springer: Cham, Switzerland, 2021; pp. 165–177.
10. Hassan, W.H.; Nile, B.K.; Mahdi, K.; Wesseling, J.; Ritsema, C. A Feasibility Assessment of Potential Artificial Recharge for Increasing Agricultural Areas in the Kerbala Desert in Iraq Using Numerical Groundwater Modeling. *Water* **2021**, *13*, 3167.
11. Tao, Z.C.; Cui, Z.N.; Yu, J.Q.; Khayatnezhad, M. Finite Difference Modelings of Groundwater Flow for Constructing Artificial Recharge Structures. *Iran. J. Sci. Technol. Trans. Civ. Eng.* **2022**, *46*, 1503–1514. [CrossRef]
12. Abraham, M.; Mohan, S. Effectiveness of artificial recharge structures in enhancing groundwater storage: A case study. *Indian J. Sci. Technol.* **2015**, *8*, 1–10.
13. Chitsazan, M.; Movahedian, A. Evaluation of artificial recharge on groundwater using MODFLOW model (case study: Gotvand plain-Iran). *J. Geosci. Environ. Prot.* **2015**, *3*, 122–132. [CrossRef]
14. Shi, X.; Jiang, S.; Xu, H.; Jiang, F.; He, Z.; Wu, J. The effects of artificial recharge of groundwater on controlling land subsidence and its influence on groundwater quality and aquifer energy storage in Shanghai, China. *Environ. Earth Sci.* **2016**, *75*, 195. [CrossRef]
15. Akhter, S.; Hossain, S. Groundwater modeling of Dhaka city and surrounding areas and evaluation of the effect of artificial recharge to aquifers. *World J. Res. Rev.* **2017**, *5*, 54–60.
16. Sashikkumar, M.C.; Selvam, S.; Kalyanasundaram, V.L.; Johnny, J.C. GIS based groundwater modeling study to assess the effect of artificial recharge: A case study from Kodaganar river basin, Dindigul district. *Tamil Nadu. J. Geol. Soc. India* **2017**, *89*, 57–64. [CrossRef]
17. Xu, Y.; Shu, L.; Zhang, Y.; Wu, P.; Eshete, A.A.; Mabedi, E.C. Physical experiment and numerical simulation of the artificial recharge effect on groundwater reservoir. *Water* **2017**, *9*, 908. [CrossRef]
18. Mohammadzadeh-Habili, J.; Khalili, D. Assessment of artificial recharge dams and improvement of their groundwater-recharge capacity. *J. Hydrol. Eng.* **2020**, *25*, 04020011. [CrossRef]

19. Kim, G.B.; Hwang, C.I.; Choi, M.R. Assessment of the need and potential for groundwater artificial recharge based on the water supply, water demand, and aquifer properties in a water shortage region of South Korea. *Environ. Earth Sci.* **2021**, *80*, 115. [CrossRef]
20. Kim, G.B.; Hwang, C.I.; Shin, H.J.; Choi, M.R. Applicability of groundwater recharge rate estimation method based on artificial neural networks in unmeasured area. *J. Geol. Soc. Korea* **2019**, *55*, 693–701. [CrossRef]
21. Hunt, R.; Feinstein, D. MODFLOW-NTW: Robust handling of dry cells using a Newton formulation of MODFLOW-2005. *Ground Water* **2012**, *50*, 659–663. [CrossRef]
22. Babac, D.; Babac, P. Wells with horizontal drains. In *Theory, Practice, Calculation Examples*; Balby International: Belgrade, Serbia, 2009; p. 530.
23. Dong, Y.; Zhao, P.; Zhou, W. Effect of artificial aquifer recharge on hydraulic conductivity using single injection well. In Proceedings of the 2011 International Symposium on Water Resource and Environmental Protection (ISWREP2011), Xi'an, China, 20–22 May 2011; pp. 2115–2117.

water

MDPI

Article

Determination of Potential Aquifer Recharge Zones Using Geospatial Techniques for Proxy Data of Gilgel Gibe Catchment, Ethiopia

Tarekegn Dejen Mengistu [1,2], Sun Woo Chang [1,2], Il-Hwan Kim [2], Min-Gyu Kim [2] and Il-Moon Chung [1,2,*]

1 Department of Civil and Environmental Engineering, University of Science and Technology (UST), Daejeon 34113, Korea; tarekegnmengistu@kict.re.kr (T.D.M.); chang@kict.re.kr (S.W.C.)
2 Department of Water Resources and River Research, Korea Institute of Civil Engineering and Building Technology, Goyang 10223, Korea; kimilhwan@kict.re.kr (I.-H.K.); kimmingyu@kict.re.kr (M.-G.K.)
* Correspondence: imchung@kict.re.kr

Abstract: The lack of valuable baseline information about groundwater availability hinders the robust decision-making process of water management in humid, arid, and semi-arid climate regions of the world. In sustainable groundwater management, identifying the spatiotemporal and extrapolative monitoring of potential zone is crucial. Thus, the present study focused on determining potential aquifer recharge zones using geospatial techniques for proxy data of the Gilgel Gibe catchment, Ethiopia. Proxy data are site information derived from satellite imageries or conventional sources that are operated as a layer attribute in the geographical information system (GIS) to identify groundwater occurrence. First, GIS and analytical hierarchy process (AHP) were applied to analyze ten groundwater recharge controlling factors: slope, lithology, topographic position index lineament density, rainfall, soil, elevation, land use/cover, topographic wetness index, and drainage density. Each layer was given relative rank priority depending on the predictive implication of groundwater potentiality. Next, the normalized weight of thematic layers was evaluated using a multi-criteria decision analysis AHP algorithm with a pairwise comparison matrix based on aquifer infiltration relative significance. Lithology, rainfall, and land use/cover were dominant factors covering a weight of 50%. The computed consistency ratio (CR = 0.092, less than 10%) and consistency index (CI = 0.1371) revealed the reliability of input proxy layers' in the analysis. Then, a GIS-based weighted overlay analysis was performed to delineate very high, high, moderate, low, and very low potential aquifer zones. The delineated map ensures very high (29%), high (25%), moderate (28%), low (13%), and very low (5%) of the total area. According to validation, most of the inventory wells are located in very high (57%), high (32), and moderate (12%) zones. The validation results realized that the method affords substantial results supportive of sustainable development and groundwater exploitation. Therefore, this study could be a vigorous input to enhance development programs to alleviate water scarcity in the study area.

Keywords: aquifer recharge; groundwater resources; GIS and remote sensing; GIS-based AHP; groundwater potential mapping; proxy data sources

Citation: Mengistu, T.D.; Chang, S.W.; Kim, I.-H.; Kim, M.-G.; Chung, I.-M. Determination of Potential Aquifer Recharge Zones Using Geospatial Techniques for Proxy Data of Gilgel Gibe Catchment, Ethiopia. *Water* **2022**, *14*, 1362. https://doi.org/10.3390/w14091362

Academic Editors: Sang Yong Chung, Gyoo-Bum Kim and Venkatramanan Senapathi

Received: 6 April 2022
Accepted: 19 April 2022
Published: 22 April 2022

1. Introduction

Assessing the continuous availability of freshwater in all climatic regions with low development costs is crucial to withstand the water-food security issues [1,2]. Nowadays, freshwater demand has become a global issue due to natural and anthropogenic influences [3–6]. Groundwater is a renewable resource that provides areas with limited access to surface water [7,8]. Groundwater is essential to improve socio-economic development in the rapidly increasing demand of rural and urban populations [9,10]. Africa has a substantial amount of groundwater storage that has not been adequately utilized and managed yet [11]. In Africa and most developing countries, such as Ethiopia, subsurface water is the leading source of crop production and drinking water supply [11–13]. Despite this, most people suffer water scarcity due to poor knowledge in identifying and utilizing water resource potentials to the limit [5,14,15]. However, most groundwater aquifers do not afford sufficient scientific evidence to assess groundwater flow for setting long-term utilization and national development goals [13,16].

The variability of groundwater recharge affects freshwater availability [17]. The spatiotemporal assessment of groundwater recharge is becoming a critical concern due to data scarcity in different climatic regions [18]. Recharging becomes effective if the infiltration is concentrated at a given time and space [19]. Groundwater recharge requires detailed assessments of scientific principles governing groundwater flow processes [20]. Delineating potential aquifer recharge zones (PARZ) in the areas of interest is vital in determining water demand and supply under hydrologic extremes [21,22]. However, several factors hinder water sector development from attaining sustainable and appropriate mitigating strategies [23–25]. The physical, hydrological, and climate variables are the key factors that govern the subsurface flow of the aquifer system [26]. Determining the governing controlling factors leads to reasonable design input and decision-making processes for complex systems. Identifying potential groundwater sites remains challenging as its exploration relies on precise analysis of hydrogeological data [27]. Recently, geospatial techniques have become essential tools in spatiotemporal predictions of groundwater [28–30].

Geospatial techniques have been used in many recharge studies to avoid dilemmas of methods that rely on groundwater flow [31,32], field surveys [13,32], and geophysical technologies [33], which are both costly and complex. The robust application of Geographic Information Systems (GIS) and Remote Sensing (RS) can manage the amount of spatially correlated governing parameters [34–37]. In addition, GIS increases the accuracy of results, minimizing bias [38]. The development of these technologies with Multi-Criteria Decision Analysis (MCDA) provided a new scientific inquiry in recent groundwater studies [39–42]. GIS and RS are effective tools that offer a fast and systematic spatiotemporal depiction of productive areas [2,14,29,43–45]. MCDA technique has been accepted worldwide for solving decision problems in driving groundwater controlling parameters [46–49]. The Analytical Hierarchy Process (AHP), advanced by Saaty [50], is the extensively used MCDA technique for complex decision-making in groundwater studies worldwide. The AHP provides a formalized approach for creating better resolutions of varied domains of highly dynamic, uncertain, and unpredictable complex decision-making problems [51,52]. Many authors comprehensively used different groundwater flow controlling parameters to delineate potential zones with GIS-based MCDA [25,30,53,54]. Various controlling factors were applied in other research conducted worldwide though the results vary from site [27]. Hence, accurate predictions of groundwater recharge zones challenged scientists as factors that control infiltration have remained unsolved [19,55].

In the Gilgel Gibe catchment, there is an alarming expansion of many satellite cities with unregulated population growth and increasing freshwater demand. Groundwater will remain the critical source of freshwater as most streams are drying up due to natural and anthropogenic activities. Groundwater has immense advantages of drought resilience [13], less treatment cost, and slow response to climate variability [56]. The area is mainly rain-fed agriculture despite the high potential for irrigable land. As rainfall is erratic, rainfall-dependent agriculture does not support sustainable crop production, resulting in food insecurity issues. Therefore, exploration of groundwater aquifers and identifying potential zones is desirable for sustainable groundwater management and selection of productive aquifers. Exploring PARZ is one of the best approaches for emerging viable water utilization for domestic, irrigation, and industrial developments [15]. PARZ can be determined from proxy data sources and has substantial advantages under data-scarce conditions [36,57–59]. Proxy data are site information produced from satellite imageries or conventional sources prepared as a geospatial layer attribute in the GIS environment. Satellite imagery affords tremendous insights into hydrologic characteristics [27,60]. However, robust application of proxy data has not been exclusively applied in the study area, which causes overexploitation and an unreliable decision-making process. Hence, the present study aims to determine potential aquifer recharge zones from proxy data sources of Gilgel Gibe catchment using geospatial techniques based on MCDA and AHP. This study used ten thematic layers to reinforce the data analysis and interpretation process. The present study will become significant baseline information to enhance aquifer management and engineering judgments.

2. Materials and Methods

2.1. Description of Study Area

The Gilgel Gibe catchment (GGC) is situated in Eastern Africa in the Omo-Gibe river basin of Southwestern Ethiopia. The watershed geographically (Figure 1) lies between (226126.133, 304388.193) North Latitudes and (811017.916, 883266.897) East Longitudes. Numerous intermittent rivers also characterize the catchment. The primary water source for the rivers is precipitation from the mountainous areas. The weather in Ethiopia is primarily managed through topography and seasonal variations in the Inter-Tropical Convergence Zone [61]. Ethiopia has a two-season tropical climate. The magnitude of rainfall varies with topography, location, and elevation. The catchment falls within the humid tropical class with a monomodal rainfall distribution. A semi-arid and humid subtropical weather is typically GGC. The cyclic wave of the intertropical convergence area and numerous panoramas mainly affect the climate conditions in Ethiopia. The GGC has an average annual rainfall and air temperature of 1405.8 mm and 18.5 °C, respectively. There is a dry winter weather season between October and April and a wet season (the summer season) between May and September. The geography of the catchment is typically rugged topography with varied mountains, hills, and plains, with upper plateaus divided by deep V-shaped valleys into the flanks and flat river terraces around the main river.

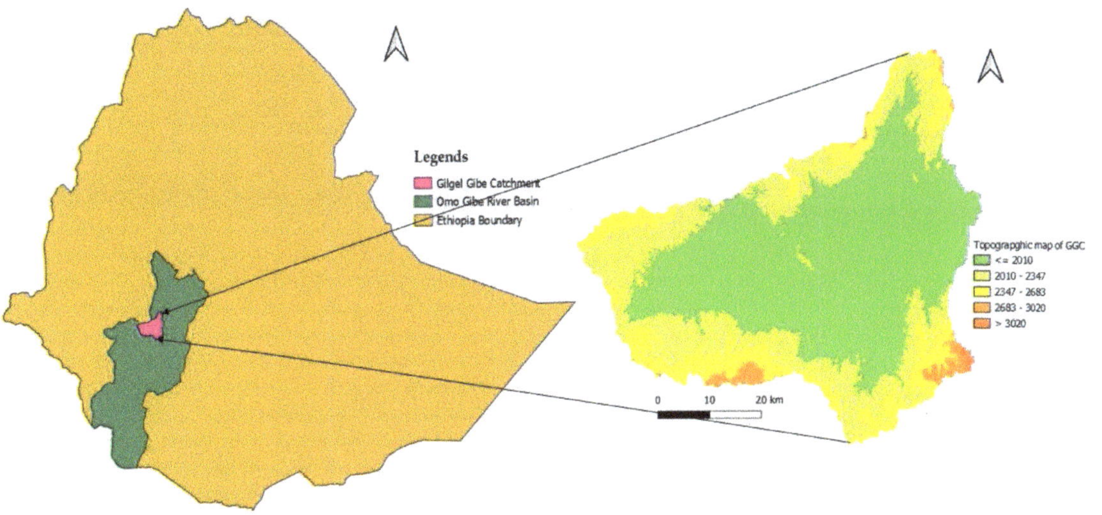

Figure 1. Location map of the study area.

2.2. Aquifers Recharge Controlling Parameters

Geospatial techniques, a synergistic effect of RS, GIS, and AHP, were employed to combine ten significant proxy data sets which affect groundwater potential and recharge rates. The proxy information was produced from satellite images using a digital image-processing algorithm. The generated proxy layers were prepared with a GIS environment to reclassify and adjust for further interpretation and spatial manipulation. Finally, a set of weights for their features were decided based on personal judgment and experts' opinion considering their relative significance for aquifer recharging potentiality in the area of interest. The comprehensive workflow approach for this study to delineate the final PARZ is shown in Figure 2.

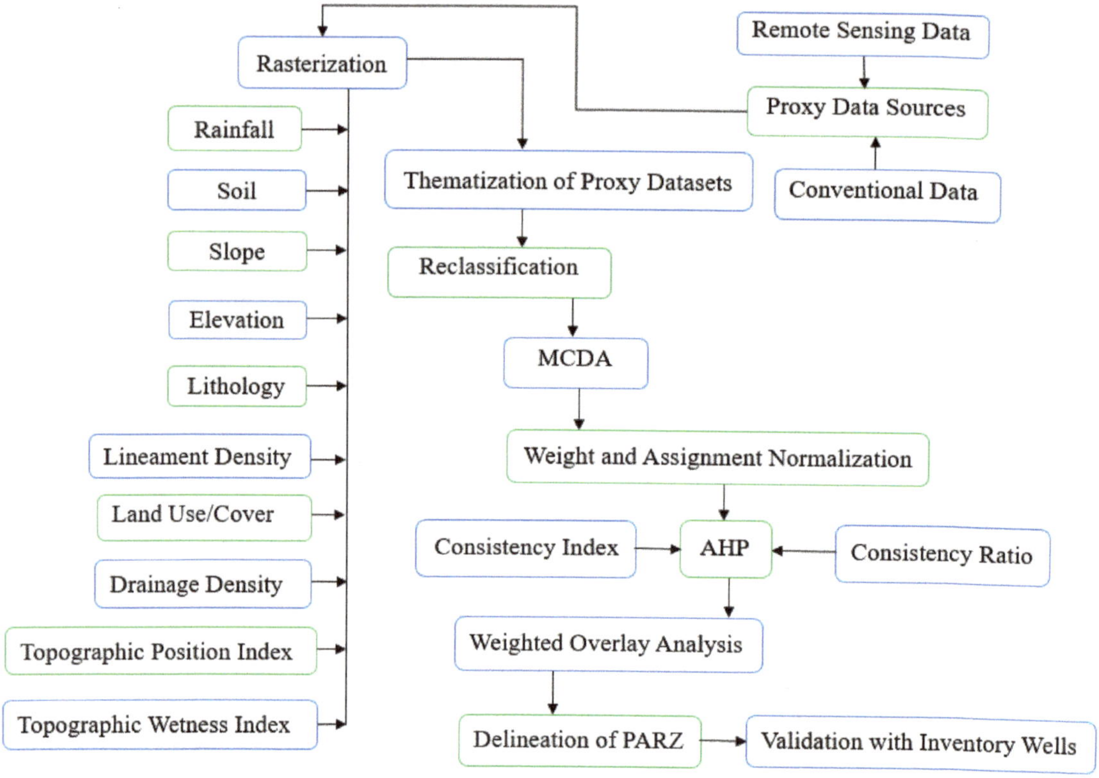

Figure 2. General flowchart of study methodology.

2.2.1. Rainfall

Rainfall is the primary basis for infiltration and recharge. Groundwater flow and storage capacity depend on the percolation rate in the aquifer [42,62,63]. The rainfall records of GGC and surrounding areas were collected from the National Meteorological Agency. Rainfall maps evolved from the historic rainfall information (1988–2018) measured from meteorological stations inside the study area's buffer zone. The rainfall map was produced using the soil and water assessment tool (SWAT) in the QGIS environment using the inverse distance weight (IDW) interpolation technique. The rainfall depth plays a significant role in infiltration; consequently, the weight rainfall classes were assigned (Figure 3a). The rainfall classes have been reclassified and ranked into five classes, mainly based on their effect on the flow and storage of groundwater.

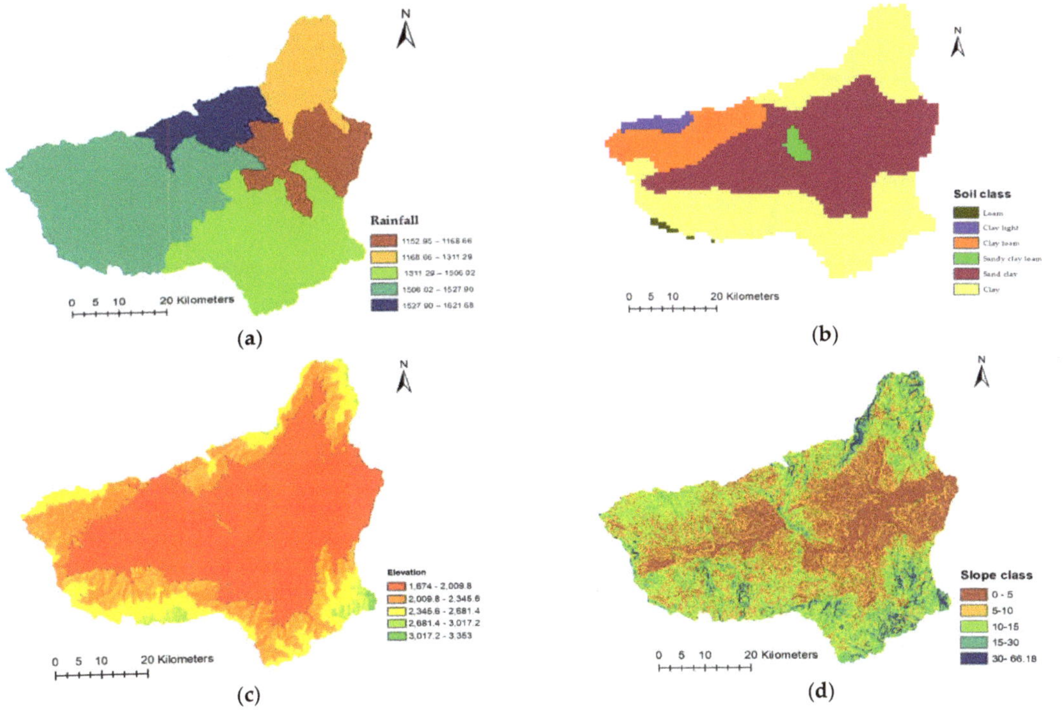

Figure 3. Thematic map of (**a**) Rainfall, (**b**) Soil, (**c**) Elevation, (**d**) Slope.

2.2.2. Elevation and Slope

The Digital Elevation Model (DEM) is a quantitative demonstration of landscape significant for science, hydrological applications, geologic analyses, and agricultural land management [64]. The required spatial data sets were projected to Adindan UTM Zone 37, the transverse Mercator projection parameter for Ethiopia. For the present study, 30-m grid resolution DEM was used and retrieved from freely available open-source product United States Geological Survey (USGS), Shuttle Radar Topography Mission (SRTM). The slope changes elevation between two locations and directly influences groundwater recharge and infiltration [65]. The water flow energy is driven by the slope [66,67]. Thus, slope and infiltration rate are inversely correlated factors [35,68,69]. This study generated the slope from SRTM DEM (Figure 3c), using the spatial analysis tool in the GIS environment. The slope was then reclassified and ranked into five classes flat, gentle, moderate, high, and steep slopes reliant on aquifers recharging potentiality, as indicated in (Figure 3d).

2.2.3. Land Use/Cover and Soil Map

The land use/cover (LULC) happens mainly due to population increase, urbanization, and anthropogenic activities [70,71], affecting infiltration, runoff, evapotranspiration rates, groundwater quantity, and recharge process [72,73]. LULC data were retrieved from World's First Global Land Cover Datasets at a 30 m Resolution, GlobeLand30 [74–76]. The LULC has been classified into inland water bodies, agricultural land, grassland, forest-mixed, range and shrubland, and settlement and urban. Groundwater recharge is subjective with LULC of the study area [35,70,77,78]. LULC is classified (Figure 4a) according to infiltration capacity suitable for aquifer potential. Soil influences groundwater recharging potential by controlling runoff and infiltration rates [36]. The catchment soil information was acquired from FAO Harmonized World Soil Database (HWSD) [79], a freely accessible open-source/www.fao.org/geonetwork/website (accessed on 23 April 2021). HWSD viewer v 1.21 was used to generate a code for soil properties. First, six dominant soil textures were distinguished, sandy clay loam, loam, clay loam, clay light, clay, and clay-heavy. Then the soil map linked to the soil database intended to hold data for all of the world's soils. Finally, the soil textures that are suitable for groundwater recharge were analyzed based on FAO classification on infiltration and water holding capacity.

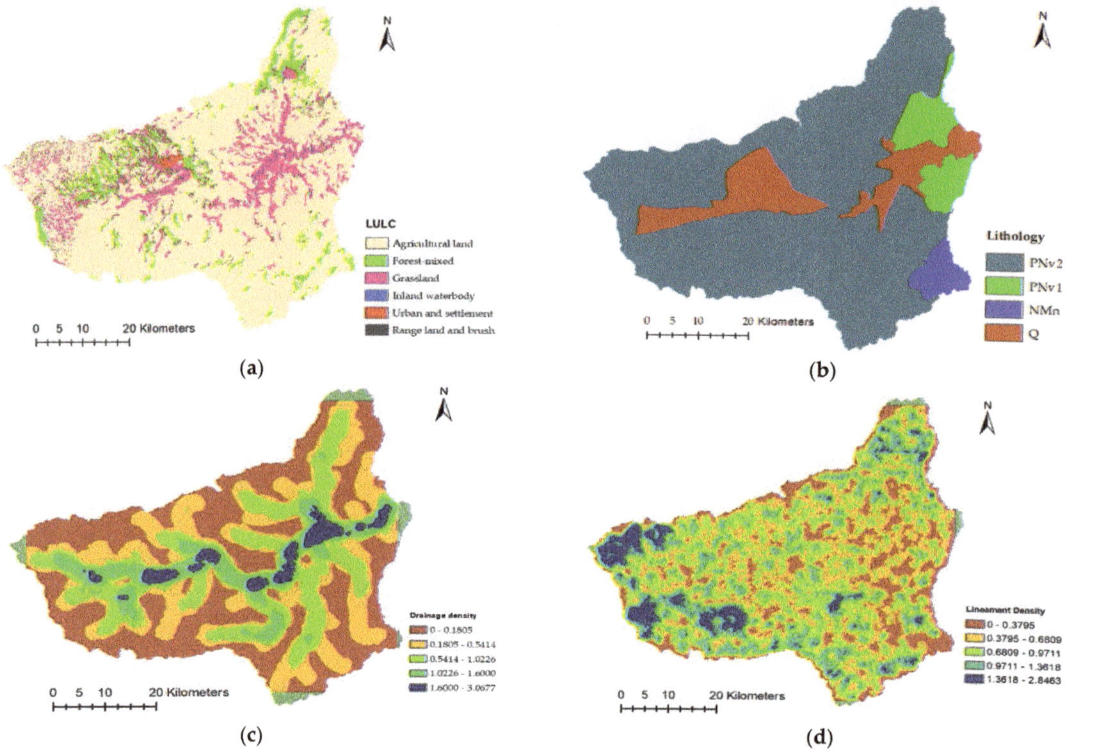

Figure 4. Thematic map of (**a**) LULC, (**b**) Lithology, (**c**) Drainage density, (**d**) Lineament density.

2.2.4. Drainage Density

Drainage density (DD) is the typical flow structure to a pooled point. DD refers to the closeness of stream networks and the total length of channel segment per unit area [69], as indicated in Equation (1). It indicates surface and subsurface formation, soil permeability, infiltration, and runoff. High DD values have high excess surface flow signifying a probability of low groundwater availability [59,80–82]. This implies that DD and permeability have an inverse correlation in determining potential aquifer sites [44,83]. In this study, DD was prepared using SRTM DEM by line density tool of the GIS environment and reclassified (Figure 4c) as appropriate for groundwater potential and recharging effects.

$$DD = \sum_{i}^{n} \frac{D_i}{A_i} \tag{1}$$

where Di is the stream's total length (km), and A is the watershed unit area (km^2).

2.2.5. Lineament Density

Lineaments are physical discontinuity of the Earth's surface and linear plan landscapes critical for infiltration [42,84,85]. They are linear features in a terrain that demonstrates a fundamental geological structure such as folds, faults, joints, or fractures as secondary porosity [68,86,87]. Lineaments characterize the weak zones which disturb groundwater flow, increasing infiltration to the subsurface [88]. The thematic layer of LD (km/km^2) is defined as the total length of all verified lineaments divided by the unit area of grid cells [68], as given in Equation (2). The lineaments increase groundwater recharging potential implicitly if linked to an aquifer [68]. The lineaments were converted to an LD map using the Line Density Tool in the Spatial Analysis Tool inverse distance to a power function of the GIS environment. The LD was interpolated with Geospatial Data Abstraction Library (GDAL) toolbox with Inverse distance weighted (IDW). Each unit was then given weights on the suitability of infiltration contributions to groundwater, as indicated in (Figure 4d). Thus, high LD areas are suitable for groundwater recharge, and low LD is ideal for groundwater recharge and discharges [81,88,89].

$$LD = \sum_{i}^{n} \frac{L_i}{A_i} \tag{2}$$

where L_i indicates lineaments' total length (km), A represents the unit area (km^2)

2.2.6. Lithology

Lithology governs the quantity and quality of groundwater occurrence and permeability of aquifer rocks [49,90]. It is a governing factor for empathetic recharge and infiltration processes [39,91,92]. Therefore, lithological characteristics should be examined to define PARZ accurately. The dominant geological materials are volcanic rocks, intercalated basalt flows with quaternary volcanic, lacustrine deposits, and alluvial sediments exposure [93]. The catchment consists of volcanic rocks (37–11 Ma) [94], the alluvial and lacustrine (sandy silt, clay, diatomite, limestone, and beach sand), Eocene and Paleocene-Pleistocene undivided lacustrine and fluvial sediments, sand, silt gravel conglomerate (Omo group), agglomerates, and basalts [20,93]. The spatial occurrence of the different geological materials is very complex and heterogeneous and not known in detail. Each lithology unit has a different impact on groundwater permeability. Hence, it was ranked as the aquifer's recharging and permeability feature (Figure 4b).

2.2.7. Topographic Position Index

The topographic position index (TPI) is an algorithm extensively applied to quantify topographic positions and systematize land surface categorizations [95]. TPI is a terrain classification method where each data point's altitude is evaluated against its neighbors [96]. In addition, it is correlated with various physical processes in the given aquifer system [97]. First, a TPI map was produced using the GDAL raster analysis tool of the QGIS environment. Then, the reclassification and ranking were conducted based on each unit's recharging suitability, as shown (Figure 5a).

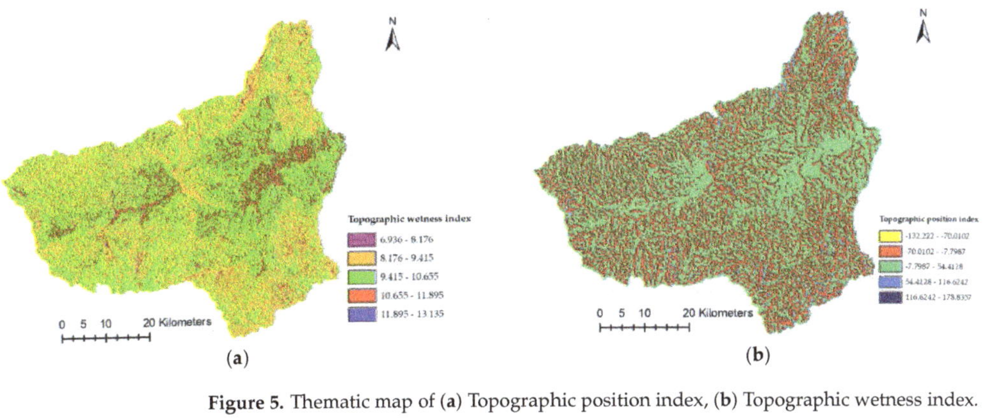

(a) (b)

Figure 5. Thematic map of (**a**) Topographic position index, (**b**) Topographic wetness index.

2.2.8. Topographic Wetness Index

The topographic wetness index (TWI) measures terrain-driven variation in soil moisture, the subsurface lateral transmissivity, and blends the upslope catchment area of the aquifer system [98–100] and downslope drainage for each cell in a DEM [82]. TWI is used to evaluate the landscape influence on hydrological processes of infiltration. TWI was prepared using the TOPMODEL toolbox [101] in the QGIS. This model simulates hydrologic fluxes of water throughout the watershed [82], SAGA terrain analysis tool. Then, reclassified and ranked (Figure 5b) depending on the groundwater recharging capabilities [81].

2.3. *Multi-Criteria Decision Analysis*

Multi-criteria decision analysis (MCDA) is a technique used for configuring numerous problems and a principle of measurement by pairwise comparison and governed by experts' decisions [42]. Data normalization is widely used in GIS-based MCDA and diminishes all decision criteria to a standard measurement scale [39,41,57,88]. In this study, MCDA was used to configure numerous pairwise comparisons matrices' (PCM) to derive priority scales for proxy data layers. The AHP is a powerful tool introduced and developed by Saaty [50] to tackle different decision-making questions providing an MCDA for measuring one unique alternative. Saaty's AHP was used to normalize hierarchical orders for the comparative effectiveness of various proxy data layers. Then, it creates a PCM based on expert opinions or judgment between the standards chosen. Saaty's [50] importance scale of intensity relative importance (Table 1) compared the thematic maps classification criteria. The weights were normalized by averaging the values in each row to the corresponding ranking, which gives the normalized weights of each parameter.

The normalized principal eigenvector was computed to find the relative importance of parameters per PCM to find a consistency ratio (CR). The influence of each factor was decided by experts' knowledge [23,35] and personal judgment. Finally, the PCM of the assigned weights to different thematic layers using [102,103] were normalized using the eigenvector approach [50,104]. The CR, which indicates the suitability of a PCM, was computed to inspect the normalized implications as presented in [50,104,105]. CR is a measurement of the consistency of the PCM and calculated using Equation (3). The comparison of the random consistency index shows how essential a layer is [50]. The consistency index (CI) designates the tolerability of the reciprocal matrix, which is computed with the mean Eigenvalue of the PCM (λmax), random consistency index (RI), and the number of applied parameters (N) [50,106] as in Equations (3) and (4). The CR of the matrix is significant if less than 10% unless the matrix should be re-evaluated based on the (CI) measure tabulated in (Table 2) [104].

$$\mathrm{CR} = \frac{\mathrm{CI}}{\mathrm{RI}} \tag{3}$$

$$\mathrm{CI} = \frac{\lambda\mathrm{max} - \mathrm{N}}{\mathrm{N} - 1} \tag{4}$$

Table 1. Saaty's 1–9 scale of relative importance [104].

Scale	Importance
1	Equal importance
2	Weak importance
3	Moderate importance
4	Moderate plus
5	Strong plus
6	Strong importance
7	Very strong importance
8	Very very strong importance
9	Extreme importance

Table 2. Tabulated standards of N for Saaty's ratio index.

N	1	2	3	4	5	6	7	8	9	10
RI	0	0	0.58	0.89	1.12	1.24	1.32	1.41	1.45	1.49

2.4. Estimating Spatial Distribution of Recharge Zones

The PCM of the given weight for proxy data classes was constructed and normalized using AHP and the eigenvector approach. Next, a comparative rank for each raster was given depending on the extrapolative implication of groundwater potentiality. Next, the thematic raster layers were overlaid based on weighted overlay analysis (WOA) in the GIS environment. Finally, after the thematic maps were integrated, an output raster map was obtained to classify and quantify the potential aquifer recharge zones.

3. Results and Discussion

3.1. Assessing the Effects of Groundwater Controlling Parameters

Numerous parameters of measurable infiltration rates mainly control the potential aquifer recharge zone map development. Hence, the groundwater movement and occurrence controlling proxy data layers were presented and discussed in the following section according to the degree of infiltration to the subsurface for the potentiality of the aquifer system.

3.1.1. Rainfall and Soil Textures

The annual catchment rainfall ranges from 1152.95–1621.68 mm. The areal rainfall distribution is classified into five as 1152.95–1168.66, 1168.66–1311.29, 1311.29–1506.02, 1506.02–1527.90, and 1527.90–1621.68 mm very low, low, moderate, high, and very high, respectively (Figure 3a). Low rainfall shows low recharge indicating low groundwater potential zones, while the excessive rainfall quantities suggest the chance of high recharge, the potentiality of high groundwater zones. The catchment zones that obtained high rainfall have a chance of getting more percolated water than with low rainfall. In contrast, the high rainfall class is given high ranks as of its significance for groundwater recharge potentiality of the aquifer system. Rainfall infiltrates if the soil has sufficient permeability or goes away as overland flow based on the aquifer's nature. Therefore, it is reasonable to account for precipitation to determine its influence on groundwater recharge. Similar studies confirmed that rainfall is a critical parameter governing the groundwater potential recharge [23,25,41,53,57].

Soil texture controls the infiltration rate into the subsurface and influences groundwater recharge. The high hydraulic conductivity of the soil allows more infiltration and limits evaporation [29,45,107,108]. Therefore, the study area soil was reclassified and comprised of six soil types (Figure 3b). The soil was characterized by very low to very high infiltration capacity. Therefore, ranks were assigned as sandy clay loam (high), loam (high), clay loam (moderate), clay light (poor), clay (poor), and clay-heavy (very poor) based on infiltration and water holding capacity as suitable for groundwater. The groundwater potential is higher in a coarse-grained structure and very low in a fine-grained structured soil type. The least weight was assigned for clayey soil because clay horizons considerably restrict percolation and have low permeability.

3.1.2. Elevation and Slope

The elevation of the study area varies from 3353 m (highest) to 1674 m (lowest). In elevated areas, runoff conditions are high, and low infiltration indicates water delays over the land surface for percolation [68]. The elevation was reclassified and ranked as suitable for groundwater recharging potential 1674–2009.8 m (very high), 2009.8–2345.6 m (high), 2345.6–2681.4 m (moderate), 2681.4–3017.2 m (low), 3017.2–3353 m (very low) as shown in (Figure 3c). The lower elevation areas were more favorable for groundwater recharge predictions than higher elevations. However, soil texture is the leading cause of percolation in highly elevated areas. Potential groundwater is high at lower elevations with sandy loam soil. The percolation rate varies from low to very low at low elevations with clay soil. Similar studies confirmed that elevation was a good decision and robust parameter for delineating PARZ [59,88,109]. The slope affects groundwater percolation during the overland and base flow [110]. The landscape slope is classified into five classes based on suitability of recharging potential and infiltration rate, then assigned weights to each slope class. The classified slope (in degrees) was 0–5 (Flat), 5–10 (gentle), 10–15 (moderate), 15–30 (hill), and 30–66.18 (steep) with very high, high, moderate, low, and very low, respectively, of recharging suitability (Figure 3d). The slope classes with flat, gentle, moderate, hilly, and steep are 15.41%, 15.61%, 16.53%, 40.77%, and 11.68%, respectively. As steepness increases, the percolation rate decreases due to the high speed of overland flow. Therefore, flat and gentle slopes indicate the presence of high groundwater recharging potentials.

3.1.3. Land Use/Cover and Lithology

The LULC substantially influences infiltration rates [71–73,111]. This study ranked LULC classes based on the capacity to infiltrate water into the ground and water holding capacity. Inland water bodies, agricultural land, grassland, forest-mixed, range and shrubland, and settlement and urban (Figure 4a) were weighted and ranked as very high, high, moderate, slightly moderate, low, and very low, respectively, as suitable for recharging potentiality. The LULC classes of GGC constitute about 77.52% (agriculture), 12.08% (grassland), 9.84% (forest-mixed), and others for water bodies, settlement and urban, and range and shrubland. Agricultural land has high groundwater potential with more porosity increasing soil water percolation [48]. An intense agricultural activity changes the hydrologic cycle by causing soil moisture conditions and recharge [70,77]. Waterbodies are the most fundamental and permanent source of groundwater recharge. Forest-mixed sites have lower groundwater recharge rates than grassland increases groundwater recharge [84] and reduce the runoff process by increasing infiltration rates, thereby augmenting groundwater recharge. On the other hand, built areas generate runoff having poor groundwater recharging potential. Other studies have agreed that land use covers are essential in delineating potential groundwater zones [30,47,59,68,86].

Infiltration is predominantly good in plateaus covered by thick alluvial sediments [20]. The dominant lithology is divided into four aquifer subclasses based on permeability and productivity potentials. The study area Lithological unit groundwater prospect classification is adapted from [20], as shown in (Figure 4b). These are very highly permeable-volcanic sand (NMn), lower felsic, volcanic, and sedimentary formation (PNv1), sandy pyroclastic sediments of the Pleistocene-Holocene volcanic group (Qv1), upper felsic volcanic (PNv2). The quaternary sediments are alluvium, river gravels, fans, and travertine [20,112]. The quaternary alluvial has good potential for deposits and structures with higher permeability and productivity [20,112]. If fractured, quaternary volcanic have to yield considerable groundwater flow [113]. Numerous studies confirmed the importance of lithological features in PARZ map development [33,49,57,59,85].

3.1.4. Drainage Density and Lineament Density

Drainage density (DD) influences the distribution of runoff and infiltration [30]. Low DD is associated with widely spaced streams due to less resistant surface materials or high infiltration. Low DD leads to a coarse texture, while high DD is favorable with a high infiltration rate [23,114]. DD (km/km^2) values are categorized into five classes as 0–0.18045 (very high), 0.18045–0.54136 (high), 0.54136–1.02256 (moderate), 1.0226–1.6000 (low), and 1.6000–3.0677 (very low) as suitable for groundwater recharge rate rankings, respectively, as revealed in (Figure 4c). High DD values indicate low groundwater availability assigned with low rank having high runoff. The lowest DD values were given the highest level for decreasing runoff in the area, as low DD is related to higher recharge and higher groundwater potentials. High DD indicates impermeable sub-surface and mountainous relief. On the other hand, the low DD of the watershed reveals that it is composed of permeable subsurface, good land cover, and more infiltration capacity. Similar studies confirmed the applicability of DD in mapping potential groundwater zone [23,25,30,47,89]. Lineament density (LD) is directly relational to the groundwater perspective [85,89]. The LD (km/km^2) was determined and reclassified into five categories, 0–0.3795 (very low), 0.3795–0.6809 (low), 0.6809–0.9711 (moderate), 0.9711–1.3618 (high), and 1.3618–2.8463 (very high) as per prospects of groundwater suitability as shown in (Figure 4d). In a very high LD, there is a high infiltration rate to the subsurface, whereas, in low LD, the rate of infiltration is low [23,25,30,47,114].

3.1.5. Topographic Positioning Index and Topographic Wetness Index

The topographic position index (TPI) is used in figuring out the landscape's upper, middle, and lower components [97]. High TPI values are located close to the ends of hills, while small TPI is close to valley bottoms [47]. The highest rating is given for low TPI due to the potentiality of groundwater prospects. Accordingly, the TPI of this study was reclassified and ranked (Figure 5a) as −133.22 to −70.0102 (very high), −70.0102 to −7.7987 (high), −7.7987 to 54.4128 (moderate), 54.4128 to 116.6242 (low), and 116.6242 to 178.8357 (very low). These studies confirmed TPI as a vigorous groundwater flow controlling parameter in delineating PARZ [47,115]. The topographic wetness index (TWI) was applied to quantify topographic control of groundwater infiltration capability due to topographical effects [84,116]. The lowest rank was given to the lowest TWI and vice versa. Then, TWI were reclassified into five categories (Figure 5b) 6.936–8.176 (very low), 8.176–9.415 (low), 9.415–10.655 (moderate), 10.655–11.895 (high), and 11.895–13.135 (very high) as suitability for infiltration of soil water into the subsurface. This indicates that TWI considerably influences runoff concentration at the soil surface. Likewise, various studies used TWI as a robust parameter in determining PARZ [47,86,117].

3.2. *Multi-Criteria Decision Analysis and Weight Normalization*

In this study, the AHP was developed hierarchically, based on expert opinions, weighing the criteria were used to normalize principal eigenvector values depending on MCDA. The MCDA has been widely used in many studies to predict potential groundwater zones [25,40,44,45,67,84]. The weights were normalized and considered by averaging values to find the normalized weights of respective parameters. According to interpretation, the normalized weights of each feature of proxy data layers for groundwater potentiality are indicated in Table 3. In the PCM, the consistency ratio value was less than 0.1, indicating that all experts' weightings are consistent and suitable for implementation. The calculated normalized principal Eigenvector value ($\lambda max = 11.371$). A PCM is consistent only if λmax is equal to or more than the examined thematic layers; otherwise, a new matrix must be created. In the present study, λmax is 11.371 and has ten thematic layers, confirming that there is no need to develop a matrix of new data layers. Hence, the principal eigenvalue (λmax) exemplifies a purpose for the matrix divergence [106]. The λmax was computed depending on the consistency ratio from the PCM of ten parameters. The consistency index was calculated to overcome the consistency ratio formula, which results in CI = 0.1371. Then, taking the RI value as 1.49 from Table 2, the computed consistency ratio (CR = 0.0920), less than 10%, and the given weights were valid for further analysis. The CI and CR were evaluated, and a PCM of ten groundwater-controlling factors was generated in the AHP process. The consistency evaluation indicates that the PARZ s map developed using MCDA is reasonably accurate.

Table 3. Normalized weights for proxy layers as suitable for groundwater recharge.

Thematic Layers	Normalized Weight (%)	Category	Ranks	Similar Efforts
Rainfall	14.5	1152.95–1168.66	Very low	[23,25,30,41,53,57,59]
		1168.66–1311.29	Low	
		1311.29–1506.02	Moderate	
		1506.02–1527.90	High	
		1527.90–1621.68	Very high	
Lithology	20.2	NMn	High	[33,49,57,59,85]
		PNv1	Moderately	
		Qv1	Moderately	
		PNv2	Poor	
Land use/cover	15.3	WATR	Very high	[25,30,59,68,80,88,89]
		AGRC	High	
		RNGE	Moderate	
		FRST	Slightly moderate	
		RNGB	Low	
		URBN	Very low	
Drainage density	7.2	0–0.18045	Very high	[23,25,30,47,89]
		0.18045–0.54136	High	
		0.54136–1.02256	Moderate	
		1.0226–1.6000	Low	
		1.6000–3.0677	Very low	
Elevation	5.1	1674–2009.8 m	Very high	[43,59,88,109]
		2009.8–2345.6 m	High	
		2345.6–2681.4 m	Moderate	
		2681.4–3017.2 m	Low	
		3017.2–3353 m	Very low	
Slope	5.4	0–5	Flat	[23,25,28,43,53,57,114]
		5–10	Gentle	
		10–15	Moderate	
		15–30	Hill	
		30–66.18	Steep	
Lineament density	8.8	0–0.3795	Very low	[23,25,30,47,114]
		0.3795–0.6809	Low	
		0.6809–0.9711	Moderate	
		0.9711–1.3618	High	
		1.3618–2.8463	Very high	
Soil	8.8	Sandy loam	High	[23,25,30,36,53,57,59]
		Loam	High	
		Clay loam	Moderate	
		Clay light	Poor	
		Clay	Poor	
		Clay-heavy	very poor	
Topographic position index	7.5	−133.22 to −70.0102	Very high	[47,84,115]
		−70.0102 to −7.7987	High	
		−7.7987 to 54.4128	Moderate	
		54.4128 to 116.6242	Low	
		116.6242 to 178.8357	Very low	
Topographic wetness index	7.3	6.936–8.176	Very low	[47,84,86,116,117]
		8.176–9.415	Low	
		9.415–10.655	Moderate	
		10.655–11.895	High	
		11.895–13.135	Very high	

3.3. Evaluating the Spatial Distribution of Potential Recharge Zones

In the present study, the importance of factors is assigned based on experts' opinions and personal judgment as there is no standard scale for a simple WOA. The thematic layer's relative weight is defined using a PCM based on the expert judgment of other similar studies [24,30,59,68,69]. These criteria were weighted based on the recharging potentiality by constructing pairwise assessments in MCDA. The maps were reclassified into five classes based on aquifer recharging potentiality to examine the areas with different stages of aquifer recharging potential. The weighted index map of the study area was developed using ten thematic layers with WOA in the GIS environment. The raster layers and their corresponding weights have been assigned to account for relative significance from a recharging perspective. The computed scale importance weighted normalization showed the value of 20.2% for lithology, 15.3% for LULC, 7.3% for topographic wetness index, 8.8% for lineament density, 5.4% for slope, 5.1% for elevation, 8.8% for soil, 14.5% for rainfall, 7.2% for drainage density, and 7.5% for topographic position index. The dominant factors governing the aquifer recharging potential were lithology, rainfall, and LULC, covering about 50% of the total weight. The catchment is delineated into five PARZ zones: very high, high, moderate, low, and very low, combining all reclassified raster layers. The developed map covers very high (29%), high (25%), moderate (28%), low (13%), and very low (5%) of the total area. The developed PARZ map with high potential zones predominantly exists in the Eastern part (Figure 6). Accordingly, most of the southwestern and northeast parts of the catchment are categorized under high and very high groundwater potentiality. As revealed from the consistency index criteria, the selected groundwater controlling factors were reliable.

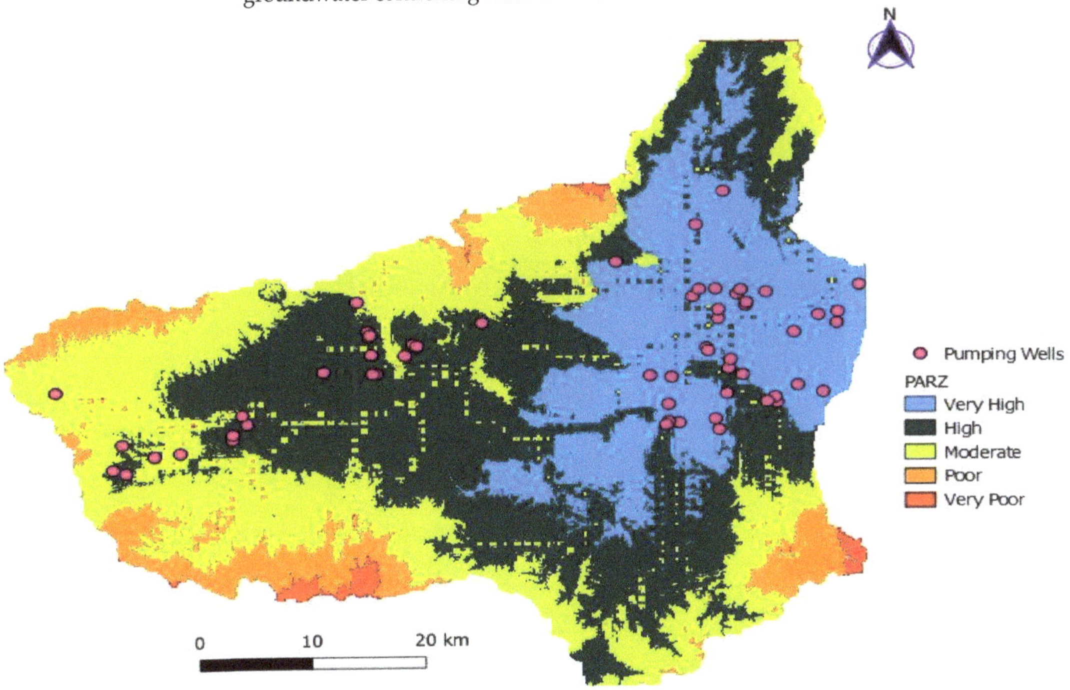

Figure 6. Spatial distribution of potential aquifer recharge zones of the study area.

The very high groundwater potential comprises quaternary alluvium deposits and quaternary volcanics and sediments. In addition, most of the tertiary volcanics and sediments, undivided quaternary alluvium, eluvium, and lacustrine sediments, have high to moderate potential zones. The very high aquifer recharging potential is found in the flat plains of the catchment with high rainfall distribution, which has a very high infiltration potential to the subsurface flow. The high potential zone is due to high lineament density with gentle slope, alluvial plains, and valleys. As a result, quaternary lacustrine sediment controls dominate low drainage density, flatter slopes, and agricultural land distribution. Loamy soil, high lineament density, and low drainage density promote groundwater aquifers' high infiltration rate. The moderate PARZ is mainly agrarian land flatter to gentle slope, with high lineament density and low drainage density. The high elevation and impermeable rock formations increase runoff resulting in low groundwater recharging. Steep slopes and the hilly regions, clay soil, and very high drainage density are the main reasons for low infiltration potential and low groundwater storage capacity.

3.4. Validation of Developed Map

The probability of aquifer recharge zone defined by geospatial techniques is usually validated by matching existing inventory data [67]. The validity of the potential aquifer recharge map was performed by overlaying the point data of inventory wells with the generated PARZ map. In addition, the result has been validated using the available inventory data or pumping wells information in the catchment (Figure 6). The validation points of the PARZ indicated that most of the production wells are located in very high (57%), high (32), and moderate (12%) delineated recharge zones with a productivity rate range of 0.35—21.6 L/s. This supports an agreement between the groundwater inventory data and PARZ defined using geospatial techniques. Based on the distribution of wells and the corresponding yield values, the developed PARZ map was complimented as a proxy for the aquifer productivity of existing pumping wells. Depending on their productivity, most African aquifers [118,119] are classified as very high (>20), high (5–20), moderate (2–5), low-moderate (0.5–2), low (0.1–0.5), very low (<0.1) in liters/second. However, from total inventory wells in the Gilgel Gibe catchment, according to the aquifer productivity classification of most African aquifers [118,119], there exists very high (25%), high (22%), moderate (28%), and low (25%) production wells. Therefore, the developed aquifer recharge map and the production wells are mismatched at some locations. This study verifies that the applicability of proxy data sources to delineate is trustworthy in groundwater development and management. This confirms that the significance of the weighted influencing parameter threshold values through GIS-based overlay analysis as credible support for delineating potential recharge zones is valuable for stakeholders to augment groundwater management and reduce the uncertainty of the decision-making process and resources allocation.

4. Conclusions

In the present study, the integrated application of GIS, RS, and MCDA were applied to identify potential aquifer recharge zones for the proxy data of the Gilgel Gibe catchment, Ethiopia. The study has proven that geospatial techniques can afford suitable proxy data analysis for viable groundwater assessments. An attempt was made to analyze ten governing data layers using GIS and AHP, and the corresponding weights of thematic layers have been assigned considering the relative implication of recharging. Each layer has no equal importance in controlling groundwater flow within the aquifer system. Accordingly, the most significant factors with a weighted value of lithology (20.2%), lineament density (8.8%), slope (5.4%), elevation (5.1%), soil (8.8%), LULC (15.3%), rainfall (14.5%), drainage density (7.2%), topographic position index (7.5%), and topographic wetness index (7.3%), respectively. Accordingly, the assessment revealed that lithology, rainfall, and LULC are dominant factors covering 50% of the controlling groundwater flow in the study area.

The computed consistency ratio (CR = 0.0920), less than 10%, verified that the established results are reasonably accurate and given valid weights. The developed map was classified into five classes: very high, high, moderate, low, and very low groundwater potentiality using weighted overlay analysis with GIS-based AHP analysis. The developed map with high potential predominantly exists in the central, western, and eastern parts. Areas with poor potential contribute to high runoff and relatively low infiltration. Therefore, a very high zone indicates the most appropriate place for productive wells allocations in practical applications. The validation indicated that most of the inventory wells are located in very high (57%), high (32), and moderate (12%) zones. The study is the first attempt to develop potential aquifer recharge zones in the Gilgel Gibe catchment. Hence, the findings are essential in providing insights to reduce complexity in decision-making processes for proper groundwater aquifer management. Furthermore, the study confirms that the applied techniques are helpful for groundwater exploration in varied climates and remote regions. The present study recommends that assessing the amount of exploitable and vulnerable groundwater in the catchment is indispensable for sustainable utilization and ecosystem management.

Author Contributions: Conceptualization, T.D.M. and I.-M.C.; writing—original draft preparation, T.D.M.; writing—review and editing, I.-M.C.; resources and data curation, M.-G.K. and I.-H.K.; visualization, M.-G.K.; supervision, I.-M.C. and S.W.C.; funding acquisition, I.-M.C. and S.W.C. All authors have read and agreed to the published version of the manuscript.

Funding: This research was supported by a grant from a Strategic Research Project (20220178-001) funded by the Korea Institute of Civil Engineering and Building Technology and partially supported by a grant from the Korea Ministry of Environment (MOE) as a Demand Responsive Water Supply Service Program (146515).

Institutional Review Board Statement: Not applicable.

Informed Consent Statement: Not applicable.

Data Availability Statement: Not applicable.

Conflicts of Interest: The authors declare no conflict of interest.

References

1. Sophocleous, M. Global and Regional Water Availability and Demand: Prospects for the Future. *Nat. Resour. Res.* **2004**, *13*, 61–75. [CrossRef]
2. Jha, M.K.; Kamii, Y.; Chikamori, K. Cost-effective Approaches for Sustainable Groundwater Management in Alluvial Aquifer Systems. *Water Resour. Manag.* **2009**, *23*, 219–233. [CrossRef]
3. Menon, S.V. Ground Water Management: Need for Sustainable Approach. MPRA (Munich Personal RePEc Archieve) Paper, 15 October, 2007. Available online: https://mpra.ub.uni-muenchen.de/6078 (accessed on 1 January 2020).
4. Pavelic, P. *Groundwater Availability and Use in Sub-Saharan Africa: A Review of 15 Countries*; International Water Management Institute (IWMI): Colombo, Sri Lanka, 2012; ISBN 9789290907589.
5. Calow, R.C.; MacDonald, A.M.; Nicol, A.L.; Robins, N.S. Ground Water Security and Drought in Africa: Linking Availability, Access, and Demand. *Ground Water* **2010**, *48*, 246–256. [CrossRef] [PubMed]
6. Milly, P.C.D.; Dunne, K.A.; Vecchia, A.V. Global pattern of trends in streamflow and water availability in a changing climate. *Nature* **2005**, *438*, 347–350. [CrossRef]
7. Foster, S.S.D.; Chilton, P.J. Groundwater: The processes and global significance of aquifer degradation. *Philos. Trans. R. Soc. London. Ser. B Biol. Sci.* **2003**, *358*, 1957–1972. [CrossRef]
8. Chowdhury, A.; Jha, M.K.; Chowdary, V.M.; Mal, B.C. Integrated remote sensing and GIS-based approach for assessing groundwater potential in West Medinipur district, West Bengal, India. *Int. J. Remote Sens.* **2009**, *30*, 231–250. [CrossRef]
9. Berhanu, B.; Seleshi, Y.; Melesse, A.M. Surface Water and Groundwater Resources of Ethiopia: Potentials and Challenges of Water Resources Development. In *Nile River Basin*; Melesse, A.M., Abtew, W., Setegn, S.G., Eds.; Springer: Cham, Switzerland, 2014; Volume 9783319027, pp. 97–117. ISBN 978-3-319-02719-7.
10. Awulachew, S.B.; Erkossa, T.; Namara, R. Irrigation potential in Ethiopia: Constraints and opportunities for enhancing the system. Research Report, International Water Management Institute, Addis Ababa. *Gates Open Res.* **2010**, *2*, 1–59.
11. MacDonald, A.M.; Bonsor, H.C.; Dochartaigh, B.É.Ó.; Taylor, R.G. Quantitative maps of groundwater resources in Africa. *Environ. Res. Lett.* **2012**, *7*, 024009. [CrossRef]
12. Famiglietti, J.S. The global groundwater crisis. *Nat. Clim. Change* **2014**, *4*, 945–948. [CrossRef]
13. Foster, S.; Tuinhof, A.; Van Steenbergen, F. Managed groundwater development for water-supply security in Sub-Saharan Africa: Investment priorities. *Water SA* **2012**, *38*, 359–366. [CrossRef]
14. Gumma, M.K.; Pavelic, P. Mapping of groundwater potential zones across Ghana using remote sensing, geographic information systems, and spatial modeling. *Environ. Monit. Assess.* **2013**, *185*, 3561–3579. [CrossRef]

15. Macdonald, A.; Dochartaigh, B.Ó. Mapping for Water Supply and Sanitation (WSS) in Ethiopia. Research-inspired Policy and Practice Learning in Ethiopia and the Nile region (RiPPLE). In *Working Paper-11, January 2009*; WaterAid Ethiopia: Addis Ababa, Ethiopia, 2009.
16. Moges, S. AgWater Solutions Project Case Study Agricultural Use of Ground Water in Ethiopia: Assessment of Potential and Analysis of Economics, Policies, Constraints and Opportunities. *Gates Open Res.* **2012**, *3*, 136. [CrossRef]
17. Dile, Y.T.; Tekleab, S.; Ayana, E.K.; Gebrehiwot, S.G.; Worqlul, A.W.; Bayabil, H.K.; Yimam, Y.T.; Tilahun, S.A.; Daggupati, P.; Karlberg, L.; et al. Advances in water resources research in the Upper Blue Nile basin and the way forward: A review. *J. Hydrol.* **2018**, *560*, 407–423. [CrossRef]
18. Chung, I.-M.; Sophocleous, M.A.; Mitiku, D.B.; Kim, N.W. Estimating groundwater recharge in the humid and semi-arid African regions: Review. *Geosci. J.* **2016**, *20*, 731–744. [CrossRef]
19. Nimmo, J.; Stonestrom, D.A.; Healy, R.W. Aquifers: Recharge. In *Fresh Water and Watersheds*, 2nd ed.; CRC Press: Boca Raton, FL, USA, 2020; pp. 11–15.
20. Kebede, S. *Groundwater in Ethiopia*; Springer: Berlin/Heidelberg, Germany, 2013; ISBN 978-3-642-30390-6.
21. Foster, S.; Kemper, K. Sustainable Groundwater Management: Management. In *GW Mate World Bank*; Springer: Berlin/Heidelberg, Germany, 2003; p. 6.
22. Tuinhof, A.; Foster, S.; Van Steenbergen, F.; Talbi, A.; Wishart, M. *Strategic Overview Series Number 5 Appropriate Groundwater Management Policy for Sub-Saharan Africa in Face of Demographic Pressure and Climatic Variability Sustainable Groundwater Management Contributions to Policy Promotion*; World Bank: Washington, DC, USA, 2011.
23. Ahmad, I.; Dar, M.A.; Andualem, T.G.; Teka, A.H. GIS-based multi-criteria evaluation of groundwater potential of the Beshilo River basin, Ethiopia. *J. Afr. Earth Sci.* **2020**, *164*, 103747. [CrossRef]
24. Fenta, A.A.; Kifle, A.; Gebreyohannes, T.; Hailu, G. Spatial analysis of groundwater potential using remote sensing and GIS-based multi-criteria evaluation in Raya Valley, northern Ethiopia. *Hydrogeol. J.* **2015**, *23*, 195–206. [CrossRef]
25. Andualem, T.G.; Demeke, G.G. Groundwater potential assessment using GIS and remote sensing: A case study of Guna tana landscape, upper blue Nile Basin, Ethiopia. *J. Hydrol. Reg. Stud.* **2019**, *24*, 100610. [CrossRef]
26. Abiy, A.Z.; Melesse, A.M. Evaluation of watershed scale changes in groundwater and soil moisture storage with the application of GRACE satellite imagery data. *Catena* **2017**, *153*, 50–60. [CrossRef]
27. Díaz-Alcaide, S.; Martínez-Santos, P. Review: Advances in groundwater potential mapping. *Hydrogeol. J.* **2019**, *27*, 2307–2324. [CrossRef]
28. Magesh, N.S.; Chandrasekar, N.; Soundranayagam, J.P. Delineation of groundwater potential zones in Theni district, Tamil Nadu, using remote sensing, GIS and MIF techniques. *Geosci. Front.* **2012**, *3*, 189–196. [CrossRef]
29. Fashae, O.A.; Tijani, M.N.; Talabi, A.O.; Adedeji, O.I. Delineation of groundwater potential zones in the crystalline basement terrain of SW-Nigeria: An integrated GIS and remote sensing approach. *Appl. Water Sci.* **2014**, *4*, 19–38. [CrossRef]
30. Tolche, A.D. Groundwater potential mapping using geospatial techniques: A case study of Dhungeta-Ramis sub-basin, Ethiopia. *Geol. Ecol. Landsc.* **2021**, *5*, 65–80. [CrossRef]
31. Medici, G.; Engdahl, N.B.; Langman, J.B. A Basin-Scale Groundwater Flow Model of the Columbia Plateau Regional Aquifer System in the Palouse (USA): Insights for Aquifer Vulnerability Assessment. *Int. J. Environ. Res.* **2021**, *15*, 299–312. [CrossRef]
32. Manna, F.; Walton, K.M.; Cherry, J.A.; Parker, B.L. Mechanisms of recharge in a fractured porous rock aquifer in a semi-arid region. *J. Hydrol.* **2017**, *555*, 869–880. [CrossRef]
33. Lentswe, G.B.; Molwalefhe, L. Delineation of potential groundwater recharge zones using analytic hierarchy process-guided GIS in the semi-arid Motloutse watershed, eastern Botswana. *J. Hydrol. Reg. Stud.* **2020**, *28*, 100674. [CrossRef]
34. Roy, A.; Keesari, T.; Sinha, U.K.; Sabarathinam, C. Delineating groundwater prospect zones in a region with extreme climatic conditions using GIS and remote sensing techniques: A case study from central India. *J. Earth Syst. Sci.* **2019**, *128*, 201. [CrossRef]
35. Singh, S.K.; Zeddies, M.; Shankar, U.; Griffiths, G.A. Potential groundwater recharge zones within New Zealand. *Geosci. Front.* **2019**, *10*, 1065–1072. [CrossRef]
36. Nigussie, W.; Hailu, B.T.; Azagegn, T. Mapping of groundwater potential zones using sentinel satellites (−1 SAR and -2A MSI) images and analytical hierarchy process in Ketar watershed, Main Ethiopian Rift. *J. Afr. Earth Sci.* **2019**, *160*, 103632. [CrossRef]
37. Abdalla, F.; Moubark, K.; Abdelkareem, M. Groundwater potential mapping using GIS, linear weighted combination techniques and geochemical processes identification, west of the Qena area, Upper Egypt. *J. Taibah Univ. Sci.* **2020**, *14*, 1350–1362. [CrossRef]
38. SRINIVASA RAO, Y.; JUGRAN, D.K. Delineation of groundwater potential zones and zones of groundwater quality suitable for domestic purposes using remote sensing and GIS. *Hydrol. Sci. J.* **2003**, *48*, 821–833. [CrossRef]
39. Malczewski, J. GIS-based multicriteria decision analysis: A survey of the literature. *Int. J. Geogr. Inf. Sci.* **2006**, *20*, 703–726. [CrossRef]
40. Adiat, K.A.N.; Nawawi, M.N.M.; Abdullah, K. Assessing the accuracy of GIS-based elementary multi criteria decision analysis as a spatial prediction tool–A case of predicting potential zones of sustainable groundwater resources. *J. Hydrol.* **2012**, *440–441*, 75–89. [CrossRef]
41. Aykut, T. Determination of groundwater potential zones using Geographical Information Systems (GIS) and Analytic Hierarchy Process (AHP) between Edirne-Kalkansogut (northwestern Turkey). *Groundw. Sustain. Dev.* **2021**, *12*, 100545. [CrossRef]
42. Agarwal, R.; Garg, P.K. Remote Sensing and GIS Based Groundwater Potential & Recharge Zones Mapping Using Multi-Criteria Decision Making Technique. *Water Resour. Manag.* **2016**, *30*, 243–260. [CrossRef]

43. Mukherjee, P.; Singh, C.K.; Mukherjee, S. Delineation of Groundwater Potential Zones in Arid Region of India-A Remote Sensing and GIS Approach. *Water Resour. Manag.* **2012**, *26*, 2643–2672. [CrossRef]
44. Chowdhury, A.; Jha, M.K.; Chowdary, V.M. Delineation of groundwater recharge zones and identification of artificial recharge sites in West Medinipur district, West Bengal, using RS, GIS and MCDM techniques. *Environ. Earth Sci.* **2010**, *59*, 1209–1222. [CrossRef]
45. Singh, L.K.; Jha, M.K.; Chowdary, V.M. Multi-criteria analysis and GIS modeling for identifying prospective water harvesting and artificial recharge sites for sustainable water supply. *J. Clean. Prod.* **2017**, *142*, 1436–1456. [CrossRef]
46. Saraf, A.K.; Choudhury, P.R. Integrated remote sensing and gis for groundwater exploration and identification of artificial recharge sites. *Int. J. Remote Sens.* **1998**, *19*, 1825–1841. [CrossRef]
47. Nair, H.C.; Padmalal, D.; Joseph, A.; Vinod, P.G. Delineation of Groundwater Potential Zones in River Basins Using Geospatial Tools—An Example from Southern Western Ghats, Kerala, India. *J. Geovisualization Spat. Anal.* **2017**, *1*, 5. [CrossRef]
48. Zghibi, A.; Mirchi, A.; Msaddek, M.H.; Merzougui, A.; Zouhri, L.; Taupin, J.-D.; Chekirbane, A.; Chenini, I.; Tarhouni, J. Using Analytical Hierarchy Process and Multi-Influencing Factors to Map Groundwater Recharge Zones in a Semi-Arid Mediterranean Coastal Aquifer. *Water* **2020**, *12*, 2525. [CrossRef]
49. Dar, T.; Rai, N.; Bhat, A. Delineation of potential groundwater recharge zones using analytical hierarchy process (AHP). *Geol. Ecol. Landsc.* **2010**, *56*, 699–711. [CrossRef]
50. Saaty, R.W. The analytic hierarchy process—what it is and how it is used. *Math. Model.* **1987**, *9*, 161–176. [CrossRef]
51. Bernasconi, M.; Choirat, C.; Seri, R. The Analytic Hierarchy Process and the Theory of Measurement. *Manag. Sci.* **2010**, *56*, 699–711. [CrossRef]
52. Bushan, N.; Rai, K. Strategic-Decision-Making-Navneet-Bhushan-and-Kanwal-Rai. Springer: London, UK, 2004; ISBN 1852337567.
53. Hussein, A.-A.; Govindu, V.; Nigusse, A.G.M. Evaluation of groundwater potential using geospatial techniques. *Appl. Water Sci.* **2017**, *7*, 2447–2461. [CrossRef]
54. Murthy, K.S.R.; Mamo, A.G. Multi-criteria decision evaluation in groundwater zones identification in Moyale-Teltele subbasin, South Ethiopia. *Int. J. Remote Sens.* **2009**, *30*, 2729–2740. [CrossRef]
55. Nimmo, J.R.; Healy, R.W.; Stonestrom, D.A. Aquifer Recharge. In *Encyclopedia of Hydrological Sciences*; John Wiley & Sons, Ltd: Chichester, UK, 2005; pp. 2229–2246. ISBN 0470848944.
56. Siebert, S.; Burke, J.; Faures, J.M.; Frenken, K.; Hoogeveen, J.; Döll, P.; Portmann, F.T. Groundwater use for irrigation—A global inventory. *Hydrol. Earth Syst. Sci.* **2010**, *14*, 1863–1880. [CrossRef]
57. Yıldırım, Ü. Identification of Groundwater Potential Zones Using GIS and Multi-Criteria Decision-Making Techniques: A Case Study Upper Coruh River Basin (NE Turkey). *ISPRS Int. J. Geo-Inf.* **2021**, *10*, 396. [CrossRef]
58. Mengistu, T.D.; Chung, I.-M.; Chang, S.W.; Yifru, B.A.; Kim, M.-G.; Lee, J.; Ware, H.H.; Kim, I.-H. Challenges and Prospects of Advancing Groundwater Research in Ethiopian Aquifers: A Review. *Sustainability* **2021**, *13*, 11500. [CrossRef]
59. Berhanu, K.G.; Hatiye, S.D. Identification of Groundwater Potential Zones Using Proxy Data: Case study of Megech Watershed, Ethiopia. *J. Hydrol. Reg. Stud.* **2020**, *28*, 100676. [CrossRef]
60. Ganapuram, S.; Kumar, G.T.V.; Krishna, I.V.M.; Kahya, E.; Demirel, M.C. Mapping of groundwater potential zones in the Musi basin using remote sensing data and GIS. *Adv. Eng. Softw.* **2009**, *40*, 506–518. [CrossRef]
61. Fazzini, M.; Bisci, C.; Billi, P. The Climate of Ethiopia. In *World Geomorphological Landscapes*; 2015; pp. 65–87. ISBN 9789401780261.
62. Selvam, S.; Dar, F.A.; Magesh, N.S.; Singaraja, C.; Venkatramanan, S.; Chung, S.Y. Application of remote sensing and GIS for delineating groundwater recharge potential zones of Kovilpatti Municipality, Tamil Nadu using IF technique. *Earth Sci. Inform.* **2016**, *9*, 137–150. [CrossRef]
63. Terzer, S.; Wassenaar, L.I.; Araguás-Araguás, L.J.; Aggarwal, P.K. Global isoscapes for $\delta^{18}O$ and $\delta^{2}H$ in precipitation: Improved prediction using regionalized climatic regression models. *Hydrol. Earth Syst. Sci.* **2013**, *17*, 4713–4728. [CrossRef]
64. Moura-Bueno, J.M.; Dalmolin, R.S.D.; Ten Caten, A.; Ruiz, L.F.C.; Ramos, P.V.; Dotto, A.C. Assessment of Digital Elevation Model for Digital Soil Mapping in a Watershed with Gently Undulating Topography. *Rev. Bras. Ciência Do Solo* **2016**, *40*, 1–15. [CrossRef]
65. Grinevskii, S.O. The effect of topography on the formation of groundwater recharge. *Mosc. Univ. Geol. Bull.* **2014**, *69*, 47–52. [CrossRef]
66. Awawdeh, M.; Obeidat, M.; Al-Mohammad, M.; Al-Qudah, K.; Jaradat, R. Integrated GIS and remote sensing for mapping groundwater potentiality in the Tulul al Ashaqif, Northeast Jordan. *Arab. J. Geosci.* **2014**, *7*, 2377–2392. [CrossRef]
67. Allafta, H.; Opp, C.; Patra, S. Identification of Groundwater Potential Zones Using Remote Sensing and GIS Techniques: A Case Study of the Shatt Al-Arab Basin. *Remote Sens.* **2020**, *13*, 112. [CrossRef]
68. Yeh, H.-F.; Cheng, Y.-S.; Lin, H.-I.; Lee, C.-H. Mapping groundwater recharge potential zone using a GIS approach in Hualian River, Taiwan. *Sustain. Environ. Res.* **2016**, *26*, 33–43. [CrossRef]
69. Rahmati, O.; Nazari Samani, A.; Mahdavi, M.; Pourghasemi, H.R.; Zeinivand, H. Groundwater potential mapping at Kurdistan region of Iran using analytic hierarchy process and GIS. *Arab. J. Geosci.* **2015**, *8*, 7059–7071. [CrossRef]
70. Riley, D.; Mieno, T.; Schoengold, K.; Brozović, N. The impact of land cover on groundwater recharge in the High Plains: An application to the Conservation Reserve Program. *Sci. Total Environ.* **2019**, *696*, 133871. [CrossRef]
71. Owuor, S.O.; Butterbach-Bahl, K.; Guzha, A.C.; Rufino, M.C.; Pelster, D.E.; Díaz-Pinés, E.; Breuer, L. Groundwater recharge rates and surface runoff response to land use and land cover changes in semi-arid environments. *Ecol. Process.* **2016**, *5*, 16. [CrossRef]

72. Jinno, K.; Tsutsumi, A.; Alkaeed, O.; Saita, S.; Berndtsson, R. Effects of land-use change on groundwater recharge model parameters. *Hydrol. Sci. J.* **2009**, *54*, 300–315. [CrossRef]
73. Adhikari, R.K.; Mohanasundaram, S.; Shrestha, S. Impacts of land-use changes on the groundwater recharge in the Ho Chi Minh city, Vietnam. *Environ. Res.* **2020**, *185*, 109440. [CrossRef] [PubMed]
74. Zhang, W.W.; Chen, J.; Liao, A.P.; Han, G.; Chen, X.H.; Chen, L.J.; Peng, S.; Wu, H.; Zhang, J. Geospatial knowledge-based verification and improvement of GlobeLand30. *Sci. China Earth Sci.* **2016**, *59*, 1709–1719. [CrossRef]
75. Chen, J.; Chen, J.; Liao, A.; Cao, X.; Chen, L.; Chen, X.; He, C.; Han, G.; Peng, S.; Lu, M.; et al. Global land cover mapping at 30m resolution: A POK-based operational approach. *ISPRS J. Photogramm. Remote Sens.* **2015**, *103*, 7–27. [CrossRef]
76. Chen, J.; Chen, J. GlobeLand30: Operational global land cover mapping and big-data analysis. *Sci. China Earth Sci.* **2018**, *61*, 1533–1534. [CrossRef]
77. Scanlon, B.R.; Reedy, R.C.; Stonestrom, D.A.; Prudic, D.E.; Dennehy, K.F. Impact of land use and land cover change on groundwater recharge and quality in the southwestern US. *Glob. Chang. Biol.* **2005**, *11*, 1577–1593. [CrossRef]
78. Razandi, Y.; Pourghasemi, H.R.; Neisani, N.S.; Rahmati, O. Application of analytical hierarchy process, frequency ratio, and certainty factor models for groundwater potential mapping using GIS. *Earth Sci. Inform.* **2015**, *8*, 867–883. [CrossRef]
79. Nachtergaele, F.; Van Velthuizen, H.; Verelst, L.; Batjes, N.; Dijkshoorn, K.; Van Engelen, V.; Fischer, G.; Jones, A.; Montanarella, L.; Petri, M.; et al. Harmonized World Soil Database. In Proceedings of the 19th World Congress of Soil Science, Soil Solutions for a Changing World, Brisbane, Australia, 1–6 August 2010.
80. Manap, M.A.; Sulaiman, W.N.A.; Ramli, M.F.; Pradhan, B.; Surip, N. A knowledge-driven GIS modeling technique for groundwater potential mapping at the Upper Langat Basin, Malaysia. *Arab. J. Geosci.* **2013**, *6*, 1621–1637. [CrossRef]
81. Arunbose, S.; Srinivas, Y.; Rajkumar, S.; Nair, N.C.; Kaliraj, S. Remote sensing, GIS and AHP techniques based investigation of groundwater potential zones in the Karumeniyar river basin, Tamil Nadu, southern India. *Groundw. Sustain. Dev.* **2021**, *14*, 100586. [CrossRef]
82. Kopecký, M.; Macek, M.; Wild, J. Topographic Wetness Index calculation guidelines based on measured soil moisture and plant species composition. *Sci. Total Environ.* **2021**, *757*, 143785. [CrossRef] [PubMed]
83. Kaliraj, S.; Chandrasekar, N.; Magesh, N.S. Identification of potential groundwater recharge zones in Vaigai upper basin, Tamil Nadu, using GIS-based analytical hierarchical process (AHP) technique. *Arab. J. Geosci.* **2014**, *7*, 1385–1401. [CrossRef]
84. Arulbalaji, P.; Padmalal, D.; Sreelash, K. GIS and AHP Techniques Based Delineation of Groundwater Potential Zones: A case study from Southern Western Ghats, India. *Sci. Rep.* **2019**, *9*, 2082. [CrossRef] [PubMed]
85. Varade, A.M.; Khare, Y.D.; Yadav, P.; Doad, A.P.; Das, S.; Kanetkar, M.; Golekar, R.B. 'Lineaments' the Potential Groundwater Zones in Hard Rock Area: A Case Study of Basaltic Terrain of WGKKC-2 Watershed from Kalmeswar Tehsil of Nagpur District, Central India. *J. Indian Soc. Remote Sens.* **2018**, *46*, 539–549. [CrossRef]
86. Abrams, W.; Ghoneim, E.; Shew, R.; LaMaskin, T.; Al-Bloushi, K.; Hussein, S.; AbuBakr, M.; Al-Mulla, E.; Al-Awar, M.; El-Baz, F. Delineation of groundwater potential (GWP) in the northern United Arab Emirates and Oman using geospatial technologies in conjunction with Simple Additive Weight (SAW), Analytical Hierarchy Process (AHP), and Probabilistic Frequency Ratio (PFR) techniques. *J. Arid Environ.* **2018**, *157*, 77–96. [CrossRef]
87. Igor, V. *Florinsky Digital Terrain Analysis in Soil Science and Geology*; Elsevier: Amsterdam, The Netherlands, 2012. ISBN 9780123850362.
88. Pinto, D.; Shrestha, S.; Babel, M.S.; Ninsawat, S. Delineation of groundwater potential zones in the Comoro watershed, Timor Leste using GIS, remote sensing and analytic hierarchy process (AHP) technique. *Appl. Water Sci.* **2017**, *7*, 503–519. [CrossRef]
89. Maity, D.K.; Mandal, S. Identification of groundwater potential zones of the Kumari river basin, India: An RS & GIS based semi-quantitative approach. *Environ. Dev. Sustain.* **2019**, *21*, 1013–1034. [CrossRef]
90. Arabameri, A.; Rezaei, K.; Cerda, A.; Lombardo, L.; Rodrigo-Comino, J. GIS-based groundwater potential mapping in Shahroud plain, Iran. A comparison among statistical (bivariate and multivariate), data mining and MCDM approaches. *Sci. Total Environ.* **2019**, *658*, 160–177. [CrossRef]
91. Gates, J.B.; Steele, G.V.; Nasta, P.; Szilagyi, J. Lithologic influences on groundwater recharge through incised glacial till from profile to regional scales: Evidence from glaciated Eastern Nebraska. *Water Resour. Res.* **2014**, *50*, 466–481. [CrossRef]
92. Saravanan, S.; Saranya, T.; Jennifer, J.J.; Singh, L.; Selvaraj, A.; Abijith, D. Delineation of groundwater potential zone using analytical hierarchy process and GIS for Gundihalla watershed, Karnataka, India. *Arab. J. Geosci.* **2020**, *13*, 695. [CrossRef]
93. Regassa, A.; Van Daele, K.; De Paepe, P.; Dumon, M.; Deckers, J.; Asrat, A.; Van Ranst, E. Characterizing weathering intensity and trends of geological materials in the Gilgel Gibe catchment, southwestern Ethiopia. *J. Afr. Earth Sci.* **2014**, *99*, 568–580. [CrossRef]
94. Bonini, M.; Corti, G.; Innocenti, F.; Manetti, P.; Mazzarini, F.; Abebe, T.; Pecskay, Z. Evolution of the Main Ethiopian Rift in the frame of Afar and Kenya rifts propagation. *Tectonics* **2005**, *24*. [CrossRef]
95. De Reu, J.; Bourgeois, J.; Bats, M.; Zwertvaegher, A.; Gelorini, V.; De Smedt, P.; Chu, W.; Antrop, M.; De Maeyer, P.; Finke, P.; et al. Application of the topographic position index to heterogeneous landscapes. *Geomorphology* **2013**, *186*, 39–49. [CrossRef]
96. Muddarisna, N.; Yuniwati, E.D.; Masruroh, H.; Oktaviansyah, A.R. An Automated Approach Using Topographic Position Index (TPI) for Landform Mapping (Case Study: Gede Watershed, Malang Regency, East Java, Indonesia). *IOP Conf. Ser. Earth Environ. Sci.* **2020**, *412*, 012027. [CrossRef]
97. Mokarram, M.; Roshan, G.; Negahban, S. Landform classification using topography position index (case study: Salt dome of Korsia-Darab plain, Iran). *Model. Earth Syst. Environ.* **2015**, *1*, 40. [CrossRef]

98. Moore, I.D.; Grayson, R.B.; Ladson, A.R. Digital terrain modelling: A review of hydrological, geomorphological, and biological applications. *Hydrol. Processes* **1991**, *5*, 3–30. [CrossRef]
99. Sørensen, R.; Zinko, U.; Seibert, J. On the calculation of the topographic wetness index: Evaluation of different methods based on field observations. *Hydrol. Earth Syst. Sci.* **2006**, *10*, 101–112. [CrossRef]
100. Beven, K.J.; Kirkby, M.J. A physically based, variable contributing area model of basin hydrology / Un modèle à base physique de zone d'appel variable de l'hydrologie du bassin versant. *Hydrol. Sci. Bull.* **1979**, *24*, 43–69. [CrossRef]
101. Beven, K. Topmodel: A critique. *Hydrol. Processes* **1997**, *11*, 1069–1085. [CrossRef]
102. Saaty, T.L. Decision making with the analytic hierarchy process. *Int. J. Serv. Sci.* **2008**, *1*, 83. [CrossRef]
103. Saaty, T.L. Highlights and critical points in the theory and application of the Analytic Hierarchy Process. *Eur. J. Oper. Res.* **1994**, *74*, 426–447. [CrossRef]
104. Saaty, T.L. How to make a decision: The analytic hierarchy process. *Eur. J. Oper. Res.* **1990**, *48*, 9–26. [CrossRef]
105. Podvezko, V. Application of AHP Technique. *J. Bus. Econ. Manag.* **2009**, *10*, 181–189. [CrossRef]
106. Brunelli, M. *Introduction to the Analytic Hierarchy Process*; SpringerBriefs in Operations Research; Springer: Cham, Switzerland, 2015; ISBN 978-3-319-12501-5.
107. Kebede, S.; Travi, Y.; Asrat, A.; Alemayehu, T.; Ayenew, T.; Tessema, Z. Groundwater origin and flow along selected transects in Ethiopian rift volcanic aquifers. *Hydrogeol. J.* **2008**, *16*, 55–73. [CrossRef]
108. Besbes, M.; De Marsily, G. From infiltration to recharge: Use of a parametric transfer function. *J. Hydrol.* **1984**, *74*, 271–293. [CrossRef]
109. Mallick, J.; Singh, C.K.; Al-Wadi, H.; Ahmed, M.; Rahman, A.; Shashtri, S.; Mukherjee, S. Geospatial and geostatistical approach for groundwater potential zone delineation. *Hydrol. Processes* **2015**, *29*, 395–418. [CrossRef]
110. Chenini, I.; Ben Mammou, A.; El May, M. Groundwater Recharge Zone Mapping Using GIS-Based Multi-criteria Analysis: A Case Study in Central Tunisia (Maknassy Basin). *Water Resour. Manag.* **2010**, *24*, 921–939. [CrossRef]
111. Cherkauer, D.S.; Ansari, S.A. Estimating Ground Water Recharge from Topography, Hydrogeology, and Land Cover. *Ground Water* **2005**, *43*, 102–112. [CrossRef]
112. Ajay Kumar, V.; Mondal, N.C.; Ahmed, S. Identification of Groundwater Potential Zones Using RS, GIS and AHP Techniques: A Case Study in a Part of Deccan Volcanic Province (DVP), Maharashtra, India. *J. Indian Soc. Remote Sens.* **2020**, *48*, 497–511. [CrossRef]
113. Medici, G.; Smeraglia, L.; Torabi, A.; Botter, C. Review of Modeling Approaches to Groundwater Flow in Deformed Carbonate Aquifers. *Groundwater* **2021**, *59*, 334–351. [CrossRef]
114. Achu, A.L.; Thomas, J.; Reghunath, R. Multi-criteria decision analysis for delineation of groundwater potential zones in a tropical river basin using remote sensing, GIS and analytical hierarchy process (AHP). *Groundw. Sustain. Dev.* **2020**, *10*, 100365. [CrossRef]
115. Lee, S.; Hyun, Y.; Lee, S.; Lee, M.-J. Groundwater Potential Mapping Using Remote Sensing and GIS-Based Machine Learning Techniques. *Remote Sens.* **2020**, *12*, 1200. [CrossRef]
116. Mallick, J.; Khan, R.A.; Ahmed, M.; Alqadhi, S.D.; Alsubih, M.; Falqi, I.; Hasan, M.A. Modeling Groundwater Potential Zone in a Semi-Arid Region of Aseer Using Fuzzy-AHP and Geoinformation Techniques. *Water* **2019**, *11*, 2656. [CrossRef]
117. Benjmel, K.; Amraoui, F.; Boutaleb, S.; Ouchchen, M.; Tahiri, A.; Touab, A. Mapping of Groundwater Potential Zones in Crystalline Terrain Using Remote Sensing, GIS Techniques, and Multicriteria Data Analysis (Case of the Ighrem Region, Western Anti-Atlas, Morocco). *Water* **2020**, *12*, 471. [CrossRef]
118. MacDonald, A.M.; Bonsor, H.C. *An Initial Estimate of Depth to Groundwater across Africa*; Groundwater Science Programme Open Report OR/11/067; British Geological Survey: Nottingham, UK, 2011; 26p.
119. Macdonald, A.; Ochartaigh, B.; Bonsor, H.; Davies, J.; Key, R. *Developing Quantitative Aquifer Maps for Africa*; British Geological Survey Internal Report IR/10/103; British Geological Survey: Nottingham, UK, 2010; 34p.

water

MDPI

Article

Appraisal of Groundwater Vulnerability Pollution Mapping Using GIS Based GOD Index in Tiruchendur, Thoothukudi District, India

Selvam Sekar [1,*], Jesuraja Kamaraj [1], Sivasubramanian Poovalingam [1], Radhika Duraisamy [2], Venkatramanan Senapathi [3] and Chung Sang Yong [4]

1 Department of Geology, V.O. Chidambaram College, Thoothukudi 628008, Tamil Nadu, India
2 Department of Zoology, V.O. Chidambaram College, Thoothukudi 628008, Tamil Nadu, India
3 Department of Disaster Management, Alagappa University, Karaikudi 630003, Tamil Nadu, India
4 Department of Earth & Environmental Sciences, Institute of Environmental Geosciences, Pukyong National University, Busan 608737, Republic of Korea
* Correspondence: geoselvam10@gmail.com; Tel.: +91-994-466-4570

Abstract: Recently, groundwater resources have become the main freshwater supply for human activities worldwide, especially in semi-arid regions, and groundwater pollution from anthropological events is one of the chief environmental problems in built-up and industrial coastal areas. Many researchers around the world have conducted studies to evaluate the impact of groundwater management. For this study, GIS based GOD vulnerability models were used to assess the intrinsic impact and risk of pollution of coastal and irrigated areas in Tiruchendur Taluk, Thoothukudi district in Tamil Nadu. Here, GOD stands for G—Groundwater hydraulic confinement, O—Overlying aquifer strata, and D—Depth to groundwater. The parameters of G, O, and D show that 70% of the study area consists of an unconfined aquifer whose central part often consists of sandstone and fine to medium clay with sand along the coast that acts as an aquifer. The recorded value was 1–28 mbgl. The map of vulnerability using the GOD method shows that 32% of the medium vulnerabilities are located in the almost northern part of the study area, where the main source of pollution is from agricultural land and anthropological activities. A total of 39 groundwater samples were collected from different types of aquifers and used to validate the pollution map, using the EC concentration (230 to 15,480 µs/cm with an average of 2758 µs/cm) and NO_3^- concentration (2 to 120 mg/L with an average of 46 mg/L) in groundwater as indicators of pollution. Finally, we measured how the EC and NO_3^- parameters represent the medium vulnerability zone of the GOD model based on the pattern of their concentrations in groundwater. Therefore, the GIS with GOD model is the best model among these models for predicting groundwater vulnerability in Tiruchendur Taluk.

Keywords: groundwater vulnerability; GOD Index; GIS; EC and NO_3^- parameters

Citation: Sekar, S.; Kamaraj, J.; Poovalingam, S.; Duraisamy, R.; Senapathi, V.; Sang Yong, C. Appraisal of Groundwater Vulnerability Pollution Mapping Using GIS Based GOD Index in Tiruchendur, Thoothukudi District, India. *Water* **2023**, *15*, 520. https://doi.org/10.3390/w15030520

Academic Editors: Paolo Fabbri and Akira Kawamura

Received: 20 November 2022
Revised: 4 January 2023
Accepted: 24 January 2023
Published: 28 January 2023

1. Introduction

Groundwater is the most important water supply in the whole world, especially in coastal areas, and is considered as important as gold [1,2]. In northern India, there are many perennial rivers with surface water, but the source of surface water in southern India is very low, and coastal residents often rely on groundwater for drinking, irrigation, and all other uses. Water demand is increasing, especially in coastal and tropical regions of the world due to problems such as population growth, industrial development, and irrigation [3], and even in developed countries, groundwater pollution, as opposed to surface water pollution, poses a greater challenge for researchers on its origin, source, and pollution levels.

Groundwater vulnerability assessment has been reported in various climatic regions of the world, such as semi-arid regions, humid tropical regions, sub-tropical regions, temperate regions, and arid regions. The assessment has also been reported in various hydrogeological environments, i.e., karst aquifers, coastal region, alluvial aquifers, and hard rock aquifers [4]. The phenomenon of groundwater vulnerability is conceptualized assuming that the physical environment might act as a protector to groundwater against natural impacts to some extent, especially regarding the entry of contaminants into the subsurface [4–6]. The scope of aquifer vulnerability refers to the possibility of contamination of the aquifer by all contaminants in the surface and groundwater systems of the Earth [7]. The effectiveness and aspect of groundwater vulnerability valuation is essential for sustainable groundwater management and plays an important role in the study of high vulnerable regions [8].

There are different types of vulnerability systems to determine the vulnerability of groundwater in groundwater quality assessment, e.g., SINTACS [9], DRASTIC [10], SI [11], and GOD [12]. Among the mentioned methods, GOD was selected for this study as it is the easiest to evaluate and interpret based on the hydrogeological parameters of GOD, and all of these parameters had already been measured by the regional government and a private water authority, so we can ensure the accuracy and the quality of the data. There is a lot of background research on groundwater vulnerability indices using the GOD method, especially the studies of Akinlalu et al., in Nigeria [13]; Boufekane & Saighi, [14] and Kerzabi et al., [15] in Algeria; Lisboa et al., in Brazil [16]; and Islam et al., in Bangladesh [17]. Groundwater vulnerability is defined as the process of calculating the potential for contaminants to occur in a given event [18]. GOD [12] is an evaluation method to assess the vulnerability to various hydrogeological parameters such as G—the aquifer, O—the overlying lithology of the aquifer, and D—the depth of the water table. The result of the method for each classification varies from minimum vulnerability (zero) to maximum vulnerability (one) in the groundwater quality study.

Tamil Nadu has a high population density and is a coastal area, and nearly 90% of the people in Thoothukudi rely on groundwater as their main source of drinking water. This study focuses on groundwater contamination and helps to classify the Tiruchendur Taluk of Thoothukudi district based on its vulnerability studies. Many researchers have adapted various works to assess water quality using the WHO (World Health Organization) and Bureau of Indian Standards-based classification, irrigation codes, water quality index for drinking water and irrigation, health risk index, lake water mixing code, correlation methods, etc. [19–26]. However, this study is a new approach to identify the groundwater vulnerability zones in the coastal aquifers of Tiruchendur Taluk in Thoothukudi district using the GIS-based GOD method. The predicting groundwater vulnerability map can be further utilized as a base map for management of groundwater and its restoration of groundwater quality.

Study Area

The study area covers an area of 8°22′ N–8°40′ N latitude and 77°14′ E–78°46′ E longitude in Tiruchendur Taluk of Tuticorin district with an area of 470 km^2 in Tamilnadu, India. Figure 1 shows the NDVI index of the region expressing water, land, shrub, and vegetation covers in Tiruchendur Taluk. Around the NW–NE, the year-round surface water source of the Thamiraparani River drains into the Bay of Bengal near the Punnakayal area, and the seasonal water source of the Karamaniyar Canal drains along the southern part of Tiruchendur. Selvam et al. [27] pointed out that the average rainfall in Tamil Nadu during the northeast monsoon is 94.3 cm, while the average rainfall in the study area is 87.7 cm, which is very low compared to the rainfall in Tamil Nadu. Wind speeds of 36 to 60 km/h were recorded during the months of June to September, while 20 to 40 km/h were observed during the months of October to December [28]. Tiruchendur Taluk is one of the major agricultural areas of Thoothukudi, and the major crops are cassava, coconut, banana, mango, cashew, sugarcane, cloves, cardamom, pepper, etc., and palm products are mainly grown in the study area [19]. In India, after Gujarat, the coastal areas of Tamilnadu around Thoothukudi and Tiruchendur are also heavily involved in the sea salt industry [28]. Most of the surveyed areas are surrounded by agricultural land, with the exception of wastelands, salt pans, and water bodies [20,24]. Groundwater and surface water in the region are affected by anthropological activities related to fishing, tourism, and industries [23].

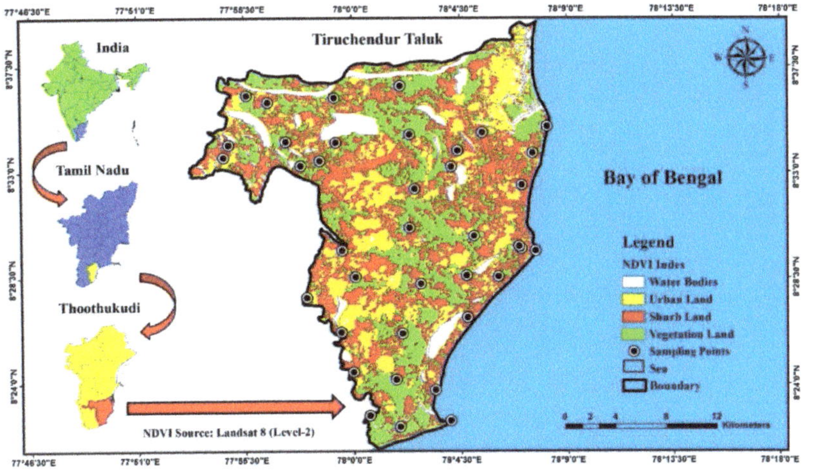

Figure 1. Study area map with NDVI Index classification.

The coastal region of Thoothukudi in Tamil Nadu has important industries such as chemicals, heavy water, petrochemicals and copper smelting (SPIC, TAC, and DCW), alkaloids, seafood, thermal power plants, port operations, and textiles. Dharangadara Chemical Works Ltd. manufactures caustic soda, liquid chlorine, trichloroethylene, refined illuminate, and PV resin. Drugs are mainly manufactured by Shantha Marine Biotechnologies (P) Ltd. Senna, and castor, corn, and cotton products are manufactured by Mother India Business Corporation. Rajashakthi & Co., a leading trader in Thoothukudi, manufactures palm jaggery, palm fruits, palm sap, palm leaves, palm candies, curry leaves, drumsticks, karnel, tari, and palmbakla. Chendhur Springs Pvt. Ltd., Manimala Water Supplier, K. Kumar Enterprise, and VVRM Water Supply are widely used for supplying quality drinking water in the study area.

2. Materials and Methods

2.1. Sampling and Analytic Techniques

Thirty-nine groundwater samples were collected in Tiruchendur Taluk, Thoothukudi district, to evaluate and verify the groundwater susceptibility system. According to the sampling guidelines, Arc-GIS 10.1 software package was used to create model stations at an equivalent distance of 1 km using GIS-based mapping and gridlines techniques. Groundwater levels from deep wells and open wells at depths ranging from 0.7 to 28 mbgl were monitored during sampling in the study area at Tiruchendur Taluk. Pre-cleaned HDPE bottles were used to collect 500 mL groundwater samples, and the water was extracted 10–15 min before sampling. Submitted water quality parameters of EC and TDS were measured during sampling using the Deluxe Water and Soil Analysis Equipment Kit multi-parameter probe (Model No. 191, Everflow Scientific Instruments, Chennai, India). All groundwater samples were then taken to a laboratory for physicochemical parameter analysis in accordance with American Public Health Association [29] standards.

2.2. Groundwater Vulnerability Assessment

In this study, the evaluation method GOD was mainly used to assess the vulnerability of groundwater in the coastal region of Tiruchendur. This is the best parametric technique that uses few parameters associated with the DRASTIC method, and this method has been accepted by many researchers [13,30,31]. The simplest and most pragmatic vulnerability assessment system for GOD was developed in England [12] by using three hydraulic and hydrological parameters such as groundwater occurrence (G), overall aquifer lithology (O), and depth to water table (D); Table 1 shows all types of each parameter (G, O, and D) with their individual estimated weights for vulnerability calculation [3]. The results of the index GOD are classified into five headings ranging from 0 to 1 [13]. The GOD index is calculated using the following equation, and GOD is defined by the multiple of the three parametric indices G (groundwater occurrence), O (overlying lithology), and D (water depth) (Equation (1)).

$$\text{GOD Index} = G_c \times O_c \times D_c \quad (1)$$

where Gc represents the weight of the aquifer occurrence parameter, Oc represents the weight of the aquifer lithology, and Dc represents the weight of the depth of the water table. In this measure, if two parameters have the value of 1, the result of the impact indicates the value of the third parameter. Finally, the results of GOD were classified into six categories, from pristine (0) to most vulnerable (1) Table 2; [31,32]. For example, if the sampling station classified under the unconfined aquifer (Gc = 1), fractured limestone acts as an overlying lithology (Oc = 1), then we can get the depth of water in between 10–20 mbgl (Dc = 0.7), and it will be classified under high vulnerable zone.

Table 1. Ratting of GOD model indicators.

Parameters	Type	Ratting
GroundwaterOccurrence (G)	No aquifer	0
	Aquifer confined and artesian	0.1
	Confined and non-artesian aquifer	0.2
	Semi-Confined Aquifer	0.3
	Aquifer with fairly permeable cover	0.4-0.6
	Unconfined aquifer	0.7–1
Overlyinglithology ofaquifer (O)	Residual soil	0.4
	Alluvial silt, clay, marl, fine limestone	0.5
	Wind, silt, tuff, igneous rock, and fractured metamorphic	0.6
	Sand and gravel, sandstone, tuff	0.7
	Gravel (colluviums)	0.8
	Limestone	0.9
	Fractured or karst limestone	1
Depth to the groundwater(D)	0–2	1
	2–5	0.9
	5–10	0.8
	10–20	0.7
	20–50	0.6
	50–100	0.5
	>100	0.4

Table 2. Vulnerability Classification.

Vulnerability Class	GOD Index
0	No Vulnerability
0–0.1	Negligible
0.1–0.3	Low Vulnerability
0.3–0.5	Medium Vulnerability
0.5–0.7	High Vulnerability
0.7–1	Very High Vulnerability

2.3. *Computation of GOD's Indicators*

Based on the GOD parameters according to Foster [12] and Foster and Hirata [33], the following parameters are used to measure vulnerability: Groundwater occurrence, class of overlying aquifer, or depth to water table. Finally, a hazard map was created that combines as well as interprets the above parameters (groundwater occurrence, class of overlying aquifer, and depth to groundwater level).

2.3.1. Groundwater Occurrence (G)

Parameter (G) refers to the types of aquifers that function as aquifers at a given depth in the study area. For this study, groundwater event data were collected from pumping test data and water level mapping reports from the CGWB and PWD departments. The IDW-based spatial map of aquatic media identifies four types of aquifers: confined (10%), unconfined (5%), semi-unconfined (70%), and semi-confined (15%) (Figure 2). Therefore, shallow bodies of water, primarily in coastal and riverine plains, have been used to extract groundwater for drinking and irrigation purposes.

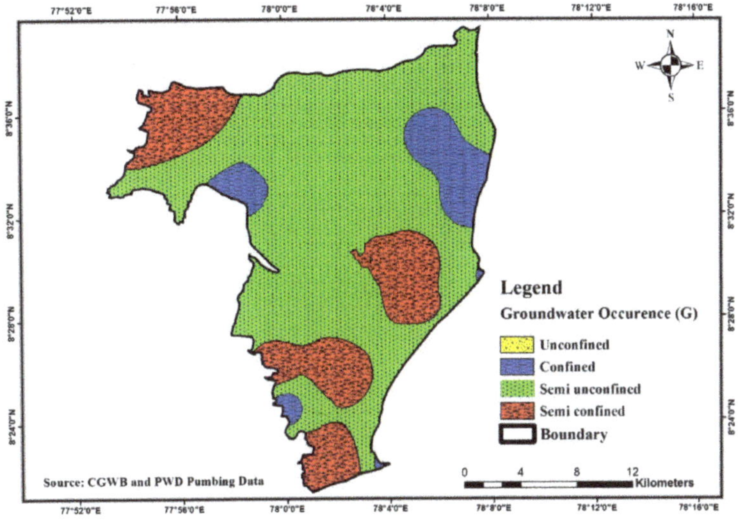

Figure 2. Groundwater occurrence spatial map in study area.

2.3.2. Overlying Aquifer Class (O)

The overlying layers of the aquifer are one of the most important factors in evaluating the groundwater's impacts because the upper layers of an aquifer can provide information about the porosity and permeability of the aquifer system [33]. As part of this study, pumping, geophysical, and lithologic well data from the CGWB and PWD fields were used to produce a spatial map of the overlying strata (Figure 3), which shows that the Tiruchendur area is mostly covered by topsoil (5%), sand with clay (13%), sandstone (20%), clay (19%), clay with shale (16%), sand (12%), and weathered granitic gneiss layers (15%). In the coastline, sandy loam and sand acted as overlying layers.

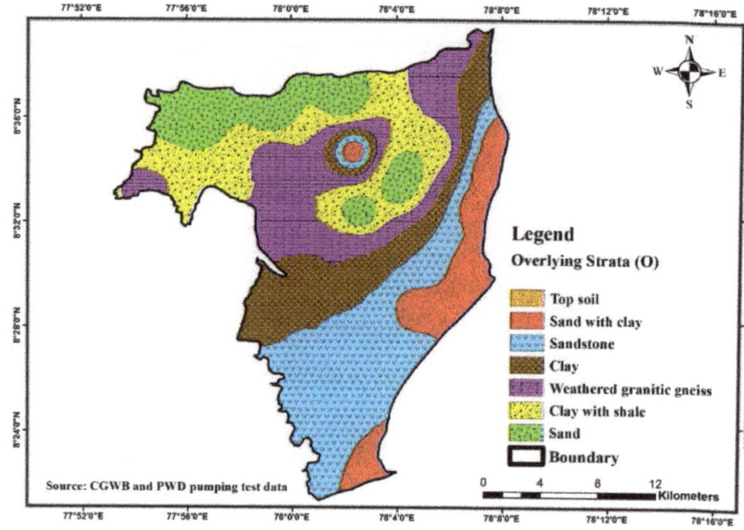

Figure 3. Overlying strata spatial map of the study area.

2.3.3. Depth to Groundwater (D)

The depth to groundwater regulates the risk of contamination as it relates to the thickness of the subsurface material. The depth to the groundwater determines the risk of contamination as it relates to the thickness of the subsurface material that must be traversed by the infiltrated water before it reaches the groundwater saturation zone. Particularly in coastal areas, the depth from the ground surface to the water level affects groundwater contamination because groundwater contamination increases with decreasing depth [34]. In the present study, this parameter was determined using 39 groundwater samples collected from open wells and drilled wells at depths ranging from 1 to 28 mbgl (Figure 4). The shallow depth of the water level (0–2 mbgl) was observed along the coast, and the deep groundwater level (20–28 mbgl) was observed beyond the coastline in the study area.

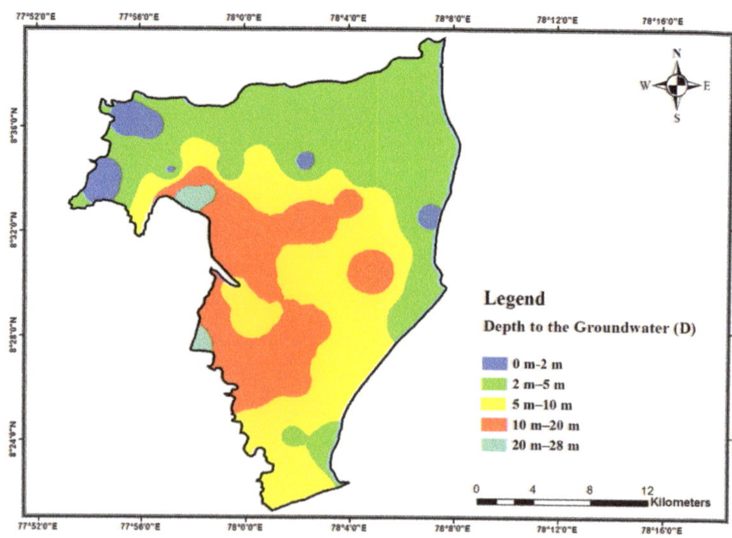

Figure 4. Spatial map of depth to the groundwater of the study area.

2.4. Method to Model Validation

Evaluating a method means that we further validate the result, which is essential for research [35,36]. No one uses a specific model for vulnerability map assessment, so most authors use the method that is most appropriate for them [37,38]. At the same time, in the field of groundwater pollution assessment, the groundwater vulnerability system is reviewed by storing data on the source of a pollutant that is abundant in the study area [35]. In the study by Jesuraja et al., [19,20] the DWQI, IWQI, and groundwater pollution index were used to determine the level of groundwater in the vicinity of Tiruchendur Taluk that is not suitable for drinking water and irrigation. Selvam et al., [22] also confirmed the presence of nitrate and fluoride pollution in the study area. Agricultural activities play an important role for the population of the study area, so in this study, it was decided to check the results of ultimate GOD susceptibility using nitrate (NO_3^-) and electrical conductivity (EC).

3. Results and Discussion

3.1. Geochemical Assessment for Drinking and Irrigation

The significant and analyzed water quality parameters of electrical conductivity (EC) and nitrate (NO_3^-) are classified according to their concentrations in groundwater (Table 3). The EC and NO_3^- concentrations were classified for drinking water [39–41] and irrigation purposes [41,42] respectively. A specific range of EC was found in groundwater, ranging from 230 to 15,480 (μs/cm) EC. According to the estimate of EC, 44% of groundwater samples were harmful for drinking and irrigation purposes (3000 μS/cm). The observed NO_3^- concentrations from open wells and deep wells in the study area ranged from 2 to 120 mg/L, and 32% of the samples were above the limits for drinking water use. NO_3^- concentration measurements are classified as useful (25–50 mg/L) in 24% of the samples and harmful for drinking water use (>50 mg/L) in 21% of the groundwater samples, and harmful for irrigation purposes (50–100 mg/L) in more than 21% of the samples.

Table 3. EC- and NO_3^--based water classification for drinking and irrigation.

Scale of Category	EC in μS/cm				NO_3^- in mg/L			
	Drinking [39–41]		Irrigation [41,42]		Drinking [39–41]		Irrigation [41,42]	
	Grade	% of Samples	Grade	% of Samples	Grade	% of Samples	Grade	% of Samples
Very good	0–180		0–250		0–10	31	0–10	31
Good	180–400	3	250–750	13	10–25	24	10–30	24
Usable	400–2000	50	750–2000	40	25–50	24	30–50	24
Usable with caution	2000–3000	3	2000–3000	3			50–100	18
Harmful	>3000	44	>3000	44	>50	21	>100	3

According to the geochemical parameters in the coastal areas, high EC concentrations were found in groundwater and abnormal NO_3^- concentrations in agricultural areas of the study area. In the southern coastal areas of Thoothukudi region in Tamil Nadu, many researchers have identified the reasons for the high number of pollutants in water quality, and the details to improve the specific parameters are given in Table 4.

Table 4. Identified groundwater pollution sources in the study area.

Year	Pollution Source	Flagged Pollutants and Parameters	Reference
1993	Over exploitation	Groundwater salinity and quality	[43]
2009	Domestic effluents	Groundwater Contamination	[44]
2011	anthropogenic contamination (likesalt pans and fertilizer)	To enhance the Na^+ and Cl^-	[45]
2011	Industrial activities	Fluoride contamination	[21]
2012	Infiltration process from sewers canals, unprotected drains and industrial effluents	To increase the TDS > 1500 mg/L	[46]
2012	Salt pans	Enhance alkaline nature	[47]
2012	Industrial effluents	Metal pollution	[45]
2012	Agricultural return flow, domestic sewage, septic tanks or other anthropogenic activities	Nitrate pollution	[46]
2013	Seawater influence or salt pan deposits or ionic exchangeprocess	Increase Na^+ in groundwater	[48]
2014	Owing to the modern day issues of sea level rise, irregular patterns of rainfall due to climate change	Increased demand of groundwater	[49]
2014	Chemical industries, salt, flower dying, copper wire, copper alloy, alkali chemicals and fertilizers, petro-chemicals & plastics industries	Overall groundwater quality	[21]
2018	Urbanization, Over exploitation and Industrialization	To decrease the groundwater quality	[50]
2019	Rural and private septic systems, sinkholes, municipal sewage systems and tourism	To increase micro plastic in groundwater	[23]
2020	Fishing industries	High organic compounds in the groundwater	[51]

According to the geochemical parameters in the coastal areas, high EC concentrations were found in groundwater and abnormal NO_3^- concentrations in agricultural areas of the study area. In the southern coastal areas of Thoothukudi region in Tamil Nadu, many researchers have identified the reasons for the high number of pollutants in water quality, and the details to improve the specific parameters are given in Table 4.

3.2. GOD Vulnerability Map

The flowchart of this study shows how the maps were preferred and the specific process used to generate the spatial distribution of each responsible parameter (Figure 5). According to the weight of vulnerability and the evaluation process, each indicator was evaluated to determine the extent of vulnerability. From the interpolation of all G, O, and D parameters, a master map of GOD vulnerability was created for each indicator using GIS software (version 10.1). The GOD index impact map was created by overlaying the water table, topsoil, groundwater depth, and soil type maps. The final map of hazard GOD shows that the study area was classified into three hazard classes from negligible to a medium hazard, with values ranging from 0.08 to 0.50. The percentage results of the studied aquifer show that the study area was classified as very low risk (3%), low risk (65%), and moderate risk (32%) (Figure 6). According to the results, most of the study areas are in the low hazard zone, while the northern part of the study area is in the high and moderate zone, especially on the Punnakayal bank of the Thamiraparani River. At the same time, no area is affected that is not classified in the highest and most sensitive zone. Shallow water levels were found in the coastal and reverie levels, which may increase the vulnerability of the aquifer compared to other lithological factors and the overlying layers of the aquifer, as shallow water is very favorable for groundwater pollution and anthropological activities [3,17,31].

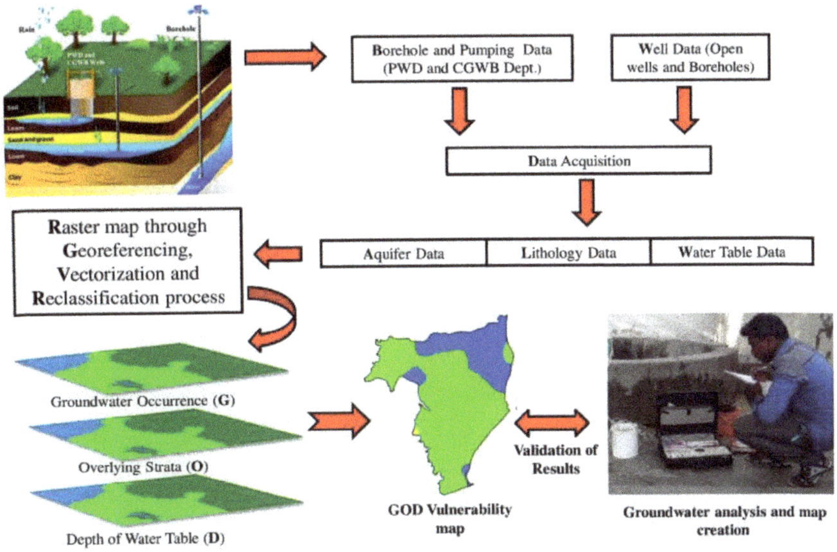

Figure 5. Flow chart of the GOD model working methodology.

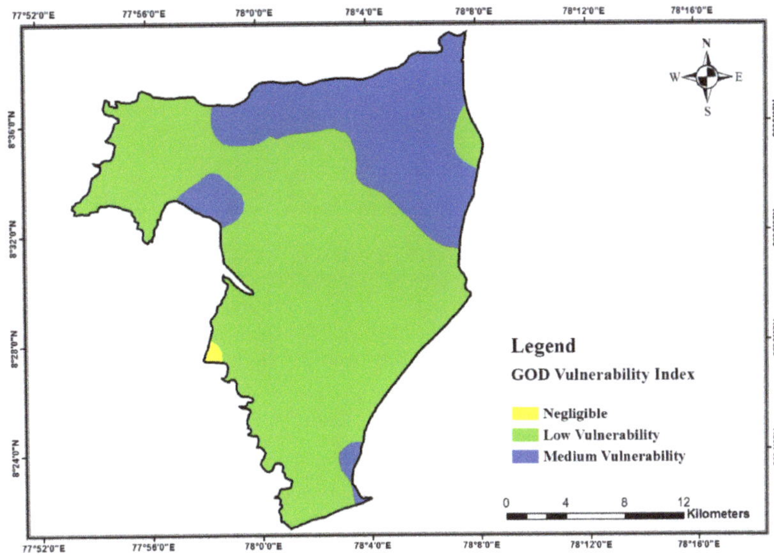

Figure 6. Spatial distribution of the GOD intrinsic vulnerability map of the study area.

3.3. GOD Model Validate with Geochemical Parameters

According to Singha et al., [52], the results of the vulnerability zones give us a little more confidence when validating the retention of the geochemical values obtained. Nitrate (NO_3^-) and EC are the most confirmed anthropogenic contamination indicators of the study area's groundwater resources, so its concentration in groundwater could correlate with the GOD risk index [21,22]. The reason for choosing NO_3^- is that the chief sources of nitrate in groundwater are numerous anthropogenic activities, such as fertilizers used in agriculture, which show the salinity and pollution of groundwater [53–57]. According to the WHO 2017, the maximum adequate nitrate concentration for human health is 50 mg/L [58], but it is known that a nitrate absorption >3 mg/L in groundwater indicates anthropogenic pollution [59]. The nitrate concentration of groundwater in the northern portion of the study area and in the agricultural area of Thamiraparani River was more than 60 mg/L.

It can be resolved that the increase in nitrate absorption in this area, particularly in the north, is likely to be related to pollution input from agricultural fertilizers in this area. The spatial correlation diagram clearly shows that the trends in NO_3^- concentrations are consistent with the indices of GOD vulnerability in the north, and that the higher NO_3^- concentrations in the study area represent the intermediate vulnerability zone. With the exception of the estuary, low NO_3^- concentrations often represent a low hazard zone because of the absence of agricultural activities compared to other areas (Figure 7).

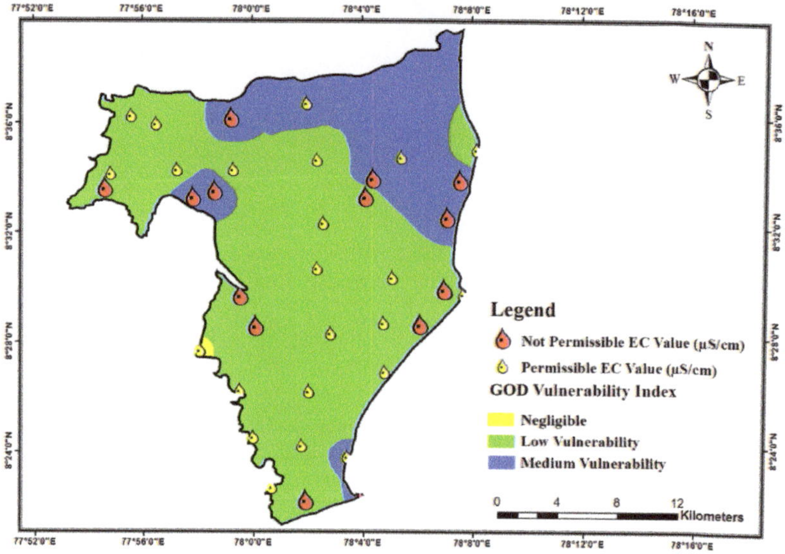

Figure 7. Spatial distribution of the GOD intrinsic vulnerability with an EC concentration.

EC concentration describes the general quality and suitability of groundwater. As with nitrate, the value of EC increases not only in river basins but also in other areas, indicating unsuitable concentration. However, high EC concentrations are observed in the moderately vulnerable areas of the NE and NW, as well as in coastal areas (Figure 8), as high EC values indicate not only agricultural mismanagement but also secondary salts and seawater leaching [59].

3.4. Mitigation of Groundwater Sources from Vulnerability

It is not always adequate to know the status of the fault memory and water quality, which is always present in the field of water conservation research. So it can be said that the next step is to do what needs to be done to improve groundwater resources so that what the public and researchers can have a positive impact on society. Therefore, the following describes ways to reduce groundwater contamination in the coastal and agricultural areas of Tiruchendur.

Proper disposal of domestic waste in the residential areas of the coastal regions of Tiruchendur and Thoothukudi is essential.

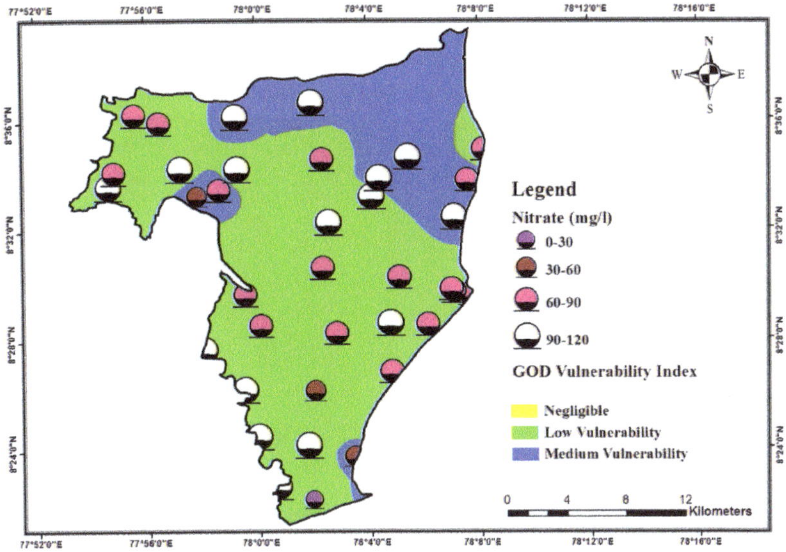

Figure 8. Spatial distribution of the GOD intrinsic vulnerability with an NO_3^- concentration.

Establishment of a special team to monitor and manage tourism, especially to mitigate the local surface water levels at the mouth of the Punnakayal Thamiraparani River.

Develop the separate guidelines based on the local environment for the companies linked to the wastewater treatment system.

Reducing fertilizer use, especially on agricultural land and informing producers of the impact.

Upgrading a large number of monitoring wells by state and central water management agencies to collect accurate data near industrial and fish factories.

Part of this is ensuring water quality for a private team and all types of private sector employees to provide safe drinking water for public schools and hospitals.

Conducting public education programs about water pollution and raising awareness about groundwater management.

Mandate rainwater collection in residential areas to reduce storm water runoff.

Sustainable water management is integral to the future of food and agriculture such as drip irrigation systems.

Sewage water treatment in high populations areas such as Tiruchendur, Punnakayal, Udangudi, Kulasai, etc.

Desalination is a process that takes away mineral components from saline water for companies not drinking.

With support from the communities, structures are built for soil and water conservation to reduce the volume and velocity of runoff for better protection against topsoil loss and to improve soil moisture retention.

Require many businesses in Thoothukudi district to set up their own coastal pollution monitoring commission, as wastewater is discharged directly into the Bay of Bengal.

4. Conclusions

In the field of groundwater chemistry, this study is the first attempt to investigate the impact on groundwater using a GIS-based GOD exposure method in TiruchendurTaluk of South Thoothukudi district. The vulnerability map of the GOD system confirms that almost 32% of the Thamiraparani River plain and northern coastal areas are classified as medium vulnerability. The remaining areas (65%) are less vulnerable, reflecting aquifer degradation in the study area. In this assessment of the GOD index, geological systems and water bodies contribute to pollution compared to the shallow layers of this region. In the zones of medium vulnerability, we can observe shallow groundwater levels of less than 5 mbgl, which is very conducive to agricultural and human pollution of groundwater. To further evaluate the susceptibility of evolution measurements, this study compares the significant geochemical parameters of EC and NO_3^- concentrations with the GOD susceptibility map. Both parameters are closely related because high concentrations were observed in the zones of medium vulnerability, which ensure that water easily combines with various geochemical components derived from noxious pesticides and their widespread use in agricultural lands and anthropological activities. Therefore, high-risk activities should not be allowed in high-risk areas to achieve economic benefits and to reduce the risk of pollution. To reduce the risk of pollution in moderate-risk areas, a precautionary measure should be taken before exploiting aquifers and before starting extensive agricultural activities in the area. The GIS-based GOD model can be used to classify areas susceptible to groundwater quality management in the Thoothukudi coastal area and serve as one of the latest databases for the government.

Author Contributions: Conceptualization, methodology, software, validation, formal analysis, investigation, resources, data accusation, writing—original draft preparation, S.S. and J.K.; writing—review and editing, S.P. and R.D.; visualization, supervision, project administration, V.S. and C.S.Y.; funding acquisition, R.D. All the authors significantly contributed to manuscript preparation. All authors have read and agreed to the published version of the manuscript.

Funding: The authors acknowledge financial support by the University Grants Commission Scheme for Trans-Disciplinary Research for India's Developing Economy (STRIDE) (Grant no: No.F.2-18/2019 (STRIDE-I)).

Data Availability Statement: Data is available on request from corresponding author.

Conflicts of Interest: The authors declare no conflict of interest.

References

1. Rahman, M.R.; Islam, A.R.M.T.; Shammi, M. Emerging trends of water quality monitoring and applications of multivariate tools. *Water Eng. Model. Math. Tools* **2021**, *14*, 271–283.
2. Rakib, M.; Sasaki, J.; Matsuda, H.; Quraishi, S.B.; Mahmud, J.; Doza, B.; Ullah, A.A.; Fatema, K.J.; Newaz, A.; Bhuiyan, M.A. Groundwater salinization and associated co-contamination risk increase severe drinking water vulnerabilities in the southwestern coast of Bangladesh. *Chemosphere* **2020**, *246*, 125646. [CrossRef]
3. Ghazavi, R.; Ebrahimi, Z. Assessing groundwater vulnerability to contamination in an arid environment using DRASTIC and GOD models. *Int. J. Environ. Sci. Technol.* **2015**, *2015*, 2909–2918. [CrossRef]
4. Bera, A.; Mukhopadhyay, B.P.; Chowdhury, P.; Ghosh, A.; Biswas, S. Groundwater vulnerability assessment using GIS-based DRASTIC model in Nangasai River Basin, India with special emphasis on agricultural contamination. *Ecotoxicol. Environ. Saf.* **2021**, *214*, 112085. [CrossRef]
5. Popescu, I.C.; Gardin, N.; Brouyére, S.; Dassargues, A. Groundwater vulnerability assessment using physically based modeling: From challenges to pragmatic solutions. In *Model CARE 2007 Proceedings, Calibration and Reliability in Ground-Water Modeling*; Refsgaard, J.C., Kovar, K., Haarder, E., Nygaard, E., Eds.; IAHS Publication: Wallingford, UK, 2008; No. 320.
6. Oroji, B. Groundwater vulnerability assessment with using GIS in Hamadan–Bahar plain, Iran. *Appl. Water Sci.* **2019**, *9*, 196. [CrossRef]
7. Agyemang, A. Vulnerability Assessment of Groundwater to NO3 Contamination Using GIS, DRASTIC Model and Geostatistical Analysis. Master's Thesis, Department of Geosciences, East Tennessee State University, Johnson City, TN, USA, 2017. Paper 3264. Available online: https://dc.etsu.edu/etd/3264 (accessed on 8 December 2017).
8. Knouz, N.; Boudhar, A.; Bachaoui, E.M.; Saadi, C. Comparative approach of three popular intrinsic vulnerability methods: Case of the Beni Amir groundwater (Morocco). *Arab. J. Geosci.* **2018**, *11*, 281. [CrossRef]
9. Civita, M. *Le Carte Della Vulnerability à Degli Acquiferi All'inquinamento: Teoria e Practica [Contamination Vulnerability Mapping of the Aquifer: Theory and Practice]*; Pitagora: Bologna, Italy, 1994; Volume 13.
10. Aller, L. *DRASTIC: A Standardized System for Evaluating Ground Water Pollution Potential Using Hydrogeologic Settings*; Robert, S., Ed.; Kerr Environmental Research Laboratory: Ada, OK, USA, 1985.
11. Ribeiro, R.J. A Sociedade Contra o Social: O Alto Custo da Vida Pública No Brasil: Ensaios. Companhia das Letras: Sao Paulo, Brazil, 2000; pp. 1–233.

12. Foster, S. Fundamental concepts in aquifer vulnerability, pollution risk and protection strategy. In: Van Duijvenbooden W, Van Waegeningh HG (eds), Vulnerability of soil and groundwater to pollutants. *Proc. Inf. TNO Comm. Hydrol. Res. Hague* **1987**, *38*, 69–86.
13. Akinlalu, A.A.; Mogaji, K.A.; Adebodun, T.S. Assessment of aquifer vulnerability using a developed "GODL" method (modified GOD model) in a schist belt environ, Southwestern Nigeria. *Environ. Monit. Assess.* **2021**, *193*, 1–27. [CrossRef] [PubMed]
14. Boufekane, A.; Saighi, O. Application of groundwater vulnerability overlay and index methods to the Jijel plain area (Algeria). *Groundwater* **2018**, *56*, 143–156. [CrossRef]
15. Kerzabi, R.; Mansour, H.; Yousfi, S.; Marín, A.I.; Navarro, B.A.; Bensefia, K.E. Contribution of remote sensing and GIS to mapping groundwater vulnerability in arid zone: Case from Amour Mountains- Algerian Saharan Atlas. *J. Afr. Earth Sci.* **2021**, *182*, 104277. [CrossRef]
16. Lisboa, É.G.; Mendes, R.L.R.; Figueiredo, M.M.P.; Bello, L.A.L. Fuzzy-Probabilistic Model for a Risk Assessment of Groundwater Contamination: Application to an Urban Zone in the City of Belém, Pará, Brazil. *Water* **2020**, *12*, 1437. [CrossRef]
17. Islam, T.; Foysol Mahmud, M.; Zafor, A. GIS based vulnerability assessment of shallow groundwater pollution in the southwest region of Bangladesh using GOD method. In *Proceedings of International Conference on Planning, Architecture & Civil Engineering*; Rajshahi University of Engineering & Technology: Rajshahi, Bangladesh, 2021.
18. Voudouris, K. Assessing groundwater pollution risk in Sarigkiol basin, NW Greece. In *River Pollution Research Progress*; Nova Science Publishers Inc.: New York, NY, USA, 2009; pp. 265–281.
19. Jesuraja, J.; Sekar, S.; Roy, P.D.; Senapathi, V.; Chung, S.Y.; Perumal, M.; Nath, A.V. Groundwater pollution index (GPI) and GIS-based appraisal of groundwater quality for drinking and irrigation in coastal aquifers of Tiruchendur, South India. *Environ. Sci. Pollut. Res.* **2021**, *28*, 29056–29074. [CrossRef]
20. Jesuraja, K.; Selvam, S.; Murugan, R. GIS-based assessment of groundwater quality index (DWQI and AWQI) in Tiruchendur Coastal City, Southern Tamil Nadu, India. *Environ. Earth Sci.* **2021**, *80*, 1–17. [CrossRef]
21. Selvam, S. Irrigational Feasibility of Groundwater and Evaluation of Hydrochemistry Facies in the SIPCOT Industrial Area, South Tamilnadu, India: A GIS Approach. *Water Qual. Expo. Heal.* **2014**, *7*, 265–284. [CrossRef]
22. Selvam, S.; Jesuraja, K.; Venkatramanan, S.; Chung, S.; Roy, P.; Muthukumar, P.; Kumar, M. Imprints of pandemic lockdown on subsurface water quality in the coastal industrial city of Tuticorin, South India: A revival perspective. *Sci. Total. Environ.* **2020**, *738*, 139848. [CrossRef]
23. Selvam, S.; Jesuraja, K.; Roy, P.D.; Venkatramanan, S.; Chung, S.; Elzain, H.E.; Muthukumar, P.; Nath, A.V.; Karthik, R. Assessment of groundwater from an industrial coastal area of south India for human health risk from consumption and irrigation suitability. *Environ. Res.* **2021**, *200*, 111461. [CrossRef]
24. Singaraja, C. Relevance of water quality index for groundwater quality evaluation: Thoothukudi District, Tamil Nadu, India. *Appl. Water Sci.* **2017**, *7*, 2157–2173. [CrossRef]
25. Singaraja, C.; Chidambaram, S.; Anandhan, P.; Prasanna, M.V.; Thivya, C.; Thilagavathi, R.; Sarathidasan, J. Determination of the utility of groundwater with respect to the geochemical parameters: A case study from Tuticorin District of Tamil Nadu (India). *Environ. Dev. Sustain.* **2013**, *16*, 689–721. [CrossRef]
26. Singaraja, C.; Chidambaram, S.; Jacob, N.; Johnson Babu, G.; Selvam, S.; Anandhan, P.; Rajeevkumar, E.; Balamurugan, K.; Tamizharasan, K. Origin of high fluoride in groundwater of the Tuticorin district, Tamil Nadu. *India Appl. Water Sci.* **2018**, *8*, 54. [CrossRef]
27. Selvam, S.; Manimaran, G.; Sivasubramanian, P.; Balasubramanian, N.; Seshunarayana, T. GIS-based Evaluation of Water Quality Index of groundwater resources around Tuticorin coastal city, south India. *Environ. Earth Sci.* **2013**, *71*, 2847–2867. [CrossRef]
28. Chandrasekar, N.; Selvakumar, S.; Srinivas, Y.; Wilson, J.S.J.; Peter, T.S.; Magesh, N.S. Hydrogeochemical assessment of groundwater quality along the coastal aquifers of southern Tamil Nadu, India. *Environ. Earth Sci.* **2013**, *71*, 4739–4750. [CrossRef]
29. APHA. *Standard Method for Examination of Water and Wastewater*, 21st ed.; APHA, AWWA, WPCF: Washington, DC, USA, 2005.
30. Feumba, R. Hydrogéologieet Evaluation de la Vulnérabilité des Nappesdanslebassin Versant de Besseke (Douala, Cameroun). Ph.D. Thesis, University of Yaoundé I Cameroon, Yaounde, Cameroon, 2015; p. 254.
31. Mfonka, Z.; Ngoupayou, J.N.; Ndjigui, P.D.; Kpoumie, A.; Zammouri, M.; Ngouh, A.N.; Rasolomanana, E.H. A GIS-based DRASTIC and GOD models for assessing alterites aquifer of three experimental watersheds in Foumban (West-ern-Cameroon). *Groundw. Sustain. Dev.* **2018**, *7*, 250–264. [CrossRef]
32. Huang, C.-C.; Yeh, H.-F.; Lin, H.-I.; Lee, S.-T.; Hsu, K.-C.; Lee, C.-H. Groundwater recharge and exploitative potential zone mapping using GIS and GOD techniques. *Environ. Earth Sci.* **2012**, *68*, 267–280. [CrossRef]
33. Foster, S.; Hirata, R. *Groundwater Risk Assessment—A Methodology Using Available Data*, 1st ed.; Technical report; Pan American Center for Sanitary Engineering and Environmental Sciences (CEPIS): Lima, Peru, 1988.
34. Thirumalaivasan, D.; Karmegam, M.; Venugopal, K. AHP-DRASTIC: Software for specific aquifer vulnerability as-sessment using DRASTIC model and GIS. *Environ. Model. Softw.* **2003**, *18*, 645–656. [CrossRef]
35. Elmeknassi, M.; El Mandour, A.; Elgettafi, M.; Himi, M.; Tijani, R.; El Khantouri, F.A.; Casas, A. A GIS-based approach for geospatial modeling of groundwater vulnerability and pollution risk mapping in Bou-Areg and Gareb aquifers, northeastern Morocco. *Environ. Sci. Pollut. Res.* **2021**, *28*, 51612–51631. [CrossRef]
36. Hasan, M.; Islam, A.; Alam, J.; Peas, M.H. Groundwater vulnerability assessment in Savarupazila of Dhaka district, Bangladesh—A GIS-based DRASTIC modeling. *Groundw. Sustain. Dev.* **2019**, *9*, 100220. [CrossRef]

37. Barbulescu, A. Assessing Groundwater Vulnerability: DRASTIC and DRASTIC-Like Methods: A Review. *Water* **2020**, *12*, 1356. [CrossRef]
38. Ravbar, N.; Goldscheider, N. Comparative application of four methods of groundwater vulnerability mapping in a Slovene karst catchment. *Hydrogeol. J.* **2009**, *17*, 725–733. [CrossRef]
39. Pusatli, O.T.; Camur, M.Z.; Yazicigil, H. Susceptibility indexing method for irrigation water management planning: Applications to K. Menderes river basin, Turkey. *J. Environ. Manag.* **2009**, *90*, 341–347. [CrossRef]
40. Saidi, S.; Bouri, S.; Ben Dhia, H.; Anselme, B. A GIS-based susceptibility indexing method for irrigation and drinking water management planning: Application to Chebba–Mellouleche Aquifer, Tunisia. *Agric. Water Manag.* **2009**, *96*, 1683–1690. [CrossRef]
41. WHO. *Guidelines for Drinking-Water Quality, Recommendations*, 3rd ed.; WHO: Geneva, Switzerland, 2014; Volume 1, p. 668.
42. Richards, L.A. *Diagnosis and Improvement of Saline and Alkalinesoils*; Salinity Laboratory Staff, Ed.; US Department of Agriculture: Washington, DC, USA, 1954.
43. Subramanian, S.; Sujatha, K.; Balasubramanian, A.; Thirugnanasambandam, R.; Radhakrishnan, V. *Hydrogeology along Tuticorin Coast, Tamil Nadu*; Groundwater Development Problems in Southern Kerala: Southern Kerala, India, 1993; Volume 1, pp. 26–34.
44. Mondal, N.C.; Singh, V.S.; Rangarajan, R. Aquifer characteristics and its modeling around an industrial complex, Tuticorin, Tamil Nadu, India: A case study. *J. Earth Syst. Sci.* **2009**, *118*, 231–244. [CrossRef]
45. Singaraja, C.; Chidambaram, S.; Anandhan, P.; Prasanna, M.V.; Thivya, C.; Thilagavathi, R.; Sarathidasan, J. Geochemical evaluation of fluoride contamination of groundwater in the Thoothukudi District of Tamilnadu, India. *Appl. Water Sci.* **2014**, *4*, 241–250. [CrossRef]
46. Selvam, S.I.J.D.; Mala, R.I.J.D.; Muthukakshmi, V. A hydrochemical analysis and evaluation of groundwater quality index in Thoothukudi district, Tamilnadu, South India. *Int. J. Adv. Eng. Appl.* **2013**, *2*, 25–37.
47. Singaraja, C.; Chidambaram, S.; Anandhan, P.; Prasanna, M.V.; Thivya, C.; Thilagavathi, R. A study on the status of saltwater intrusion in the coastal hard rock aquifer of South India. *Environ. Dev. Sustain.* **2014**, *17*, 443–475. [CrossRef]
48. Sivakumar, K.; Priya, J.; Muthusamy, S.; Saravanan, P.; Jayaprakash, M. Spatial diversity of major ionic absorptions in groundwater: Recent study from the industrial region of Tuticorin, Tamil, Nadu, India. *Enviro. Geo. Chem. Acta* **2016**, *3*, 138–147.
49. Viveka, B.; Arunkumar, V.; Vasanthi, D. Assessment of Groundwater Quality in Coastal Areas of Thoothukudi District, Tamil Nadu. *Madras Agric. J.* **2019**, *106*, 1. [CrossRef]
50. Selvam, S.; Jesuraja, K.; Venkatramanan, S.; Roy, P.D.; Kumari, V.J. Hazardous microplastic characteristics and its role as a vector of heavy metal in groundwater and surface water of coastal south India. *J. Hazard. Mater.* **2020**, *402*, 123786. [CrossRef] [PubMed]
51. Singha, S.S.; Pasupuleti, S.; Singha, S.; Singh, R.; Venkatesh, A.S. A GIS-based modified DRASTIC approach for geo-spatial modeling of groundwater vulnerability and pollution risk mapping in Korba district, Central India. *Environ. Earth Sci.* **2019**, *78*, 1–19. [CrossRef]
52. Abbasnia, A.; Yousefi, N.; Mahvi, A.H.; Nabizadeh, R.; Radfard, M.; Yousefi, M.; Alimohammadi, M. Evaluation of groundwater quality using water quality index and its suitability for assessing water for drinking and irrigation purposes: Case study of Sistan and Baluchistan province (Iran). *Hum. Ecol. Risk Assess. Int. J.* **2019**, *25*, 988–1005. [CrossRef]
53. Gopinath, S.; Srinivasamoorthy, K.; Saravanan, K.; Prakash, R.; Karunanidhi, D. Characterizing groundwater quality and seawater intrusion in coastal aquifers of Nagapattinam and Karaikal, South India using hydrogeochemistry and modeling techniques. *Hum. Ecol. Risk Assess. Int. J.* **2019**, *25*, 314–334. [CrossRef]
54. Rabeiy, R.E. Assessment and modeling of groundwater quality using WQI and GIS in Upper Egypt area. *Environ. Sci. Pollut. Res.* **2017**, *25*, 30808–30817. [CrossRef]
55. Sajil Kumar, P.; James, E. Identification of hydrogeochemical processes in the Coimbatore district, Tamil Nadu, India. *Hydrol. Sci. J.* **2016**, *61*, 719–731. [CrossRef]
56. Selvam, S.; Manimaran, G.; Sivasubramanian, P. Hydrochemical characteristics and GIS-based assessment of groundwater quality in the coastal aquifers of Tuticorin corporation, Tamilnadu, India. *Appl. Water Sci.* **2012**, *3*, 145–159. [CrossRef]
57. WHO. *Guidelines for Drinking Water Quality: Fourth Edition Incorporating the First Addendum*; World Health Organization: Geneva, Switzerland, 2017.
58. Madison, R.J.; Brunett, J.O. Overview of the occurrence of nitrate in ground water of the United States. In *National Water Summary 1984-Hydrologic Events, Selected Water-Quality Trends, and Ground-Water Resources*; Water-Supply Paper 2275; U.S. Geological Survey: Reston, VA, USA, 1985; pp. 93–105. [CrossRef]
59. Satheeskumar, V.; Subramani, T.; Lakshumanan, C.; Roy, P.D.; Karunanidhi, D. Groundwater chemistry and demarcation of seawater intrusion zones in the Thamirabarani delta of south India based on geochemical signatures. *Environ. Geochem. Heal.* **2020**, *43*, 757–770. [CrossRef] [PubMed]

MDPI
St. Alban-Anlage 66
4052 Basel
Switzerland
Tel. +41 61 683 77 34
Fax +41 61 302 89 18
www.mdpi.com

Water Editorial Office
E-mail: water@mdpi.com
www.mdpi.com/journal/water

www.ingramcontent.com/pod-product-compliance
Lightning Source LLC
Chambersburg PA
CBHW061508210926
51850CB00008B/1850
* 9 7 8 3 0 3 6 5 7 8 5 9 0 *